Testing of Materials and Elements in Civil Engineering (4th Edition)

Testing of Materials and Elements in Civil Engineering (4th Edition)

Guest Editor

Krzysztof Schabowicz

MDPI

Basel • Beijing • Wuhan • Barcelona • Belgrade • Novi Sad • Cluj • Manchester

Guest Editor
Krzysztof Schabowicz
Faculty of Civil Engineering
Wrocław University of
Science and Technology
Wroclaw
Poland

Editorial Office
MDPI AG
Grosspeteranlage 5
4052 Basel, Switzerland

This is a reprint of the Special Issue, published open access by the journal *Materials* (ISSN 1996-1944), freely accessible at: https://www.mdpi.com/journal/materials/special_issues/7988X5DEM9.

For citation purposes, cite each article independently as indicated on the article page online and as indicated below:

Lastname, A.A.; Lastname, B.B. Article Title. *Journal Name* **Year**, *Volume Number*, Page Range.

ISBN 978-3-7258-8353-0 (Hbk)
ISBN 978-3-7258-8354-7 (PDF)
https://doi.org/10.3390/books978-3-7258-8354-7

Contents

About the Editor

Krzysztof Schabowicz

Krzysztof Schabowicz is a Full Professor in the Faculty of Civil Engineering at the Wroclaw University of Science and Technology (WUST), and the Head of the Building Engineering Departament. His research interests include concrete, fibre-cement, ultrasonic tomography, impact echo, impulse response, GPR and other nondestructive tests, and artificial intelligence. He is best known for developing a testing methodology of concrete structures, especially the unilaterally accessible ones, using modern non-destructive acoustic methods. He is also among the early contributors to integrating artificial intelligence with non-destructive testing for reliable, data-driven diagnostics of concrete structures. He is the innovator of eco-friendly cellulose fibre cement boards production and the diagnostics and detection of imperfections in fibre cement boards during both their production and service life. He is an author and co-author of 7 books, over 500 publications, and 10 patents. He has more than 1600 citations in Scopus. He serves as an Editor of *Materials* (MDPI) and an Editorial Board Member of *Civil Engineering and Architecture* (HRPUB), and *Nondestructive Testing and Diagnostics* (SIMP). He has developed more than 300 reviews of journal and conference articles. He is the co-author of 10 patents and 2 patent applications, and a member of the Polish Association of Civil Engineers and Technicians (PZITB) and the Polish Association of Building Mycology (PSMB).

Preface

We have published three successful editions of the Special Issue titled 'Testing of Materials and Elements in Civil Engineering', with over 130 innovative papers published. The field of materials testing in civil engineering is broad and interesting from engineering and scientific perspectives.

This new Special Issue is proposed and organized to present recent developments in the field of materials testing in civil engineering. The articles highlighted in this Special Issue relate to different aspects of testing of different materials in civil engineering, from building materials and elements to building structures. The current trend in the development of materials testing in civil engineering primarily focuses on detecting flaws and defects in elements and structures using destructive, semi-destructive, and nondestructive testing.

This Special Issue is mainly focused on novel testing approaches, the development of single and hybrid measurement techniques, and advanced signal analysis.

Topics include the following:

- Testing of materials and elements in civil engineering;
- Testing of structures made of novel materials;
- Condition assessment of civil materials and elements;
- Detecting defects that are invisible on the surface;
- Damage detection and damage imaging;
- Diagnostics of cultural heritage monuments;
- Structural health monitoring systems;
- Modeling and numerical analysis;
- Nondestructive testing methods;
- Advanced signal processing for nondestructive testing.

Krzysztof Schabowicz
Guest Editor

materials

MDPI

Article

Artificial Weathering Test Methods of Waterborne Acrylic Coatings for Steel Structure Corrosion Protection

Łukasz Ładosz [1,*], Ewa Sudoł [1], Ewelina Kozikowska [1] and Emilia Choińska [2]

[1] Construction Materials Engineering Department, Instytut Techniki Budowlanej, 00-611 Warsaw, Poland; e.sudol@itb.pl (E.S.); e.kozikowska@itb.pl (E.K.)
[2] Faculty of Materials Science and Engineering, Warsaw University of Technology, 02-507 Warsaw, Poland; emilia.choinska@pw.edu.pl
* Correspondence: l.ladosz@itb.pl; Tel.: +48-22-57-96-138

Abstract: Corrosion protection technologies based on waterborne paints have become increasingly popular as steel structure protection, which implies the need to determine relevant assessment methods considering the conditions of use and product-specific characteristics. This study attempts to evaluate the fitness of standard corrosion protection weathering methods and an original cyclic test for verifying the resistance of waterborne acrylic coatings to environmental conditions. Changes to the properties of artificially weathered coatings were analysed with reference to those observed during exposure in natural conditions. The degree of coating degradation after exposure to neutral salt spray and condensation humidity was determined to significantly exceed the changes observed in natural conditions. An original cyclic test caused changes in the appearance, microstructure, FT-IR spectrum and utility properties of the coatings, such as thickness, colour, hardness, adhesion and impedance, similar to those observed in the natural environment. The results confirm that the programming direction of waterborne coatings artificial weathering tests is adequate and promising.

Keywords: tests methods; weathering; artificial tests; eco-friendly; waterborne coatings; original cyclic test; change of properties

1. Introduction

Despite many advantages, structural steel is susceptible to corrosion in most environments. Corrosion is estimated to be responsible for losses amounting to 3–4% of the global gross domestic product [1,2]. Owing to adequate corrosion protection, the costs can be reduced by up to 35%, but, most importantly, the risk of construction disasters can be limited [3]. Steel corrosion protection involves applying organic coatings as well as using inhibitors or electrochemical protection [4]. Organic coatings are the most popular. This study focuses on eco-friendly waterborne organic coatings. They are not as popular as solvent-based coatings in construction but have become increasingly widespread due to limitations in the use of pigmented paints, e.g., containing lead tetroxide or chromium compounds with oxidation state +6 [3,5].

The main advantage of waterborne coatings compared to solvent-based ones is an up to 95% lower content of volatile organic compounds (VOCs) and lower price [3]. Still, they are less prevalent than solvent-based coatings because of their higher application regime. The surface needs to be better prepared. Nevertheless, the prognosis is promising, as through the polymerization of monomers in aggregated form, improvements in wettability have been achieved [6]. Environmental conditions should be observed more strictly during application as well. They are less efficient, have limited coating thickness, and have much longer drying times [3,4].

The results of previous studies revealed that waterborne coatings are less durable than solvent-based ones [7–9]. Nevertheless, eco-friendly waterborne systems have developed dynamically in striving for sustainable development. Waterborne and durable solid slippery and superhydrophobic coatings, resistant to abrasion, have been developed [10]. Coating manufacturers aim to achieve solutions with utility properties that are at least on par with those of solvent-based coatings [5,11]. The changes in the corrosion protection sector, resulting in innovative waterborne paints, entail improving research methods to consider the products' specificity [12].

An analysis of the literature data revealed that the currently used accelerated weathering methods, indispensable to verify the durability of corrosion protection systems, were designed for standard solvent-based paints. For different compositions of organic solvent-based and waterborne products, the previous accelerated weathering methods may be inadequate. Weathering tests are typically based on intensive moisture impact, often, to an extent, exceeding natural impacts [13,14], which may lead to misevaluation, considering the sensitivity of waterborne paints to water.

In natural conditions, coatings are exposed to many environmental factors of varied individual intensity. The main degradation factors for polymer coatings include temperature, humidity, solar radiation, wind, atmospheric precipitation, and air pollution (SO_2, NOx, and O_3) [8,15–20]. Simulating the impacts in laboratory conditions is possible but with numerous technical limitations for their combination. This might result in a lack of correlation between accelerated test results and natural effects [7]. Davalos-Monterio et al. [21] demonstrated, by examining powder coating corrosion due to scratching, a lack of correlation between natural and artificial weathering for the referenced feature. Some of the coating systems they tested, following four years of exposure in natural conditions, were characterised by significantly larger corrosion areas versus the samples subject to weathering for 4–6 months with the use of a cyclic test; for some systems, the corrosion area was larger after the cyclic test versus natural weathering. Looking for a correlation between the corrosion of incised coatings subject to accelerated and natural weathering in marine conditions, Krystel Pélissier et al. [22] demonstrated that in order to observe correlations between the impacts of natural and accelerated weathering, longer (even 5–7 years) exposure in natural conditions is required. Ramdé et al. [23] demonstrated that air humidity and temperature significantly contribute to the colour change in acrylic binder-based waterborne paint samples. Scrinzi et al. [19] investigated the colour change in samples subject to natural and accelerated weathering (cycle: UV test + NSS test), demonstrating quite a good correlation for the parameter. Scientists from Laval University [16] observed that the colour change after tests in weathering chambers featured with fluorescent lamps does not correspond to the actual colour change in natural conditions. Even though UV light is the main contributor to the colour change, other factors, such as exposure angle, additional snowfall, hail and other environmental factors, can considerably affect the parameter. Valverde and Moya presented similar conclusions in their study [24], paying attention to the fact that it is hard to set the parameters of artificial weathering in UV chambers to simulate natural weathering in a specific geographical area. Solar radiation, temperature and air humidity are highly volatile in natural conditions, whereas we can strictly control the parameters in weathering chambers. Juan and Roger [24] also indicated that, in addition to the variability in climate conditions, there are other natural factors, e.g., living organisms, which accelerate coating degradation, including colour change.

Considering the above, planning laboratory tests concurrently simulating many environmental factors seem necessary. Cyclic tests are an example of such a solution, and their execution can help researchers simulate the durability and corrosion resistance of paint systems considering the target environment's conditions of use. The task poses a challenge due to the diversity and variability in environmental conditions, even in the same country. Further research is indispensable to improve accelerated weathering methods, including but not limited to new groups of corrosion protection coatings and the correlation of accelerated and natural weathering [25]. The literature presents many results of studies demonstrating a correlation between natural and artificial degradation for specific coating parameters, e.g., electrochemical impedance spectroscopy (EIS) and Fourier-transform infrared spectroscopy (FTIR). Still, some results contradict this thesis [7,17,19,22,25,26].

A major problem in obtaining a correlation between natural and artificial weathering is related to limited amounts of data. Even accelerated corrosion tests are time-consuming and costly, so they do not generate many samples, and, consequently, such methods as neural networks cannot be applied [2]. Scrinzi et al. [19] demonstrated that in order to understand coating degradation in natural conditions, it is essential to record environmental parameters—next to exposure time—such as solar radiation, atmospheric precipitation, temperature, humidity and pollution, since the parameters do not act individually but degrade organic coatings in synergy [27].

Methods for testing the resistance of waterborne coatings to environmental conditions have not been determined so far. That is why this study attempts to assess the fitness of standard accelerated weathering methods for corrosion protection in the new category of the coatings mentioned above, which has not been analysed so far. In addition, the unprecedented original cyclic test was developed that simulates the impact of a broad spectrum of environmental factors: UV radiation, condensation humidity, neutral salt spray, negative and elevated temperature, and sulphur dioxide. An analysis was performed of the changes in waterborne acrylic coating characteristics under the influence of artificial weathering carried out with different techniques. The changes observed in accelerated tests were those observed during exposure in natural conditions. Macroscopic and microscopic evaluations were carried out, and thermal properties and spectra were analysed with FTIR. Moreover, the utility properties of the coatings were evaluated for their thickness, colour, adhesion and impedance. A flowchart of works versus the whole experiment is summarised in Figure 1.

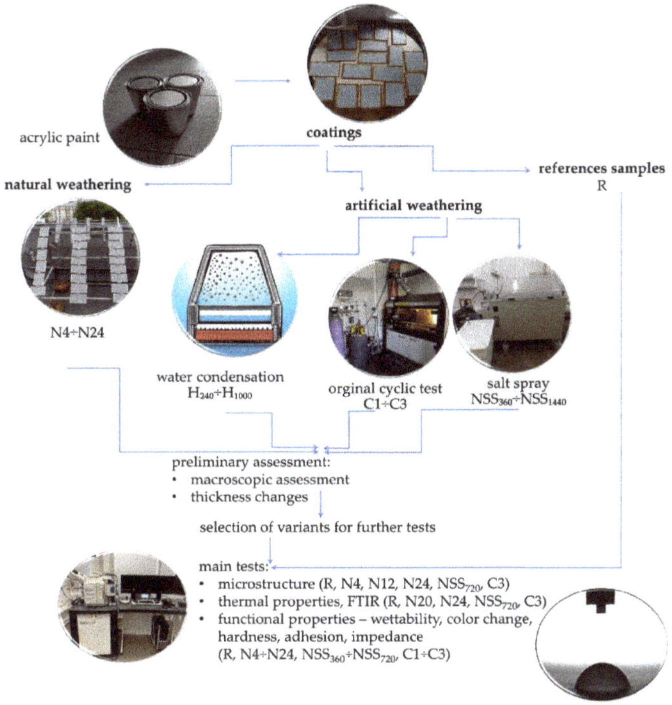

Figure 1. Experimental flowchart with photos.

2. Materials and Methods

2.1. Coatings

A coating system consisting of two layers of one-component acrylic waterborne paint, whose main components are presented in Table 1, was the test subject. Table 2 summarises the physical characteristics of the paint. The paint system used in the tests is intended for corrosion protection of steel constructions in traffic structures, e.g., bridges and flyovers.

Table 1. Composition of acrylic paint.

Ingredient Name	%
titanium dioxide	≥10–≤25
trizinc bis(orthophosphate)	≥5–≤10
2-butoxyethanol	≥1–≤3
zinc oxide	≤0.9
trimethylolpropane	≤0.3
ammonia	≤0.3

Table 2. Acrylic paint characteristics.

Density (g/cm^3)	Apparent Viscosity (Pa·s)	Content of Non-Volatile Ingredients (%)	Surface Drying Time (min)
1.308–1.339	0.69–0.71	61.2–61.8	30–35 (50 μm)

The paint was applied to S235JR structural steel panels, prepared to Sa 2 ½ grade according to ISO 8501-1 [28]. The application was performed for all samples under the same conditions using the airless spray method. It was executed by an experienced company specialising in structural steel corrosion protection services. The coatings were 200–210 μm thick.

One hundred 150 × 100 × 5 mm panels and one hundred 150 × 75 × 5 mm panels were prepared for the tests. The panel edges were protected with an extra coating layer and epoxy resin/paraffin to intensify corrosion protection of sensitive areas. A general view of selected panels is shown in Figure 2. On some panels meant for scratch corrosion evaluation tests, 100 mm long and 1 mm wide X-shaped scribes (incisions) were made. The coating was incised up to the steel substrate depth. The panels were conditioned under laboratory conditions (temp. 23 ± 2 °C, 50 ± 5% RH) for thirty days from paint application.

Figure 2. Test samples—steel panels with acrylic paint coating.

2.2. Exposure to Natural Weathering

A half of the panels were subject to natural weathering. This involved exposure on a test site (Figure 3a) in a C3 corrosion-category environment [29]. The test site was situated on the flat roof of the Building Research Institute, on the fourth floor, with geographical coordinates 52°13′ N 21°0′ E. The panels were fixed to frames (display stands) made of stainless steel and wood. The samples were positioned at 60° against the surface. A weather station Meteo LB-490, (LAB-EL, Reguły, Poland) was placed in the direct vicinity of the samples. The weather station was supposed to record environmental conditions, i.e., temperature, air relative humidity, rainfall, solar radiation intensity and NO_2 and SO_2 content. The data were recorded in ten-minute intervals. The exposure on the test site and registration of environmental conditions were carried out for twenty-four months, from June 2021 to June 2023. The weather station data are summarised in Figure 3b. All parameters except for precipitation are presented as a median. The median of the solar radiation intensity covered only the period with intensity values > 0. Atmospheric precipitation is presented as a total for the given period. The test series are marked as N4, N8... and N24, where the numerical component stands for natural weathering time in months, i.e., 4, 8... and 24 months, respectively.

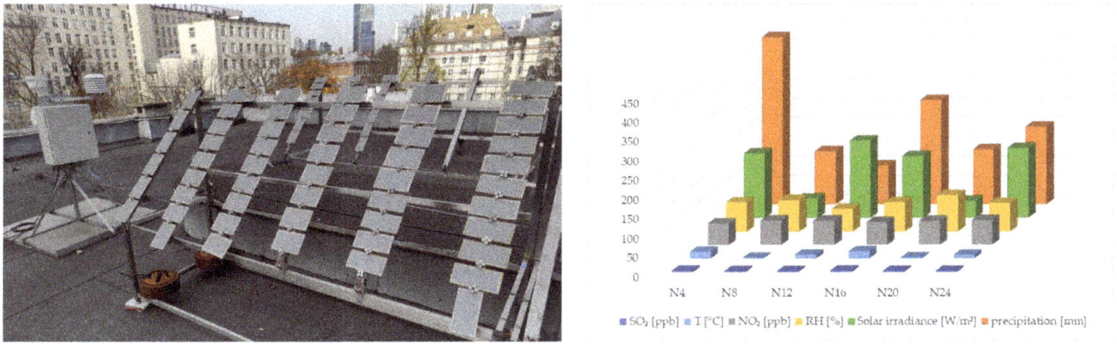

(**a**) (**b**)

Figure 3. Natural weathering: (**a**) test site with test panel frames and weather station, (**b**) environmental conditions recorded by the weather station. The data cover 24-month exposure and are presented either as a sum (precipitation) or median (other), recorded during 4 months (N4), 8 months (N8) and 24 months (N24).

At each exposure stage, macroscopic evaluation and thickness measurements were performed, and series were selected for performance testing. The microstructure was analysed after selected stages (N4, N12 and N24). A thermal analysis and FTIR were performed in the final exposure stage (N20 and N24). Non-destructive tests—thickness, colour and contact angle measurements—were carried out after each exposure stage for the same samples. The samples were cleaned with demineralised water and then conditioned for six hours under laboratory conditions. After the tests, the samples were placed back in the display stands. Samples for other tests were removed in groups, cleaned with water, conditioned under laboratory conditions for at least seven days and tested.

2.3. Exposure to Artificial Weathering

Accelerated (artificial) weathering was carried out simultaneously. The exposure is standard for solvent-based paints and concerns water condensation impact according to ISO 6270-1 [30] and neutral salt spray atmosphere according to ISO 9227 [31]. Moreover, an original cyclic test was programmed, considering a combination of a broad range of factors used in artificial weathering polymer-based coating products [17,20]. In all cases, the exposure was executed in stages, as summarised in Table 3.

Table 3. Artificial weathering exposure.

Exposure Type	Water Condensation (H)				Neutral Salt Spray (NSS)			Original Cyclic Test (C)		
Exposure time, h	240	480	720	1000	360	720	1440	360 (1 cycle)	720 (2 cycles)	1080 (3 cycles)
Series marking	H_{240}	H_{480}	H_{720}	H_{1000}	NSS_{360}	NSS_{720}	NSS_{1440}	C1	C2	C3

Exposure to condensation (H) was executed in a condensation chamber (Operations Technology Institute, Radom, Poland). It involved continuous exposure to water condensate at 38 ± 2 °C. The samples were exposed at 20° against the substrate level. The conductivity of the mineralised water used amounted to 10 µS/cm.

The NSS test was performed in the CCT 1000-TL-D chamber (VLM GmbH, Bielefeld, Germany). It involved creating a salt spray atmosphere in the chamber's working space using 5% NaCl. The samples were exposed at 70° against the substrate. The temperature in the chamber during the test amounted to 35 ± 2 °C.

The original cyclic test (C) considered a combination of a broad range of factors used in the artificial weathering of polymer-based coating products, i.e., UV radiation, temperature, humidity and neutral salt spray atmosphere [17,20]. The test was completed with the factor representing environmental pollution, which is rarely used in testing. The pollution was imitated by using a humid atmosphere containing SO_2, causing significant coating degradation [15,20] and occurring in natural conditions. The original cyclic test was composed of three stages, each consisting of 120 h of exposure to UV irradiation and water condensation according to ISO 16474-3 [32], 168 h of NSS test according to ISO 9227 [31], 24 h of exposure to negative temperature and two cycles of Kesternich test according to ISO 3231 [33]. Table 4 summarises the details of the original cyclic test. The exposure was first performed in the salt chamber mentioned above and then in the QUV spray chamber (Q-LAB, Westlake, OH, USA) at 60° against the substrate. Exposure to negative temperature was carried out in a climate chamber (Weisstechnik GmbH, Reiskirchen-Lindenstruth, Germany) by placing the samples in the chamber vertically. The exposure to a humid atmosphere containing SO_2 was performed in the CON 300 FL chamber (VLM GmbH, Bielefeld, Germany). The samples were positioned at 70° against the substrate.

Table 4. A course of a single cycle in the original cyclic test (C).

Days 1–5 (120 h) UV—ISO 16474-3	Days 6–12 (168 h) NSS Test—ISO 9227	Day 13 (24 h) Freezing	Days 14–15 (2 × 24 h) Kesternich Test
One cycle included: - 4 h irradiation by lamp UVA-340 (0.83 W/m², 60 ± 3 °C) - 4 h water condensation (50 ± 3 °C)	air temperature in the chamber: 35 ± 2 °C salt solution concentration: 5% NaCl (50 g NaCl/1 L distilled water)	temperature −20 ± 2 °C	One cycle included: - 1 L of sulphur dioxide - 8 h heating at 40 ± 3 °C, 100% RH - 16 h cooling with ventilation at 18–28 °C, <100% RH

Each exposure stage (H, NSS and C) was followed by macroscopic evaluation and thickness measurement. Moreover, some series were selected for performance tests. Considering the steel substrate's corrosion grade and almost complete degradation of the polymer coating, further testing of the NSS_{1440} series samples was abandoned. At the final stage of exposure in the salt spray atmosphere (NSS_{720}) and according to the original cyclic test (C3), the microstructure was analysed and thermal analysis and FTIR performed. Non-destructive tests—thickness, colour and contact angle measurements—were carried out after each exposure stage. The samples were cleaned with demineralised water and conditioned for 6 h under laboratory conditions. The samples for other tests were removed in groups, cleaned with water, conditioned under laboratory conditions for at least seven days and then tested. The course of the works is presented schematically against the whole experiment in Figure 1.

2.4. Macroscopic Evaluation

After each stage of natural and artificial weathering, the coating was evaluated macroscopically in daylight, with the naked eye or using a magnifying glass (magnification ×10), if necessary. The quantity and size of blistering, corrosion, flaking, cracking, chalking and corrosion from the scribe were determined. The results are expressed on a scale according to the series of ISO 4628 [34] standards, sheets from 2 to 6 and 8. In reference to blistering, cracking and chalking, a homogenous system of describing the changes as XS(Y) was used, where X stands for quantity, while Y represents the size of the changes. In both cases, a numerical scale from 0 to 5 was applied. Referring to the quantity, 0 means no changes; 1—very few changes; 2—a few changes; 3—a moderate number of changes; 4—a considerable number of changes; and 5—a dense pattern of damage. Referring to the size, 0 means changes not visible at magnification ×10; 1—visible at magnification ×10; 2—barely noticeable with the naked eye; 3—up to 0.5 mm; 4—from 0.5 mm to 5 mm; and 5—larger than 5 mm.

2.5. Thickness

The coating thickness was measured after each stage of weathering exposure, under both natural and artificial conditions. The Phascope PMP10 Duplex (Helmut Fischer, Sindelfingen, Germany) meter was used with the magnetic induction method. Six measurements in uniformly distributed areas, at least 10 mm from the edge, were performed for each sample.

2.6. Microstructure Analysis

The microstructure of the coating surface was investigated using a Sigma 500 VP scanning electron microscope with a cold-field emission (Carl Zeiss Microscopy GmbH, Köln, Germany). The microstructure surface of the coatings was examined at an accelerating voltage of 5 KeV excitation electron beam, using an SE detector, on samples sprayed with a coat of gold. The observations were carried out at a magnification of x5000. The microstructure of the coating surface was observed in the initial state (R) after exposure in natural conditions for four months (N4), twelve months (N12) and twenty-four months (N24). In addition, SEM observations were carried out on samples after artificial weathering using salt spray for 720 h (NSS_{720}) and after subjecting to the custom test covering three cycles with a total exposure of 1080 h (C3).

2.7. Thermal Properties

Thermal properties were tested with thermogravimetry (TGA) and differential scanning calorimetry (DSC). TGA measurements were performed with the Q5000IR apparatus (TA Instruments, New Castle, DE, USA). The samples were placed on platinum pans and heated up to 750 °C, at 10 °C/min heating rate and nitrogen flow rate amounting to 25 mL/min. DSC measurements used Q2000 apparatus (TA Instruments, New Castle, DE, USA). The samples were crimped in Tzero aluminium pans and tested according to a heat/cool/heat procedure in a temperature range between −50 °C and 180 °C, with a heating/cooling rate of 10 °C/min and nitrogen flow rate of 50 mL/min. The TGA and DSC results were processed and analysed with Universal Analysis software (version 4.5A, TA Instruments, New Castle, DE, USA). The samples used for the tests included non-weathered coatings (R), coatings weathered as a result of natural exposure for twenty and twenty-four months (N_{20} and N_{24}, respectively), artificially weathered under neutral salt spray for 720 h (NSS_{720}) and according to original accelerated weathering cyclic test—three cycles and total exposure time of 1080 h (C3).

2.8. Fourier-Transform Infrared Spectroscopy

The FTIR measurements were made with a Nicolet 8700 spectroscope (ThermoScientific, Madison, WI, USA) within an attenuated total reflectance (ATR) mode and diamond crystal. The measurements for all coatings were performed in three different zones. Sixty-four scans with a resolution of 4 cm^{-1} in the 400–4000 cm^{-1} range were collected and averaged. The spectra were recorded, processed and analysed with dedicated Omnic software (version 8.2.0.387, ThermoScientific, Madison, WI, USA). The samples used for the tests included non-weathered coatings (R), coatings weathered as a result of natural exposure for twenty and twenty-four months (N_{20} and N_{24}, respectively), artificially weathered under neutral salt spray for 720 h (NSS_{720}) and according to original accelerated weathering cyclic test—three cycles and total exposure time of 1080 h (C3).

2.9. Performance

After each stage of weathering exposure, both under natural and artificial conditions, parameters, such as adhesion, hardness, contact angle, colour change and impedance, were verified.

An adhesion test was performed with a DY-206 pull-off tester (Proceq, Schwerzenbach, Switzerland) according to ISO 4624 met. B [35]. Six measurements were performed for each series using 20 mm diameter dollies at a pull-off rate of 0.5 MPa/s.

Hardness was tested using a Buchholtz (TQC Sheen, Capelle aan den Ijssel, The Netherlands) hardness meter. Fifteen measurements were made for each series according to ISO 2815 [36].

The contact angle was measured with a Fibro PGX+ goniometer (TQC Sheen, Capelle aan den Ijssel, The Netherlands) by performing six measurements for each series in random areas. A 4 μL demineralised water droplet was used for the test.

The colour change was measured with SP62 spectrophotometer (X-Rite, Grand Rapids, MI, USA). D65 light was used as the light source, 10° calorimetric observer and d8° measurement geometry. Three measurements were performed for each sample in order to record the change in the a* b* colour parameters and L* psychometric brightness. The total colour change was determined as ΔE^*_{ab} according to Equation (1).

$$\Delta E^*_{ab} = \sqrt{(\Delta L)^2 + (\Delta a)^2 + (\Delta b)^2} \quad (1)$$

where: ΔE^*_{ab}—total colour change; ΔL—difference between L* before and after weathering; Δa—difference between a* before and after weathering; Δb—difference between b* before and after weathering.

Electrochemical impedance spectroscopy (EIS) was carried out with the ATLAS 0441 (ATLAS-SOLLICH, Rębiechowo, Poland) impedance analyser in a standard system with three electrodes. The working electrode was the steel sample with an applied coating in a measurement cell with a 12.57 cm^2 area. The reference electrode was made of Ag/AgCl, while a platinum wire with a mesh was the counter electrode. The measurement cells were filled with 3% NaCl solution in demineralised water. Thirty minutes after their filling with electrolyte, the coatings' resistance parameters were recorded. The frequency range in the test was between 10^5 and 0.1 Hz, and the amplitude value amounted to 100 mV.

All performance tests were carried out on the coating under laboratory conditions (23 ± 2 °C, 50 ± 5%RH).

3. Results and Discussion

3.1. Macroscopic Evaluation

The results of the macroscopic evaluation of coatings at different stages of both natural and artificial weathering are summarised in Table 5. For no test series, at no observation stage, degradation changes in the form of the coating flaking, cracking or chalking were observed, even at magnification ×10, corresponding to grade 0(S0).

Blistering was the dominant type of change. It shall be highlighted that they were not observed at any exposure stage in natural conditions. The N4–N24 series samples revealed no traces of blistering, even at magnification ×10, which corresponded to grade 0(S0). Blistering was visible during artificial weathering but occurred at different stages of selected exposures and with varied intensity. Blistering was not observed at the initial stage of water condensation exposure. A minimal number of changes, noticeable only at magnification ×10—grade 1S(1)—was observed only after 1000 h of exposure (H_{1000}). Nevertheless, it shall be emphasised that coating swelling was observed right from the first exposure stage (as presented in Figure 4a and discussed in detail in Section 3.2), making blistering evaluation difficult. The exposure to neutral salt spray caused coating blistering at the very first stage of exposure, whereby for the NSS_{360} series with no scribes, minor blistering was reported, visible only at magnification ×10—2S(1). In the case of incised samples, the damage pattern was dense but barely noticeable with the naked eye—5S(2). The longer the weathering time was, both for samples with and without a scribe, the blistering intensified, respectively, up to grade 3(S2), moderate quantity of blistering barely noticeable with the naked eye, and 5(S3), dense blistering pattern up to 0.5 mm in size. The coating behaved completely differently during exposure in the original cyclic test. Non-incised samples had no traces of blistering throughout the entire exposure period, similar to the C1 series with scribes. A significant quantity of blistering was reported for the incised samples after C2 and C3, barely noticeable with the naked eye, 4(S2). The samples with and without scribes are presented in Figure 5a,b.

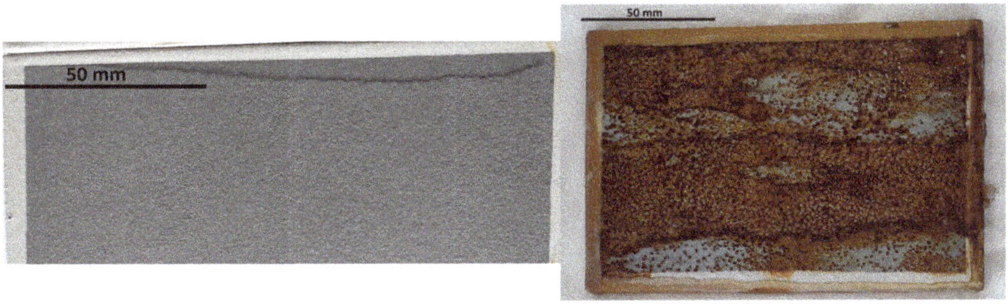

(a) (b)

Figure 4. General view of the artificially weathered sample: (**a**) coating swelling after 1000 h of water condensation exposure, (**b**) corrosion after 1440 h exposure to neutral salt spray atmosphere (NSS_{1440}).

(a) (b)

Figure 5. General view of the artificially weathered samples after 3 cycles of original test exposure: (**a**) coating without scribes, (**b**) coating with 1 mm scribes.

Table 5. Results of macroscopic evaluation of the coating after weathering.

Series Designation	Blistering ISO 4628-2	Rusting ISO 4628-3	Cracking ISO 4628-4	Flaking ISO 4628-5	Chalking ISO 4628-6	Corrosion around a Scribe, mm ISO 4628-8
N4, N8, N12, N16, N20, N24	0S(0)	Ri0	0S(0)	0S(0)	0S(0)	0
H_{240}, H_{480}, H_{7200}	0S(0)	Ri0	0S(0)	0S(0)	0S(0)	not evaluated
H_{1000}	1S(1)	Ri0	0S(0)	0S(0)	0S(0)	not evaluated
NSS_{360}	2(S1)/5(S2)	Ri3	0S(0)	0S(0)	0S(0)	0.5
NSS_{720}	3(S2)	Ri4	0S(0)	0S(0)	0S(0)	0.5
NSS_{1440}	5(S3)	Ri5	0S(0)	0S(0)	0S(0)	2.0
C1	0S(0)	Ri0	0S(0)	0S(0)	0S(0)	0
C2	0S(0)/4(S2)	Ri0	0S(0)	0S(0)	0S(0)	1.0
C3	0S(0)/4(S2)	Ri1	0S(0)	0S(0)	0S(0)	1.0

The results are expressed in degrees according to ISO 4628 [34] parts 2–6 and 8. The first component in the XS(Y) symbol, i.e., X stands for quantity, S means size and Y represents the size value (according to the description in item 2.4). If the same result was observed for all series in the group, it was presented once. The component after the slash ("/") presents the results for incised samples but only when the result for the incised samples was different than that for non-incised ones.

Rusting was also observed. The substrate corrosion area for naturally weathered samples (N4–N24 series) and after water condensation exposure (H_{240}–H_{1000}) equalled 0, Ri0 grade. When observing the samples subject to neutral salt spray, substrate corrosion was reported, which enlarged with longer exposure time. The corrosion area for the NSS_{360} series samples was 1%, Ri3 grade; for the NSS_{720} samples, it was 8%, Ri4, and 40–50%; for the NSS_{1440} samples, Ri5 grade (Figure 4b). For the samples subjected to the original cyclic test, substrate corrosion was observed only for the C3 series—Ri1 grade.

The width of corrosion around the scribe for the N4–N24 samples was 0 mm. Corrosion around the scribe was not evaluated for H_{240}–H_{1000} water condensation exposure. The corrosion around the scribe for the NSS_{360} and NSS_{720} samples was 0.5 mm, while for the NSS_{1440} samples, it amounted to 2 mm. After the cyclic test on the C1 series samples, no corrosion around the scribe was observed, while the corrosion of the C2 and C3 series samples was 1 mm wide.

The analysis of the coating macroscopic observations conducted at different weather stages, both natural and artificial, indicates that exposure to neutral salt spray was the most destructive for the acrylic coating. Blistering and rusting were observed already at the initial exposure stage (NSS240), and they intensified significantly during successive exposure stages. The degree of rusting observed in the final stage (NSS_{1440}) prevented further coating examination. It should be emphasised that such intensive changes, noticeable from the early exposure stage, were not observed for any other weathering. Significant similarity to the coating's behaviour in natural conditions (N4–N24) was, however, observed in the behaviour of the non-incised coating during the original cyclic test (C1–C3) and weathering with the use of condensation water (H_{240}–H_{1000}). In all cases, no degradation changes noticeable with the naked eye were observed. The scribe of the coating increased its susceptibility to degradation. A higher degree of changes was observed both for the series weathered with a neutral salt spray atmosphere (NSS_{360}) and according to the original cyclic test (C2, C3). It is most likely related to the electrolyte penetration running from the scribe deep into the sample, which promotes osmosis through the coating [37].

3.2. Thickness

An analysis of the coating thickness measurement results reveals that both natural and artificial weathering cause changes that aggravate with the increasing exposure duration (Figure 6). For the coating exposed in natural conditions (N4–N24), exposed to a neutral salt spray atmosphere ($NSSs_{360}$–NSS_{720}) and subjected to the original cyclic test (C1–C2), reduced thickness was observed, which matches previous study results [11,26,38]. The reduction did not exceed 10% of the original value, which amounted to ca. 200 μm. The decrease in the polymer coating thickness may be related to erosion caused by precipitation and wind-transferred mechanical pollution [39,40], additionally accelerated by the impact of UV radiation [41].

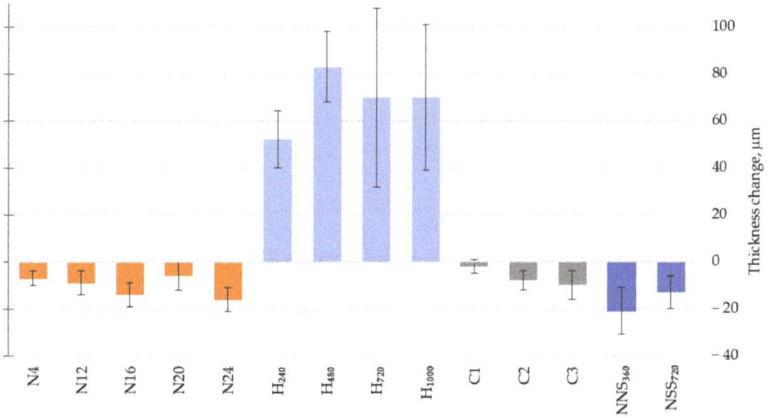

Figure 6. Change in the coating thickness after ▪ natural weathering for a period between 0 and 24 months, including intermediate evaluation after 4, 12, 16, 20 and 24 months (N4–N24); ▪ artificial weathering by water condensation for a period between 0 and 1000 h, including intermediate evaluation after 240 h, 480 h and 720 h (H_{240}–H_{1000}); ▪ after artificial weathering according to the original cyclic test between 1 (C1) and 3 (C3) cycles; ▪ artificial weathering in a neutral salt mist for a period between 0 and 720 h, including intermediate evaluation after 360 h (NSS_{360}–NSS_{720}); the error bars represent standard deviation.

The coating thickness being higher for the N20 series than for N24 is most likely related to different weather conditions during the measurement. The highest relative humidity median amounting to 93% at a relatively low temperature of ca. 3 °C was observed during the N20 series measurements (Figure 3b). A significant thickness decrease was reported for the N24 series samples, corresponding to the trend observed throughout the exposure period of 0–24 months. It should be highlighted that during the N24 measurements, the median of air relative humidity was ca. 70% RH, and the air temperature was ca. 10 °C (Figure 3b). Considering that waterborne paint systems tend to absorb water [13,14], which might cause their blistering [37], a higher thickness after exposure to humid conditions seems justified. An acrylic coating's susceptibility to swelling due to a long-term water impact was demonstrated by its thickness, higher by 50–80 µm after artificial weathering using condensation water (H_{240}–H_{1000}). A thickness increase of 25–40% against the initial state was observed in this test but not for any other exposures. Considering the significant difference against other exposures, especially in comparison with natural conditions, and simultaneously taking into account some technical constraints in performing tests on such a swollen coating, a decision was made to abandon further testing on the H_{240}–H_{1000} series. Based on the observations above, a weathering test involving long-term exposure to condensation humidity required a separate, in-depth analysis of its fitness for evaluating acrylic coatings. The impact of time and samples' seasoning conditions between the end of exposure and making the measurement is among the factors to be analysed. It seems that under adequate conditions, water evaporates from the coating, and the observed thickness increase after condensation tests can be temporary [13].

3.3. *Microstructure Analysis*

Waterborne coatings degrade during use and exposure to a corrosive environment, which is accompanied by changes in microstructure. The impact of factors characteristic of the conditions of use of corrosion protection coatings for steel structures includes mechanical loads, climate conditions (UV radiation, temperature fluctuations, water), the corrosive impact of the atmosphere and microorganisms contribute to the degradation of polymer coatings due to polymer chain breaking [8,34]. That is why this study evaluates the surface quality and verifies surface damage by analysing the coating's morphology in the initial state (R), after natural weathering (N4, N12, N24), weathered artificially in the salt spray atmosphere (NSS_{720}) and according to the original cyclic test (C3).

An analysis of the coating's surface microstructure in the initial state revealed a uniform and continuous coating (Figure 7a). The images show only minor contamination of the coating and particles of mineral fillers. Exposing the coating to natural environmental conditions for four months (N4) resulted in slight changes in the microstructure. The SEM revealed small pits with sizes around 1 μm (Figure 7b). The surface morphology analysis, carried out after twelve months of exposure (N12), demonstrated pits with sizes ranging from 5 μm to 10 μm and smaller pits in a size range between 1 μm and 2 μm (Figure 8a). The presence of these pits clearly shows polymer degradation [42,43]. Moreover, significant contamination of the material surface can be observed in the images. The change in the coating's microstructure under an environmental impact can harm the performance of the analysed materials [40,43].

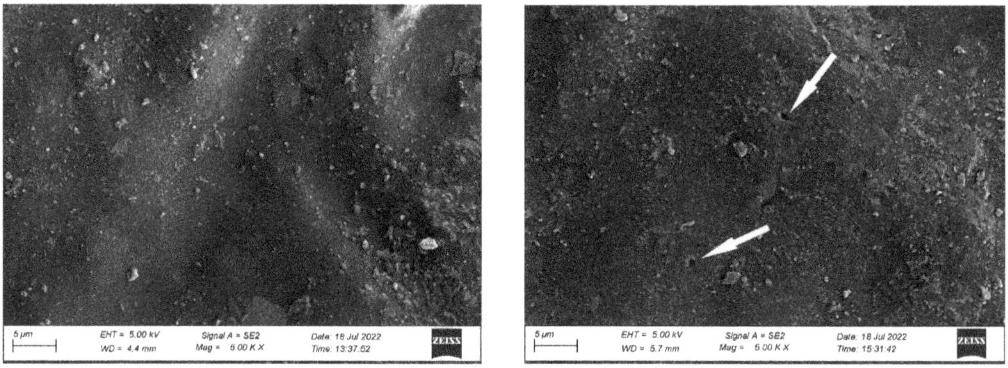

(a) (b)

Figure 7. Microstructure of the coating's surface: (**a**) initial state (R), (**b**) after natural weathering for 4 months (N4); pitting is marked by an arrow; magnification ×5000.

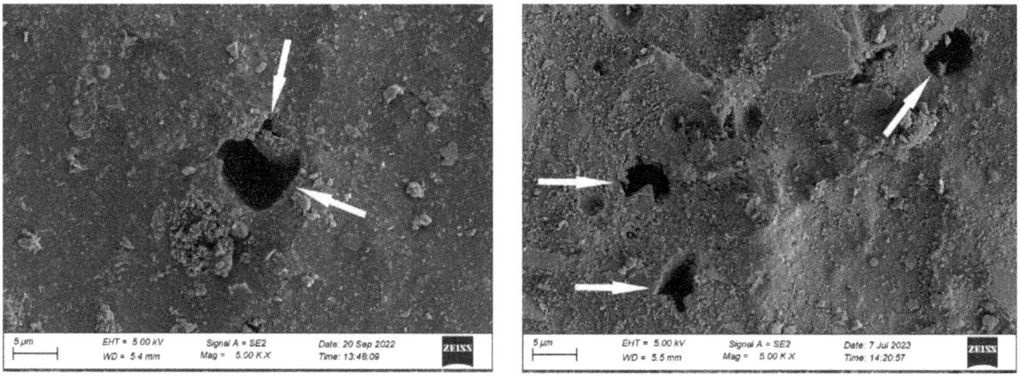

(a) (b)

Figure 8. Microstructure of the coating's surface after natural weathering for (**a**) 12 months (N12), (**b**) 24 months (N24); pitting is marked by an arrow; magnification ×5000.

Further exposure to natural conditions for twenty-four months (N24) did not intensify the coating's surface degradation. The images show pits with sizes ranging from 1 μm to 6 μm (Figure 8b), i.e., smaller ones than after twelve months of exposure. Smaller sizes of the coating defects can result from the difference in the atmospheric conditions during sampling. For the samples weathered on the test site, the microstructure change can be attributed to the impact of UV radiation, erosion due to rain and hail, and wind-transferred mechanical pollution [40]. Microscopic observations confirm the results of measurements of utility properties presented further in this study and show a correlation between surface degradation and material properties. The samples subjected to natural weathering reveal degradation changes such as contamination anchored in the material and the presence of voids and larger pits.

In the next stage of work, the coating's microstructure after exposure to natural conditions was compared with the microstructure after artificial weathering, including exposure to neutral salt spray (NSS_{720}) and according to the original cyclic test (C3). The results of microscopic observations reveal a significantly higher degradation rate of the coating after exposure to salt spray. The images show voids with sizes ranging from 15 μm to 20 μm (Figure 9a). The NSS_{720} series coating degradation was more significant than after the cyclic test, where the pit sizes ranged from 1 μm to 5 μm (Figure 9b).

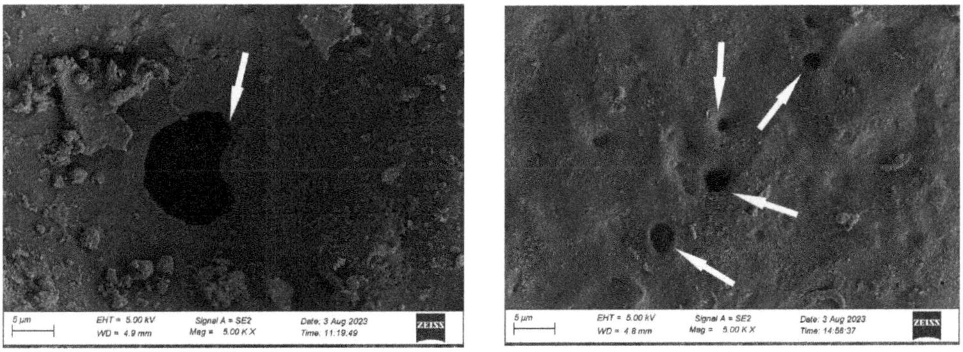

(**a**) (**b**)

Figure 9. Microstructure of the coating's surface after (**a**) exposure to neutral salt spray (NSS_{720}), (**b**) original cyclic test (C3); pitting is marked by an arrow; magnification ×5000.

The coating's microstructure changes after artificial weathering according to the original cyclic test C3 (Figure 9b) were similar to those reported for the N_{12} series samples naturally weathered for twelve months (Figure 8a) and N_{24} series samples weathered for twenty-four months (Figure 8b).

3.4. Thermal Characteristics

The TGA measurements were made to evaluate the potential changes both in the examined coatings' thermal stability and the inorganic filler content. Macroscopic evaluations revealed that changes may have occurred in the materials, which was suggested by the presence of pores on their surface. The results of the analyses are summarised in Figure 10 and Table 6.

Three stages of mass loss were observed for all coatings. The first stage, progressing most intensively for temperatures of ca. 100 °C, is related to the dehydration of the coatings and the potential evaporation of additives with low thermal stability, e.g., 2-butoxyethanol (cf. Table 1). The mass loss corresponding to the process was minor, within a 1.6–1.9% range. Then, a further slow mass loss was observed, related to the formation of anhydrides from carboxylic acids in the polymer or the loss of hydrogen atoms from the main chain [44]. The second significant decomposition stage occurs within a temperature range of ca. 350–450 °C and is related to the primary degradation of the acrylic matrix. The third decomposition stage, characterised by mass losses of 1.4–1.7%, occurring within a temperature range of ca. 525–600 °C, is the result of matrix residue carbonisation [44]. The total mass loss related to polymer degradation (marked as m_2 in the Table), i.e., covering anhydrite formation, hydrogen atom loss from the primary chain and main pyrolysis, was most significant for non-weathered samples (R) and alleviated for naturally weathered (N_{20} and N_{24}) and artificially weathered samples, both in the salt spray atmosphere (NSS_{720}) and according to the original cyclic test (C3). Consequently, the share of inorganic fillers changed, which was identified after pyrolysis carried out up to 750 °C. The observed reduction in the polymer fraction content suggests that the erosion rate of the polymer coating matrix is much higher than that of the inorganic filler.

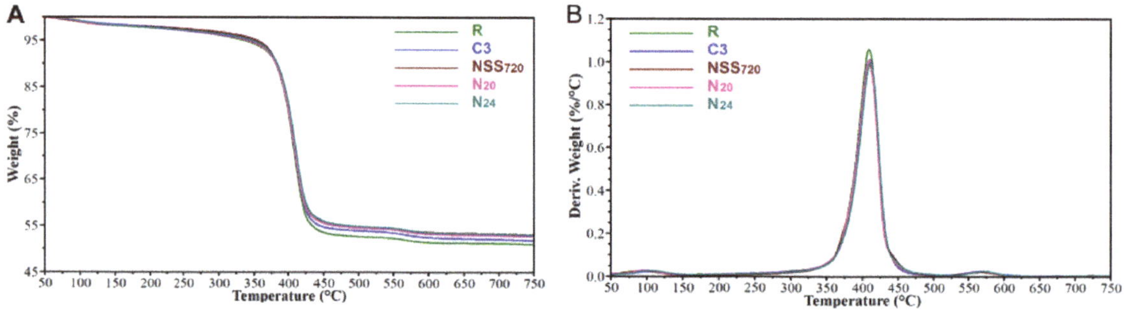

Figure 10. TGA (**A**) and DTGA (**B**) curves recorded for the tested coatings.

Table 6. Coating TGA results.

No.	Sample	m_1 (%)	m_2 (%)	$T_{d5\%}$ (°C)	T_{max} (°C)	Res_{750} (%)
1	R	1.9	45.5	337	409	51.0
2	N_{20}	1.8	43.7	341	409	52.7
3	N_{24}	1.7	43.4	347	411	53.0
4	NSS_{720}	1.7	43.4	354	409	53.0
5	C3	1.6	44.5	350	410	51.8

m_1, m_2—mass losses corresponding to coating dehydration and decomposition of the coating polymer matrix, respectively; $T_{d5\%}$—decomposition temperature corresponding to 5% mass loss; T_{max}—temperature corresponding to the fastest decomposition kinetics, determined based on the derivative signal (dW/dT(%/°C)); Res_{750}—pyrolysis residues determined for 750 °C.

It can be noticed that as a result of long-lasting natural and artificial weathering, the thermal stability of the analysed coatings did not change significantly from an application point of view. Some discrepancies in the identified $T_{d5\%}$ values can be attributed to different water contents in the coatings.

The results of the DSC analysis are summarised in Table 7 and Figure 11. For all analysed coating states, in the first heating cycle, a transition from a vitreous to a viscoelastic state (vitrification) was observed, accompanied by a broad endothermic peak of a more or less noticeable but always bimodal nature. Since it occurs only in the first cycle, it is most likely caused by water evaporation and also by evaporation or melting of the additives (e.g., 2-butoxyethanol). The vitrification temperature was observed to increase slightly due to coating weathering. This suggests that a coating-softening ingredient is washed out. The identified temperatures were slightly lower in the first heating cycle, which can be explained by the presence of water and/or other additives acting like plasticisers. The effect observed for the non-weathered sample was the opposite, i.e., T_g was slightly lower in the second cycle. From an application point of view, the results do not reveal any significant differences between the materials.

Table 7. Summary of the determined glass transition temperature, melting point * and melting enthalpy *.

Test Series	Tg [1] (°C)	Tg [2] (°C)	Tm [1] (°C)	ΔH [1] (J/g)
R	19.2	17.0	117.2	29.5
N_{20}	21.1	21.2	115.8	27.4
N_{24}	21.8	22.5	113.7	25.3
NSS_{720}	20.9	22.2	129.2	31.5
C3	23.3	23.6	137.1	20.5

* the transition is of an endothermal nature, but evaporation of water and/or 2-butoxyethanol is more likely than the melting of additive; indices [1] and [2] stand for the first and second heating cycle, respectively.

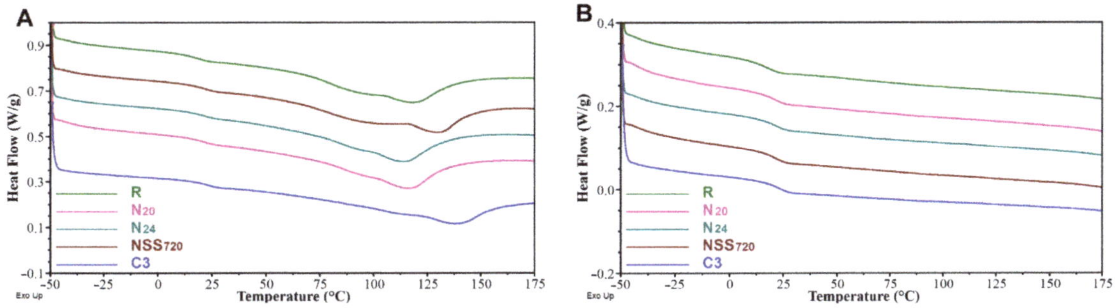

Figure 11. DSC curves recorded for 1st (**A**) and 2nd (**B**) heating cycles of non-weathered and artificially weathered coatings.

3.5. FTIR

The coatings were also tested using Fourier-transform infrared spectroscopy (FTIR). The spectra of inorganic pyrolysis residues were additionally recorded (Figure 12). Titanium dioxide should be the primary inorganic filler (Table 1). Nonetheless, the recorded spectrum is not typical of the referenced compound [45,46] talc spectrum [47,48]. The most intensive band with a peak at 1004 cm^{-1}, similar to the peak at 666 cm^{-1}, originates from the Si-O stretching vibrations, whereas the band at 3675 cm^{-1} originates from stretching of O-H surrounded by the MgMgMg triplet [49,50].

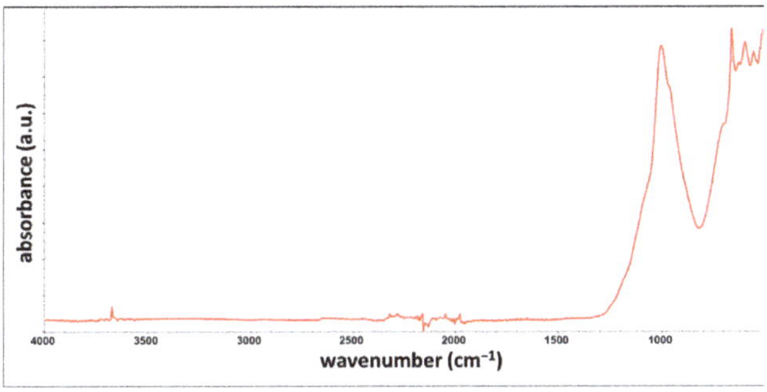

Figure 12. Spectrum recorded for residues of pyrolysed non-weathered coating.

The spectra recorded for the non-weathered coating and coatings weathered in natural conditions (N20 and N24) and artificially weathered, in the salt spray atmosphere (NSS720) and according to the original cyclic test (C3) (Figure 13), did not reveal significant differences between samples of different states. The positions of the main characteristic bands (Table 8) did not change. Nonetheless, a thorough analysis showed that as a result of coating weathering, the intensity ratio of bands characteristic for the filler (~1011 cm^{-1} and 3675 cm^{-1}) to the intensity of the band characteristic for the polyacrylic matrix (i.e., 1725 cm^{-1}, 1452 cm^{-1}, 1158 cm^{-1}, and 698 cm^{-1}) rises. This is testimony for the reduced share of the polymer phase in the coating, which complies with the TGA results. Most importantly, for the naturally weathered coatings and those exposed to artificial weathering according to the original cyclic test (C3), broadening was observed at the peak base from the C=O stretching vibrations, as shown in Figure 13. The effect is caused by the formation of photodegradation products under UV radiation [51]. The effect was not observed for the coating weathered in a salt spray environment (NSS_{720}).

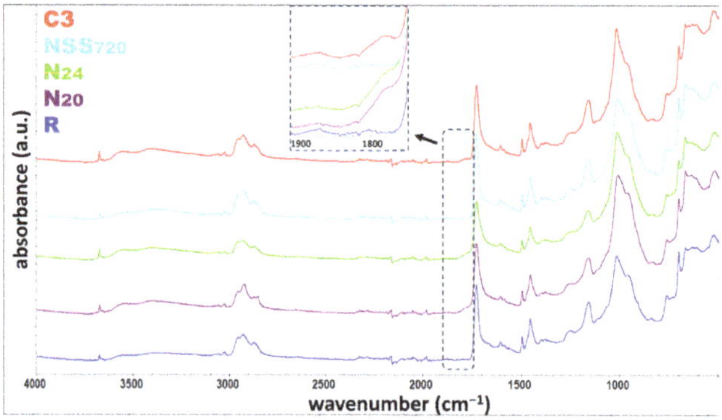

Figure 13. ATR-FTIR spectra recorded for non-weathered and artificially weathered coatings.

Table 8. ATR-FTIR absorption band positions.

Band/Material	R	N_{20}	N_{24}	NSS_{720}	C3
* O-H stretching	3676 cm^{-1}	3675 cm^{-1}	3676 cm^{-1}	3675 cm^{-1}	3676 cm^{-1}
C-H stretching	2956 cm^{-1} 2928 cm^{-1} 2873 cm^{-1}	2956 cm^{-1} 2924 cm^{-1} 2872 cm^{-1}	2957 cm^{-1} 2928 cm^{-1} 2873 cm^{-1}	2956 cm^{-1} 2928 cm^{-1} 2872 cm^{-1}	2956 cm^{-1} 2928 cm^{-1} 2872 cm^{-1}
C=O stretching	1726 cm^{-1}	1725 cm^{-1}	1725 cm^{-1}	1726 cm^{-1}	1725 cm^{-1}
C=C stretching	1603 cm^{-1} 1494 cm^{-1}	1603 cm^{-1} 1494 cm^{-1}	1603 cm^{-1} 1494 cm^{-1}	1603 cm^{-1} 1494 cm^{-1}	1603 cm^{-1} 1494 cm^{-1}
C-H bending	1453 cm^{-1} 698 cm^{-1}	1453 cm^{-1} 698 cm^{-1}	1452 cm^{-1} 698 cm^{-1}	1453 cm^{-1} 698 cm^{-1}	1453 cm^{-1} 698 cm^{-1}
C-O stretching	1158 cm^{-1}	1159 cm^{-1}	1158 cm^{-1}	1158 cm^{-1}	1158 cm^{-1}
C-H rock	840 cm^{-1}	831 cm^{-1}	830 cm^{-1}	840 cm^{-1}	836 cm^{-1}
* Si-O stretching	1014 cm^{-1} 665 cm^{-1}	1008 cm^{-1} 665 cm^{-1}	1011 cm^{-1} 666 cm^{-1}	1011 cm^{-1} 665 cm^{-1}	1015 cm^{-1} 666 cm^{-1}

* bands characteristic of inorganic filler.

Bearing in mind the material characteristics described above, the most relevant information on the possibility of representing natural weathering on the test site in laboratory tests is obtained from the FTIR analysis. It unequivocally reveals that applying a cyclic process triggers material changes identical to those in samples exposed to natural external conditions.

3.6. *Performance*

Natural weathering resulted in a contact angle change (Figure 14). Data analysis for the N4–N24 sample series shows a declining trend. The contact angle for N24 samples dropped by 11° compared to the reference sample R. A similar trend was observed for the C1–C3 series samples. The NSS_{720} samples were characterised by contact angles similar to those of N_{24} and C3. The decrease in the contact angle value in the function of time suggests an increase in the coating's hydrophilic nature due to weathering exposure. The coating's increasing wettability means its higher susceptibility to absorb moisture [3], consequently reducing its corrosion protection capability due to a larger contact area with corrosive media. The contact angle value slightly deviating from the general trend for the N12 series could be related to the atmospheric conditions during the measurement. The air relative humidity median reported for the measurement period was the lowest for the entire exposure duration, i.e., ~58% RH, whereas the solar radiation intensity was the highest ~200 W/m^2; moreover, the total precipitation in the referenced period was the lowest—~100 mm (Figure 3b). Analysing the contact angle values and changes in the coating thickness against the environmental conditions recorded by the weather station, one can observe the relationship between the measurement results and temporary conditions on the test site. The observations confirm the changing nature described in the literature, hampering test programming with accelerated methods [27].

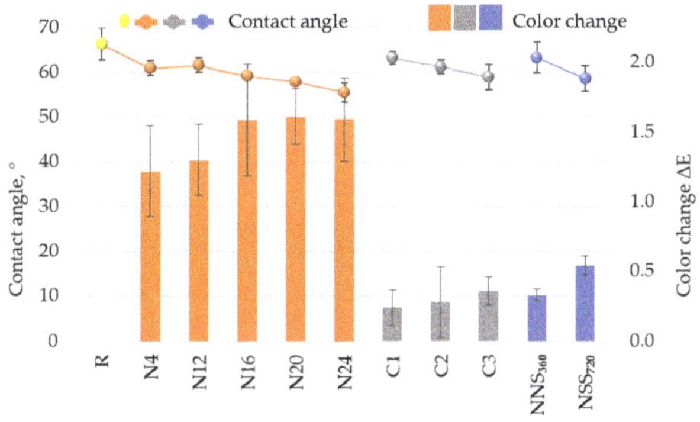

Figure 14. Change in the colour and contact angle ▪ reference samples, ▪ natural weathering for a period between 0 and 24 months, including intermediate evaluation after 4, 12, 16, 20 and 24 months (N4–N24), ▪ artificial weathering according to the original cyclic test—between 1 (C1) and 3 (C3) cycles, ▪ artificial weathering in a neutral salt mist for a period between 0 and 720 h, including intermediate evaluation after 360 h (NSS_{360}–NSS_{720}); the error bars represent standard deviation.

The coating colour change observed in this study was adversely correlated with the changes in the contact angle (Figure 14). For the N4–N24 series samples, the total colour change ΔE^*_{ab} amounted to 1.2–1.5, which is considered a very low value. A change of fewer than five units is unnoticeable to the human eye [19,26]. Under artificial conditions, a gradual colour change was observed with longer exposure times for the C1–C3 and NSS_{360}–NSS_{720} series samples, but it was over three-times lower than the change for the N4–N24 series samples. Considering the above, it can be stated that the set exposure under artificial conditions, despite taking into account the factors primarily responsible for the colour change in waterborne paint samples, i.e., UV radiation, elevated temperature and humidity [37,52], did not exert an impact similar to that during natural weathering, even in comparison with the first four months of exposure. Exposure to neutral salt spray, for which a correlation with natural weathering was reported [19] for the colour change, did not bring the expected convergence. The authors of other studies [16] indicated that the colour change after tests in weathering chambers featured with fluorescent lamps may not correspond to the actual colour change in natural conditions. Even though UV radiation is essential to colour changes, a different exposure angle, snowfall and other environmental factors can significantly contribute to the parameter change. The colour change measurement results collected during this study suggest a need to consider future modifications to the cyclic weathering test, e.g., by more prolonged exposure to UV radiation and humidity. Due to high fluctuations in solar radiation intensity in natural conditions (Figure 3b), accelerated weathering involving UV radiation is hard to simulate in a laboratory. The authors of [20] reached similar conclusions, pointing out that accelerated weathering parameters are hard to select in UV chambers to simulate natural weathering in a specific geographical area. In natural conditions, solar radiation, temperature and air humidity are highly variable, whereas the parameters can be strictly controlled in weathering chambers. It was also highlighted that in addition to variable climate conditions, other factors that accelerate colour degradation occur naturally, e.g., living organisms.

An analysis of hardness test results (Figure 15), expressing the coating's susceptibility to denting, reveals that exposure in natural conditions increases the hardness, which is illustrated by the results obtained for the N4–N24 series. For the C1–C3 series samples, a slight decrease in hardness was observed with every cycle. Still, it should be highlighted that the results obtained in the original test were similar to those reported for the N20 and N24 series. The coating hardness for the NSS_{360} and NSS_{720} series samples increased twice against the reference sample, reaching values considerably higher than those for N20–N24.

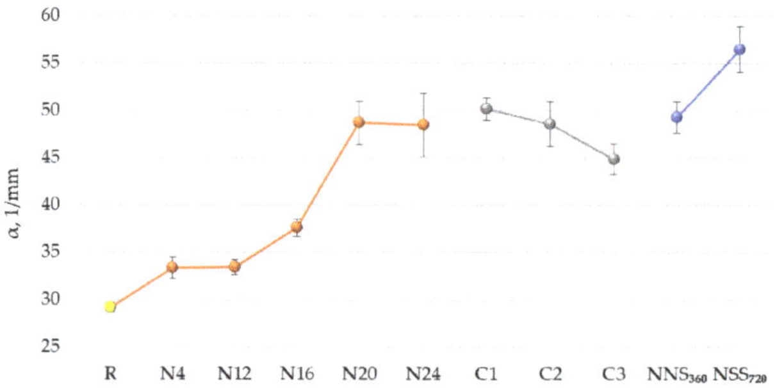

Figure 15. Coating hardness α ■ reference samples, ■ natural weathering for a period between 0 and 24 months, including intermediate evaluation after 4, 12, 16, 20 and 24 months (N4–N24), ■ artificial weathering according to the original cyclic test—between 1 (C1) and 3 (C3) cycles, ■ artificial weathering in a neutral salt mist for a period between 0 and 720 h, including intermediate evaluation after 360 h (NSS_{360}–NSS_{720}); the error bars represent standard deviation.

The hardness increases and decreases observed for the N4–24 sample series are most likely related to momentary atmospheric conditions on the test site when the samples were collected. A noticeable hardness drop observed during the cyclic test was caused by the impact of a low concentration of sulphurous acid formed in a humid atmosphere during the SO_2 test ($H_2O + SO_2 \rightarrow H_2SO_3$). The sulphurous acid concentration in the accelerated test was much higher than in natural conditions (Figure 3b) and contributed to accelerated coating oxidation, decreasing the coating hardness [53]. The above suggests a need to consider future modifications in the original cyclic test by reducing the SO_2 concentration.

The adhesion of coatings after natural weathering ranged from 5.9 MPa to 7.3 MPa (Figure 16). During exposure in natural conditions (N4–N24), no decrease in the bond stress was reported compared to the initial state (R) values. After artificial weathering, the bond stress for the NSS_{360} and NSS_{720} series samples ranged from 5.6 MPa to 6.9 MPa, while after natural weathering for the C1–C3 series, it was between 6.9 MPa and 7.5 MPa, i.e., at a level not lower than in the initial state.

The nature of the samples' failure in the adhesion test is particularly interesting. In the initial state (R), after exposure to natural conditions (N4–N24) and after accelerated weathering according to the original cycle (C1–C3), cohesive failure was observed, though it occurred in different layers (Figure 16). For the initial state, failure in the base coat (B) dominated (85%). After N4–N24 weathering, top-coat (C) failure prevailed (70–95%) but with a noticeable share of the base coat (B). After C1–C3, the failure share shifted significantly towards the top coat (C), reaching values over 90%. The breaking nature of the NSS_{360}–NSS_{720} samples differed significantly. Cohesive–adhesive failure was observed. Adhesive failure dominated for the NSS_{360} samples, 25% between the base coat and the steel substrate (A/B) and 50% between the base coat and top coat (B/C). The remaining 25% was cohesive failure within the top coat (C). After extending the exposure time for the NSS_{720} samples, a significant (35%) share was observed of the base coat separation from the substrate with steel corrosion exposure and a high (75%) share of the top-coat (C) failure.

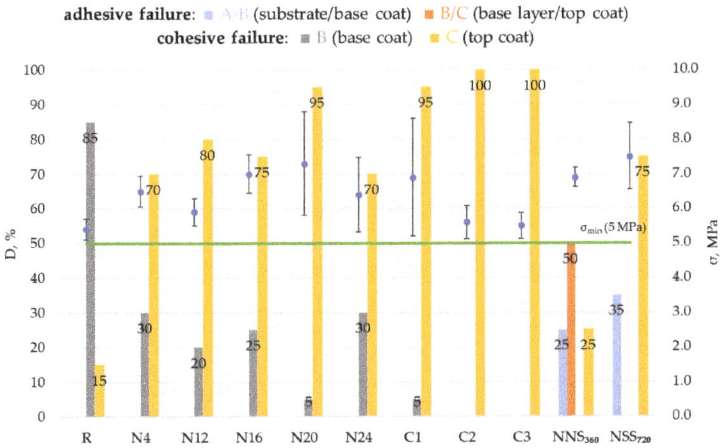

Figure 16. Results of adhesion tests expressed by the bond stress level (σ) and nature of failure (D) obtained for samples weathered naturally for a period between 0 and 24 months, including intermediate evaluation after 4, 12, 16, 20 and 24 months (N4–N24), artificially weathered according to the original cyclic test for 1 (C1) to 3 (C3) cycles, artificially weathered in neutral salt mist for a period between 0 and 720 h, including intermediate evaluation after 360 h (NSS_{360}–NSS_{720}); the error bars represent standard deviation.

By analysing the adhesion test results, one can observe the impact of environmental conditions on the periodic changes of this property. The lowest air humidity medians and the highest solar radiation intensity medians were recorded for the N12 and N24 periods (Figure 3b), which could contribute to tiny drops in the coating's adhesion to the substrate. The adhesion values obtained for the samples' exposure to weathering factors under artificial conditions (C1–C3, NSS_{360}–NSS_{720}) differed only slightly from the reference samples. It should be emphasised that the stress values for all tested variants exceeded the 5 MPa threshold considered satisfactory for steel structure corrosion protection coatings [4]. Moreover, analysing the nature of the coating failure after the adhesion test, it can be noticed that it was similar for N4–N24 natural weathering and C1–C2 accelerated weathering in the cyclic test. Cohesive failure (breaking) of the samples occurred in the base coat or top coat. Adhesive failure of the coating was, in turn, observed for the samples after a neutral salt spray test (NSS_{360}–NSS_{720}).

The coating's impedance parameter results are shown in Figure 17a,b as Bode plots [54]. The impedance modules and phase angle suggest that the impedance of the naturally weathered samples (N4–N24) remained at a similarly high level over 1×10^8 Ω cm^2, and the initial value of the phase angle was 80°. The coating's resistance to ion penetration is reached when the impedance module value at low frequencies exceeds 10^8 Ω cm^2 [7]. Only a minor impedance drop was observed at a low frequency ($|Z|_{0.01Hz}$) for the N4–N24 samples as weathering time progressed. In laboratory tests, the impedance module and phase angle values for the C1–C3 series samples were highly similar to those obtained for the N4–N24 series samples. The samples exposed to neutral salt spray were characterised by lower impedance values. For the NSS_{360} samples, at frequencies < 1 Hz, the curve flattening was observed at 10^7 Ω cm^{2,} as well as a significant decrease in the phase angle value up to 20°. The measurement results of the impedance module for the NSS_{720} samples revealed a loss in the coating's barrier function, even at high frequencies, which was demonstrated by the impedance module values of 1×10^4 Ω cm^2 and a violent drop in the phase angle.

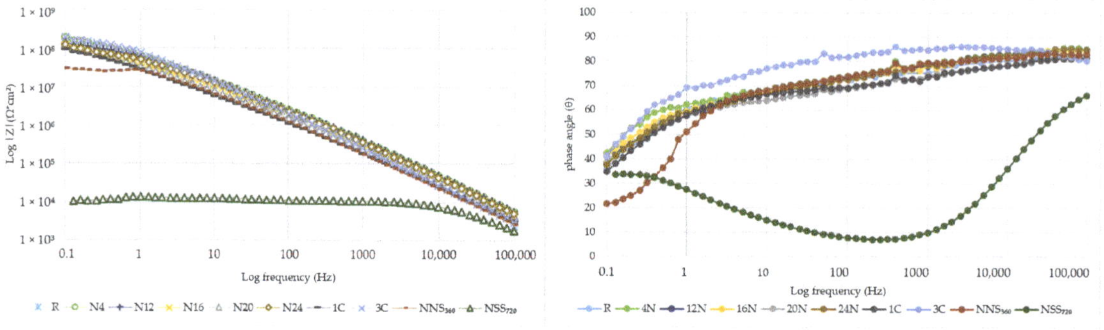

(**a**) (**b**)

Figure 17. Coating impedance parameters for naturally weathered samples for a period between 0 and 24 months, including intermediate evaluation after 4, 12, 16, 20 and 24 months (N4–N24), artificially weathered according to the original cyclic test—from 1 (C1) to 3 (C3) cycles, artificial weathering in a neutral salt mist for a period between 0 and 720 h, including intermediate evaluation after 360 h (NSS_{360}–NSS_{720}); Bode plot; impedance module log$|Z|$ (**a**), phase angle (**b**).

A slight decrease in the impedance module at 0.01 Hz for the N4–N24 naturally weathered samples is a testament to slow degradation changes occurring in the coating's barrier mechanism, e.g., as a result of electrolyte penetration and enlargement of the pores [55], which was observed in microstructure tests (Figure 8a,b). The impedance values $|Z|_{0.01Hz}$ correspond to the resistance generated by the coating due to electrolyte flow to the substrate; the higher the values, the better the coating's barrier effect is. Consequently, when the electrolyte penetrates towards the substrate, the value decreases [23]. The drop in the impedance module and phase angle for the NSS_{360} and NSS_{720} samples confirm macroscopic observations, where the first traces of the steel substrate corrosion occurred due to the presence of aggressive ions that accelerated corrosion; the coating's microstructure (Figure 8b) revealed pore sizes higher than for other series. Analysing the test results, it can be concluded that similarities occur in $|Z|_{0.01Hz}$ for the N4–N24 naturally weathered samples and after the cyclic test (C1–C3). Moreover, the EIS test results reveal that the NSS causes too severe a coating degradation. Similar conclusions were drawn by the researchers, who discovered that the salt spray test caused more considerable coating degradation than natural weathering [19]. A similarity was observed only for exposure in marine environments [17,56].

4. Conclusions

Analysis of the study results leads to the conclusion that the degradation of acrylic paint coatings after artificial weathering, performed using methods established for solvent-based coats, differs significantly from that observed in natural conditions. Exposing waterborne coatings to the long-lasting impact of salt spray and condensation humidity caused blistering, rusting and swelling, respectively, much more significant than those observed during exposure on the test site. The above challenges the fitness of the methods for evaluating waterborne paint coatings.

An alternative weathering method was developed in this study. An original cyclic test simulated the impact of a broad spectrum of environmental factors, including UV radiation, condensation humidity, neutral salt spray, negative and elevated temperature, and sulphur dioxide. For waterborne coatings, the test triggered changes similar to those observed in the natural environment. In none of the cases was degradation in the form of flaking, cracking, blistering or chalking confirmed through visual evaluation. Minor rusting was observed only around the scribe. Surface morphology analysis, performed with scanning electron microscopy, revealed the formation of up to 5 μm pits in the coating. At the same time, the FTIR demonstrated spectra convergence involving broadening at the peak base from C=O stretching vibrations. Moreover, the coating's colour change and hardness level were similar in both cases. An identical, cohesive failure was observed in the top coat in the adhesion test. Comparable values of the impedance module and phase angle were recorded. The extent of the coating change after the original cyclic test corresponds to two years of exposure on the test site, which was not observed after any other accelerated tests used in this study.

To summarize, the following can be stated:

- Accelerated weathering test methods established for solvent-based paints caused significantly different degradations of acrylic coatings than observed under natural conditions.
- An original cyclic weathering test for waterborne coatings was developed.
- After an original cyclic weathering test, degradations and changes were observed to be similar to that after two years of exposure on the test site for the following properties: surface morphology, colour, hardness, nature of failure and the impedance.

Considering the above, the programming direction of artificial weathering tests on waterborne coatings followed in this study can be considered adequate and prospective. There are plans to develop it in future studies. Modification of the original cyclic weathering by more prolonged exposure to UV radiation, humidity and reducing the SO_2 concentration is being considered. A broader range of waterborne paints intended for steel structure corrosion protection, longer exposure time in natural conditions and variability and seasonality of various climate factors will also be taken into consideration.

Author Contributions: Conceptualization, Ł.Ł. and E.S.; methodology, Ł.Ł., E.S., E.K. and E.C.; formal analysis, Ł.Ł., E.S., E.K. and E.C.; investigation, Ł.Ł., E.S., E.K. and E.C.; writing—original draft preparation, Ł.Ł., E.S., E.K. and E.C.; writing—review and editing, Ł.Ł. and E.S., visualization, Ł.Ł., E.S., E.K. and E.C. All authors have read and agreed to the published version of the manuscript.

Funding: This research was funded by the Ministry of Education and Science as part of the projects: NZM-063/2023 and NZK-080/2023.

Institutional Review Board Statement: Not applicable.

Informed Consent Statement: Not applicable.

Data Availability Statement: The data presented in this study are available on request from the corresponding author (due to privacy).

Conflicts of Interest: The authors declare no conflicts of interest. The funders had no role in the study's design; in the collection, analyses, or interpretation of data; in the writing of the manuscript; or in the decision to publish the results.

References

1. Nash, W.; Zheng, L.; Birbilis, N. Deep Learning Corrosion Detection with Confidence. *Npj Mater. Degrad.* **2022**, *6*, 26. [CrossRef]
2. Yin, B.; Considine, T.A.; Emdad, F.; Kelley, J.V.; Jensen, R.E.; Rundensteiner, E.A. Corrosion Assessment: Data Mining for Quantifying Associations between Indoor Accelerated and Outdoor Natural Tests. In Proceedings of the 2020 IEEE International Conference on Big Data (Big Data), Atlanta, GA, USA, 10–13 December 2020; pp. 2929–2936.
3. Wan Mohamad Kamaruzzaman, W.M.I.; Shaifudin, M.S.; Mohd Nasir, N.A.; Syafiq Mohd Hamidi, N.A.; Yusof, N.; Adnan, A.; Jew, L.O.; Norsani Wan Nik, W.M.; Mohd Ghazali, M.S. Eggshells Biowaste Filler for Improving the Anticorrosive Behaviour of Waterborne Polyurethane Coatings on Mild Steel in Artificial Seawater. *J. Mater. Res. Technol.* **2022**, *21*, 3815–3827. [CrossRef]
4. Stojanović, I.; Cindrić, I.; Turkalj, L.; Kurtela, M.; Rakela-Ristevski, D. Durability and Corrosion Properties of Waterborne Coating Systems on Mild Steel Dried under Atmospheric Conditions and by Infrared Radiation. *Materials* **2022**, *15*, 8001. [CrossRef] [PubMed]
5. Ecco, L.G.; Fedel, M.; Deflorian, F.; Becker, J.; Iversen, B.B.; Mamakhel, A. Waterborne Acrylic Paint System Based on Nanoceria for Corrosion Protection of Steel. *Prog. Org. Coat.* **2016**, *96*, 19–25. [CrossRef]

6. Dhar, M.; Mishra, C.; Das, A.; Manna, U. Polymerization of Monomer Aggregates for Tailoring and Patterning Water Wettability. *Chem. Commun.* **2024**, *60*, 444–447. [CrossRef] [PubMed]
7. Roselli, S.N.; Romagnoli, R.; Deyá, C. The Anti-Corrosion Performance of Water-Borne Paints in Long Term Tests. *Prog. Org. Coat.* **2017**, *109*, 172–178. [CrossRef]
8. Nguyen, T.V.; Le, X.H.; Dao, P.H.; Decker, C.; Nguyen-Tri, P. Stability of Acrylic Polyurethane Coatings under Accelerated Aging Tests and Natural Outdoor Exposure: The Critical Role of the Used Photo-Stabilizers. *Prog. Org. Coat.* **2018**, *124*, 137–146. [CrossRef]
9. Rodrigues Peruchi, A.B.; Zuchinali, F.F.; Bernardin, A.M. Development of a Water-Based Acrylic Paint with Resistance to Efflorescence and Test Method to Determine the Appearance of Stains. *J. Build. Eng.* **2021**, *35*, 102005. [CrossRef]
10. Kumar, S.; Dhar, M.; Prusty, B.M.; Sarkar, D.; Das, A.; Manna, D.; Manna, U. Amidation Reaction to Derive Waterborne, Tolerant, and Optically Transparent Solid Slippery and Superhydrophobic Coatings. *Chem. Eng. J.* **2023**, *465*, 142776. [CrossRef]
11. Nguyen, T.V.; Dao, P.H.; Nguyen, T.A.; Dang, V.H.; Ha, M.N.; Nguyen, T.T.T.; Vu, Q.T.; Nguyen, N.L.; Dang, T.C.; Nguyen-Tri, P.; et al. Photocatalytic Degradation and Heat Reflectance Recovery of Waterborne Acrylic Polymer/ZnO Nanocomposite Coating. *J. Appl. Polym. Sci.* **2020**, *137*, 49116. [CrossRef]
12. Zhou, Z.; Meng, Q.; Wang, J.; Ren, P.; Li, C.; Wang, Z.; Tan, H. Impact of Marine Atmospheric Corrosion on the Thermophysical Properties of Building Coatings. *J. Build. Eng.* **2024**, *85*, 108704. [CrossRef]
13. Meng, Y.; Gao, Y.; Li, J.; Liu, J.; Wang, X.; Yu, F.; Wang, T.; Gao, K.; Zhang, Z. Preparation and Characterization of Cross-Linked Waterborne Acrylic/PTFE Composite Coating with Good Hydrophobicity and Anticorrosion Properties. *Colloids Surf. A Physicochem. Eng. Asp.* **2022**, *653*, 129872. [CrossRef]
14. Liu, M.; Mao, X.; Zhu, H.; Lin, A.; Wang, D. Water and Corrosion Resistance of Epoxy-Acrylic-Amine Waterborne Coatings: Effects of Resin Molecular Weight, Polar Group and Hydrophobic Segment. *Corros. Sci.* **2013**, *75*, 106–113. [CrossRef]
15. Pagnin, L.; Calvini, R.; Wiesinger, R.; Schreiner, M. SO_{2-} and NO_{x-} Initiated Atmospheric Degradation of Polymeric Films: Morphological and Chemical Changes, Influence of Relative Humidity and Inorganic Pigments. *Microchem. J.* **2021**, *164*, 106087. [CrossRef]
16. Cogulet, A.; Blanchet, P.; Landry, V. Evaluation of the Impacts of Four Weathering Methods on Two Acrylic Paints: Showcasing Distinctions and Particularities. *Coatings* **2019**, *9*, 121. [CrossRef]
17. Deflorian, F.; Rossi, S.; Fedel, M. Organic Coatings Degradation: Comparison between Natural and Artificial Weathering. *Corros. Sci.* **2008**, *50*, 2360–2366. [CrossRef]
18. Fedel, M.; Rossi, S.; Deflorian, F. Comparison between Natural and Artificial Weathering of E-Coated Galvanized Steel Panels. *Prog. Org. Coat.* **2013**, *76*, 194–203. [CrossRef]
19. Scrinzi, E.; Rossi, S.; Orian, F.D. Influence of Natural and Artifi Cial Weathering on Aesthetic and Protective Properties of Organic Coatings. *Corros. Rev.* **2011**, *29*, 275–285. [CrossRef]
20. Jacques, L.F.E. Accelerated and Outdoor/Natural Exposure Testing of Coatings. *Prog. Polym. Sci.* **2000**, *25*, 1337–1362. [CrossRef]
21. Davalos-Monteiro, R.; D'Ambrosio, G.; Zhou, X.; Gibbon, S.; Curioni, M. Relationship between Natural Exposure Testing and Cyclic Corrosion Testing ISO 20340 for the Assessment of the Durability of Powder-Coated Steel. *Corros. Eng. Sci. Technol.* **2021**, *56*, 742–752. [CrossRef]
22. Pélissier, K.; Le Bozec, N.; Thierry, D.; Larché, N. Evaluation of the Long-Term Performance of Marine and Offshore Coatings System Exposed on a Traditional Stationary Site and an Operating Ship and Its Correlation to Accelerated Test. *Coatings* **2022**, *12*, 1758. [CrossRef]
23. Ramdé, T.; Ecco, L.G.; Rossi, S. Visual Appearance Durability as Function of Natural and Accelerated Ageing of Electrophoretic Styrene-Acrylic Coatings: Influence of Yellow Pigment Concentration. *Prog. Org. Coat.* **2017**, *103*, 23–32. [CrossRef]
24. Valverde, J.C.; Moya, R. Correlation and Modeling between Color Variation and Quality of the Surface between Accelerated and Natural Tropical Weathering in Acacia Mangium, Cedrela Odorata and Tectona Grandis Wood with Two Coating. *Color Res. Appl.* **2014**, *39*, 519–529. [CrossRef]
25. Cristoforetti, A.; Rossi, S.; Deflorian, F.; Fedel, M. Comparative Study between Natural and Artificial Weathering of Acrylic-Coated Steel, Aluminum, and Galvanized Steel. *Mater. Corros.* **2023**, *74*, 1429–1438. [CrossRef]
26. Shang, B.; Zhang, L.; Zhu, Y.; Wu, S.; Teng, J.; He, Q.; Su, Y. Corrosion Degradation of Two Coating Systems Exposed for Three Years in a Tropical Oceanic Atmospheric Environment. *Int. J. Electrochem. Sci.* **2021**, *16*, 151010. [CrossRef]
27. Deflorian, F.; Rossi, S.; Fedrizzi, L.; Zanella, C. Comparison of Organic Coating Accelerated Tests and Natural Weathering Considering Meteorological Data. *Prog. Org. Coat.* **2007**, *59*, 244–250. [CrossRef]
28. *ISO 8501-1;* Preparation of Steel Substrates before Application of Paints and Related Products Visual Assessment of Surface Cleanliness. Part 1: Rust Grades and Preparation Grades of Uncoated Steel Substrates and of Steel Substrates after Overall Removal of Previous Coatings. International Organization for Standardization ISO: Geneva, Switzerland, 2007.
29. *ISO 12944-2;* Paints and Varnishes Corrosion Protection of Steel Structures by Protective Paint Systems—Part 2: Classification of Environments. International Organization for Standardization ISO: Geneva, Switzerland, 2017.
30. *ISO 6270-1;* Paints and Varnishes Determination of Resistance to Humidity—Part 1: Condensation (Single-Sided Exposure). International Organization for Standardization ISO: Geneva, Switzerland, 2017.
31. *ISO 9227;* Corrosion Tests in Artificial Atmospheres Salt Spray Tests. International Organization for Standardization ISO: Geneva, Switzerland, 2017.

32. *ISO 16474-3*; Paints and Varnishes Methods of Exposure to Laboratory Light Sources—Part 3: Fluorescent UV Lamps. International Organization for Standardization ISO: Geneva, Switzerland, 2021.
33. *ISO 3231*; Paints and Varnishes Determination of Resistance to Humid Atmospheres Containing Sulfur Dioxide. International Organization for Standardization ISO:: Geneva, Switzerland, 1993.
34. *ISO 4628*; Paints and Varnishes Evaluation of Degradation of Coatings—Designation of Quantity and Size of Defects, and Intensity of Uniform Changes in Appearance. International Organization for Standardization: Geneva, Switzerland, 2016.
35. *ISO 4624*; Paints and Varnishes Pull-Off Test for Adhesion. International Organization for Standardization ISO: Geneva, Switzerland, 2016.
36. *ISO 2815*; Paints and Varnishes Buchholz Indentation Test. International Organization for Standardization ISO: Geneva, Switzerland, 2003.
37. Effendy, S.; Zhou, T.; Eichman, H.; Petr, M.; Bazant, M.Z. Blistering Failure of Elastic Coatings with Applications to Corrosion Resistance. *Soft Matter* **2021**, *17*, 9480–9498. [CrossRef]
38. Gao, J.; Li, C.; Lv, Z.; Wang, R.; Wu, D.; Li, X. Correlation between the Surface Aging of Acrylic Polyurethane Coatings and Environmental Factors. *Prog. Org. Coat.* **2019**, *133*, 98–105. [CrossRef]
39. Bauer, F.; Decker, U.; Naumov, S.; Riedel, C. Photoinitiator-Free UV Curing and Matting of Acrylate-Based Nanocomposite Coatings: Part 3. *Prog. Org. Coat.* **2014**, *77*, 1085–1094. [CrossRef]
40. Kotnarowska, D.; Stanisławek, D. Influence of Climatic Ageing on Acrylic Coatings Adhesion. *Autobusy* **2018**, *6*, 518–522. [CrossRef]
41. Hurley, C.J.; Zahavi, J.; Schmitt, G.F. UV Radiation Exposure Effects on the Rain Erosion Resistance of Coated Monolithic Poly(Carbonate) Transparencies. *Wear* **1983**, *87*, 9–28. [CrossRef]
42. Schulz, U.; Trubiroha, P.; Schernau, U.; Baumgart, H. The Effects of Acid Rain on the Appearance of Automotive Paint Systems Studied Outdoors and in a New Artificial Weathering Test. *Prog. Org. Coat.* **2000**, *40*, 151–165. [CrossRef]
43. Pan, T.; Yu, Q. Long-Term Anti-Corrosion Performance of a Conducting Polymer-Based Coating System for Steels. *J. Mater. Eng. Perform.* **2016**, *25*, 2384–2394. [CrossRef]
44. Fyfe, C.A.; McKinnon, M.S. Investigation of the Thermal Degradation of Poly(Acrylic Acid) and Poly(Methacrylic Acid) by High-Resolution Carbon-13 CP/MAS NMR Spectroscopy. *Macromolecules* **1986**, *19*, 1909–1912. [CrossRef]
45. Gohari, G.; Mohammadi, A.; Akbari, A.; Panahirad, S.; Dadpour, M.R.; Fotopoulos, V.; Kimura, S. Titanium Dioxide Nanoparticles (TiO2 NPs) Promote Growth and Ameliorate Salinity Stress Effects on Essential Oil Profile and Biochemical Attributes of Dracocephalum Moldavica. *Sci. Rep.* **2020**, *10*, 912. [CrossRef] [PubMed]
46. Guzmán-Aponte, L.; Mejia, R.; Maury-Ramirez, A. Metakaolin-Based Geopolymer with Added TiO 2 Particles: Physicomechanical Characteristics. *Coatings* **2017**, *7*, 233. [CrossRef]
47. Mai, Q.; Zhou, H.; Ou, L. Flotation Separation of Chalcopyrite and Talc Using Calcium Ions and Calcium Lignosulfonate as a Combined Depressant. *Metals* **2021**, *11*, 651. [CrossRef]
48. Shameli, K.; Ahmad, M.B.; Yunus, W.Z.W.; Ibrahim, N.A.; Darroudi, M. Synthesis and Characterization of Silver/Talc Nanocomposites Using the Wet Chemical Reduction Method. *Int. J. Nanomed.* **2010**, *5*, 743–751. [CrossRef] [PubMed]
49. Djangang, C.N.; Mbey, J.A.; Ekani, C.J.; Tiam, S.T.; Blanchart, P.; Njopwouo, D. Improved Microstructure and Free Efflorescence Geopolymer Binders. *SN Appl. Sci.* **2020**, *2*, 2167. [CrossRef]
50. Corona, J.C.; Jenkins, D.M.; Dyar, M.D. The Experimental Incorporation of Fe into Talc: A Study Using X-Ray Diffraction, Fourier Transform Infrared Spectroscopy, and Mössbauer Spectroscopy. *Contrib. Mineral. Petrol.* **2015**, *170*, 29. [CrossRef]
51. Pintus, V.; Wei, S.; Schreiner, M. Accelerated UV Ageing Studies of Acrylic, Alkyd, and Polyvinyl Acetate Paints: Influence of Inorganic Pigments. *Microchem. J.* **2016**, *124*, 949–961. [CrossRef]
52. Özgenç, Ö.; Durmaz, S.; Şahin, S.; Boyaci, İ.H. Evaluation of the Weathering Resistance of Waterborne Acrylic- and Alkyd-Based Coatings Containing HALS, UV Absorber, and Bark Extracts on Wood Surfaces. *J. Coat. Technol. Res.* **2020**, *17*, 461–475. [CrossRef]
53. Kotnarowska, D. Epoxy Coating Destruction as a Result of Sulphuric Acid Aqueous Solution Action. *Prog. Org. Coat.* **2010**, *67*, 324–328. [CrossRef]
54. Sun, W.; Xing, C.; Tang, X.; Zuo, Y.; Tang, Y.; Zhao, X. Comparative Study on the Degradation of a Zinc-Rich Epoxy Primer/Acrylic Polyurethane Coating in Different Simulated Atmospheric Solutions. *J. Coat. Technol. Res.* **2021**, *18*, 397–413. [CrossRef]
55. Zhang, H.; Dun, Y.; Tang, Y.; Zuo, Y.; Zhao, X. Correlation between Natural Exposure and Artificial Ageing Test for Typical Marine Coating Systems. *J. Appl. Polym. Sci.* **2016**, *133*. [CrossRef]
56. Shao, V.; Wang, X.; Xing, W.; Zhang, G. Correlation Analysis of Natural Environment Test and Laboratory Accelerated Test. In Proceedings of the 2017 Second International Conference on Reliability Systems Engineering (ICRSE), Beijing, China, 10–12 July 2017; ISBN 978 1 5386 0918 7.

materials

MDPI

Article

Mixed-Curve Model for Evaluating the Carbonation Depth of Concrete at Different Ages

Xinhao Wang [1,2], Qiuwei Yang [1,2,*], Hongfei Cao [1,2] and Fengjiang Qin [3,4,*]

1 School of Civil and Transportation Engineering, Ningbo University of Technology, Ningbo 315211, China; wangxinhao0709@163.com (X.W.); caohongfei@nbut.edu.cn (H.C.)
2 Engineering Research Center of Industrial Construction in Civil Engineering of Zhejiang, Ningbo University of Technology, Ningbo 315211, China
3 Key Laboratory of New Technology for Construction of Cities in Mountain Area, Ministry of Education, Chongqing University, Chongqing 400045, China
4 School of Civil Engineering, Chongqing University, Chongqing 400045, China
* Correspondence: yangqiuwei79@gmail.com (Q.Y.); qinfengjiang@cqu.edu.cn (F.Q.)

Abstract: To accurately quantify the variation in concrete carbonation depth, selecting an appropriate mathematical curve model is crucial. Currently prevalent models, such as the Fick model and exponential models, confront limitations in prediction accuracy and range of application. Given that a single curve model struggles to precisely describe the pattern of concrete carbonation, this work introduces a mixed-curve-based prediction model for carbonation depth, effectively integrating the Fick model with a hyperbolic model. Compared to the Fick model, the additional term in the mixed-curve model can be viewed as a reasonable correction to better adapt to the complex and varied conditions of concrete carbonation. This hybrid model transcends the limitations of individual models, enhancing fitting precision and broadening the scope of applicability. The new model boasts a concise structure with only two fitting parameters, facilitating ease of application. To validate its superiority, rigorous comparisons were conducted between the proposed model and existing ones, leveraging experimental data from 10 distinct concrete carbonation scenarios. By comparing the average error, standard deviation, and coefficient of determination across these cases, the new model demonstrates a clear advantage over the Fick model and the exponential model. In terms of fitting errors, the average error and standard deviation of the new model are notably lower than those of the other two models. In terms of the coefficient of determination, the values achieved by the new model in all examples are closer to 1 than those of both the Fick model and the exponential model, underscoring the new model's superior fitting quality and remarkable stability. This research indicates that the combined model presented in this paper holds promising prospects for widespread application in predicting concrete carbonation depth.

Keywords: concrete; elastic modulus; compressive strength; two-parameter curve model; three-parameter curve model

1. Introduction

Concrete carbonation occurs when atmospheric CO_2 permeates concrete pores and cracks, chemically reacting with alkaline components to decrease alkalinity. Key reactions are detailed in Equations (1)–(4) [1], with pore alterations depicted in Figure 1 [2]. Note that Equations (2)–(4) are verified only for high CO_2 concentrations (>5%). This process lowers pH near reinforcing steel, accelerating corrosion, damaging structures, and reducing their lifespan [3,4]. The assessment of concrete carbonation depth is of great significance in ensuring structural safety and estimating service life [5–12].

$$\mathrm{Ca(OH)_2} + \mathrm{CO_2} \overset{\mathrm{H_2O}}{\rightarrow} \mathrm{CaCO_3} + \mathrm{H_2O} \tag{1}$$

$$(3CaO{\cdot}SiO_2{\cdot}3H_2O) + 3CO_2 \rightarrow (3CaCO_3{\cdot}2SiO_2{\cdot}3H_2O) \quad (2)$$

$$3CaO{\cdot}SiO_2 + 3CO_2 + \gamma H_2O \rightarrow SiO_2{\cdot}\gamma H_2O + 3CaCO_3 \quad (3)$$

$$2CaO{\cdot}SiO_2 + 3CO_2 + \gamma H_2O \rightarrow SiO_2{\cdot}\gamma H_2O + 2CaCO_2 \quad (4)$$

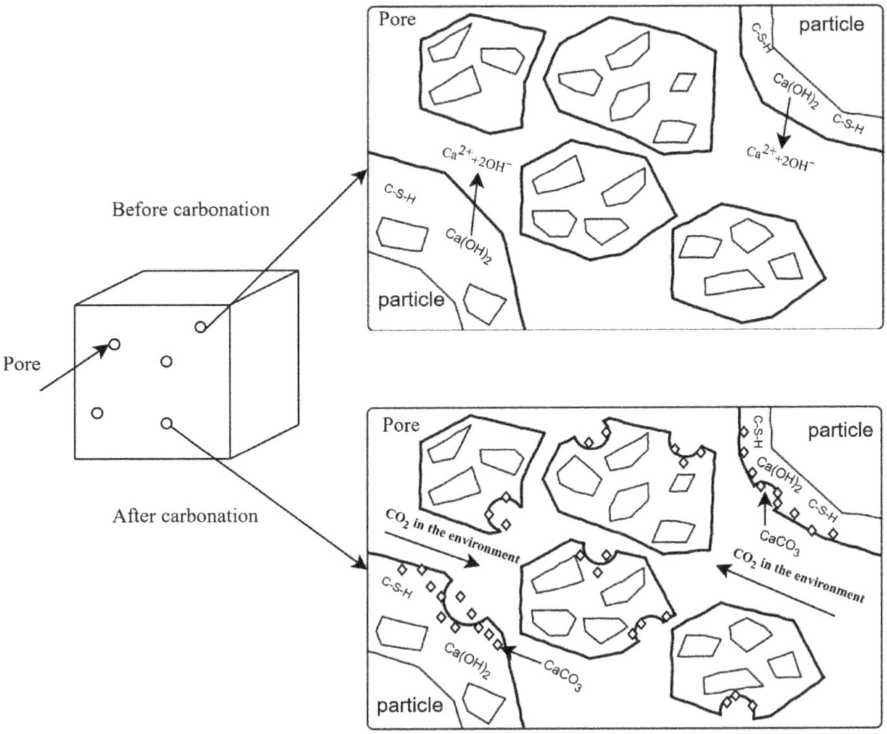

Figure 1. Carbonation reaction mechanism.

Currently, there are two main methods for evaluating concrete carbonation depth (CD): machine learning techniques and mathematical curve fitting. Among the machine learning approaches, various models are constructed, including artificial neural networks (ANN), support vector machines (SVM), and decision trees (DT), which are subsequently trained on extensive datasets to accurately evaluate the CD of concrete. This reliance on large data sets allows for more nuanced and precise assessments compared to traditional methods. Majlesi et al. [13] trained an ANN model for predicting the CD of reinforced concrete under different natural and environmental conditions. Tran [14] devised an ANN model to forecast CD, utilizing 300 experimental data points for validation. Felix et al. [2] employed an artificial neural network (ANN) coupled with a backpropagation algorithm for the purpose of predicting the CD of concrete that incorporates fly ash additives. Concha [15] created a new CD prediction model using ANN, analyzing 445 experimental data points for enhanced accuracy. Akpinar and Uwanuakwa [16] refined carbonation simulation and prediction accuracy with an ANN model utilizing a scale conjugate gradient (SCG) function. Zhang et al. [17] introduced a hybrid prediction framework merging the least squares support vector machine (LSSVM) with a metaheuristic algorithm to precisely forecast the CD of fly ash-blended concrete. Huang et al. [18] compared ANN, radial product function neural networks (RBF), random forests (RF), and SVM for CD prediction under varying conditions. Çevik et al. [19] reviewed SVM's applications in structural engineering, citing case studies. Li et al. [20] applied SVM to forecast concrete carbonation, discovering its predictive accuracy surpassed Back Propagation (BP) neural networks. Ruan [21] introduced the Support Vector Regression (SVR) to predict concrete CD, validated with real case data of CD. Londhe et al. [22] introduced Model Tree (MT), RF, and Multi Gene Genetic Programming (MGGP) for predicting concrete carbonation coefficients. Wang et al. [23] comprehensively evaluated eight ML algorithms (DT, RF, AdaBoost, CatBoost, XGBoost, K-Nearest Neighbors (KNN), SVR, and Multilayer Perceptron (MLP)) for predicting CD in concrete with solid waste. Amirhossein et al. [24] crafted artificial bee colony expression programming (ABCEP) models, rigorously analyzed, and compared the best ones with previous models. Lee et al. [25] utilized a state-of-the-art deep learning model to forecast concrete carbonation, outperforming the Architectural Institute Japanese (AIJ) Model and the Finite Element Method (FEM) analysis. Kumar et al. [26] introduced an adaptive neural fuzzy inference system (ANFIS) for estimating CD in fly ash concrete.

The model evaluated by combining mathematical curves mainly considers the influence of concrete composition on the CD of concrete. The fundamental concept behind this approach involves establishing the functional relationship between the mechanical properties of concrete and the various influencing factors based on initial experimental data. This information is then leveraged to apply data-fitting techniques, allowing for the calculation of concrete's mechanical properties under a range of different influencing conditions. Through this process, a clearer understanding of how varying factors affect performance can be achieved, leading to more accurate predictions and analyses. At present, the main models include the evaluation model based on the water–cement ratio [27,28], the evaluation model based on concrete compressive strength [29,30], and the evaluation model considering multiple parameters [31,32]. The original model was based on Fick's diffusion law to predict the depth of carbonation, and the original model was called the Fick model [33]. Other scholars have conducted in-depth research based on the Fick model and extended this model. For example, Possan et al. [34] proposed a mathematical model for estimating the CD of concrete and predicting the service life of concrete structures under carbon dioxide, which was validated using multiple sets of data. The results indicate that the model has the potential to predict the CD of concrete under boundary conditions that guide its development. Ekolu [35] proposed a mathematical model for predicting the effects of concrete composition and environmental factors on natural carbonation. Multiple datasets were used for model development and calibration, and the proposed model was compared with the fib model. The results indicate that the proposed model is relatively accurate. Liang and Lin [36] developed a new one-dimensional mathematical model. A one-dimensional linear partial differential equation was derived based on the principle of mass balance and convection dispersion equation, and an analytical solution was found through mathematical methods. Various parameters determine the numerical results of concrete carbonation. The numerical results indicate the practical application of the model. These results indicate that the proposed model can describe the carbonation process of concrete chemically and physically. The advantage of mathematical curve models lies in their ability to directly establish functional relationships between concrete CD and a multitude of influencing variables, such as the water-to-cement proportion, material temperature, pressure, and the chemical makeup of the constituents [2].

In general, while existing evaluation models are capable of predicting the mechanical properties of concrete to some degree, they exhibit certain limitations in specific areas. The drawbacks of machine learning-based evaluation models can primarily be categorized into two main issues: first, these models require a training phase before they can be effectively utilized for assessments; second, the complexity of the parameters considered by machine learning evaluation models can make their application in practical engineering challenging. This complexity can hinder their ease of use in real-world scenarios. The limitations of current mathematical curve models mainly lie in their limited applicability. For example, multi-parameter models must consider multiple factors and require a large amount of experimental data for fitting, which can lead to defects such as difficulty in obtaining variable data. To overcome the problems of insufficient early data and difficulty obtaining various variables in practical engineering, this study aims to develop a new dual-parameter curve model by integrating the advantages of existing curve models. In comparison to pre-existing curve formulations, the present model offers improved precision in assessing the interplay between the CD and the temporal evolution of concrete carbonation. Drawing upon the outcomes of this model's evaluations, a more nuanced comprehension of the CD within the component is achievable, thereby facilitating the anticipation and expeditious adoption of preventive actions. Therefore, this article first analyzes the advantages and disadvantages of the original Fick model and exponential model and then proposes a new digital curve model that overcomes the shortcomings of existing curve models. The improvement of the new model is mainly reflected in two aspects: firstly, the mathematical formula of the new model is concise, only considering the fitting of two parameters, which is convenient for practical engineering applications; second, the curve used in the new model has been improved based on the Fick model, resulting in higher fitting accuracy than existing models. This article mainly studies the application of a new model in the assessment of CD in concrete. This study engages in a comparative analysis utilizing ten sets of experimental data sourced from the established literature, aiming to evaluate the computational accuracy of a new model against existing ones. The objective is to highlight the advancements and improved performance of the new model over its predecessors.

2. Mathematical Models for Predicting the Carbonation Depth of Concrete

2.1. Analysis of Existing Models

The existing evaluation models that describe the relationship between the age of concrete and CD primarily consist of power function, exponential, and logarithmic models. The foundational power function model is the Fick model, and the corresponding curve equation is as follows [33]:

$$z(t) = a\sqrt{t} \tag{5}$$

Among them, $z(t)$ represents the quantified CD of the concrete at the specific time denoted by t; a denotes an unknown non-negative parameter that must be determined using experimental data. As indicated in Equation (5), a key advantage of the Fick model is that it requires fewer parameters for fitting, which facilitates its application in practical engineering scenarios. However, the Fick model also has its limitations. The primary drawback is that it incorporates only one parameter for fitting; thus, when dealing with experimental data that exhibit considerable variability, the fitting accuracy may be reduced, leading to substantial errors.

The curve equation of the exponential model is expressed as [37]:

$$z(t) = a + b \cdot e^{-t/c} \tag{6}$$

In Equation (6), a, b, and c are the three fitting parameters that need to be obtained through curve fitting based on experimental data. The advantage of this model is that the model function is a monotonic function, which is consistent with the increasing trend of concrete CD over time in practical engineering. And when $t \to 0$, $z(t)$ tends to converge. However, due to the need to fit three parameters in this model, more experimental data are required for fitting, making it less convenient for practical engineering applications.

2.2. The Proposed Curve Model

Due to the numerous factors influencing concrete carbonation, such as concrete strength and porosity, it is challenging to precisely capture the variation pattern of concrete carbonation depth using a single curve model. This is also the underlying reason for the poor fitting accuracy of traditional models. A promising approach is to combine various curves into a mixed model to adapt to the complexity and variability of concrete carbonation conditions. In view of this, this work proposes a new mixed-curve model, which improves the accuracy of describing the variation in concrete carbonation depth over time in practical engineering applications. The mathematical expression of this mixed-curve model is provided below:

$$z(t) = a\sqrt{t} + b\cdot\frac{\sqrt{t}}{\sqrt{t}+1} \quad (7)$$

In Equation (7), t and $z(t)$ represent the time and the corresponding concrete carbonation depth at that time, consistent with the definitions provided in previous equations. The two unknown parameters a and b are determined by fitting experimental data. Comparing Equation (7) with Equation (5), it can be seen that this new model is a hybrid of the Fick model and the hyperbolic model, where the additional term can be regarded as a reasonable modification to the Fick model to adapt to different concrete carbonation conditions. The newly added term $\frac{\sqrt{t}}{\sqrt{t+1}}$ exhibits the following characteristics: when $t = 0$, $\frac{\sqrt{t}}{\sqrt{t+1}} = 0$, and as t approaches infinity, $\frac{\sqrt{t}}{\sqrt{t+1}}$ approaches 1. This indicates that the added term $\frac{\sqrt{t}}{\sqrt{t+1}}$ is a bounded function that can, to a certain extent, compensate for the shortcomings of the Fick model in adapting to complex and variable carbonation scenarios, thereby enhancing the fitting accuracy. Compared to other mathematical curve models that involve multiple influencing parameters, the proposed combined curve model boasts a simple and clear structure, containing only two unknown fitting parameters, making it more suitable for engineering applications.

2.3. Solving Process of Fitting Parameters

In this section, we will further discuss the method for calculating the fitting parameters in the previously proposed mathematical curve model. To begin, we will create a linear equation system that relates CD to age, using the actual data obtained from experiments. Subsequently, we will derive the corresponding fitting parameters. Through the employment of diverse collections of experimental data, the linear system pertaining to Equation (7) can be formulated in the manner depicted below:

$$y = C\cdot x \quad (8)$$

$$y = \begin{Bmatrix} z(t_1) \\ \vdots \\ z(t_n) \end{Bmatrix}, \quad (9)$$

$$x = \begin{Bmatrix} a \\ b \end{Bmatrix} \quad (10)$$

$$C = \begin{bmatrix} \sqrt{t_1} & \frac{\sqrt{t_1}}{\sqrt{t_1}+1} \\ \vdots & \vdots \\ \sqrt{t_n} & \frac{\sqrt{t_n}}{\sqrt{t_n}+1} \end{bmatrix} \quad (11)$$

where $z(t_n)$ denotes the CD of the concrete specimen measured at time t_n. It is evident that both the coefficient matrices and vectors in Equation (8) can be derived from the experimental data. Consequently, the least squares method can be employed to determine the vector of undetermined fitting coefficients x. Thus, Equation (8) can be reformulated as follows:

$$C^T y = (C^T C)\cdot x \quad (12)$$

Pursuant to Equation (11), the representation of the solution for x is formulated as

$$\hat{x} = (C^T C)^{-1} C^T y \quad (13)$$

The computed value of $\hat{x}$ represents the least squares estimation of x. After obtaining the fitting parameters using the above formulas, the specific curve model is correspondingly established for predicting the carbonation depth. Different fitting parameters can be obtained by employing the aforementioned curve-fitting method for different concrete strengths, porosities, and other relevant factors. These fitting parameters, in turn, enable the generation of unique prediction curves for carbonation depth that precisely correspond to each specific concrete carbonation condition. In other words, the proposed model precisely adapts to changes in the properties of concrete materials and environmental conditions by adjusting the fitting parameters. This is a common characteristic shared by predictive models of this kind, which is to accommodate changes in the properties of the research object by adjusting the built-in parameters of the model. Figure 2 provides a more intuitive illustration of the operational process of the proposed model for predicting the carbonation depth of concrete.

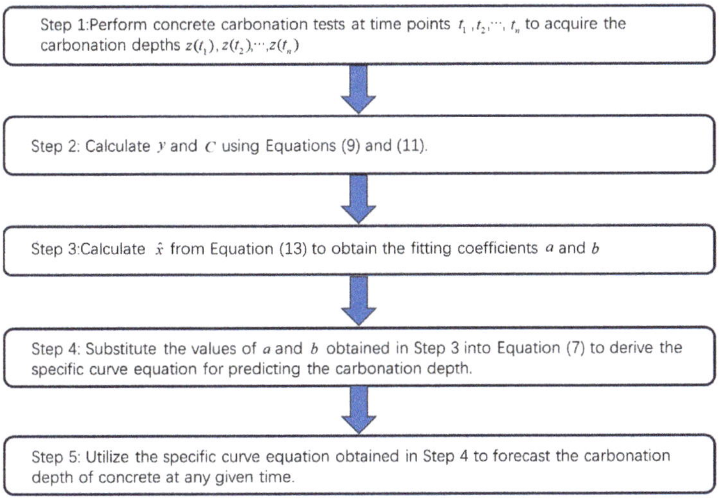

Figure 2. Schematic diagram of the process for predicting the depth of carbonation in concrete.

To assess the accuracy of these model predictions, we can compare the predicted values against the experimental values. The quantitative divergence between the predicted outcomes and the experimental measurements is defined as follows:

$$\{\delta\} = |C \cdot \hat{x} - y| \tag{14}$$

where $\{\delta\} = (\delta_1, \cdots, \delta_n)^T$ is referred to as the residual vector. In particular, δ_i represents the residual error between the ith experimental data point and its corresponding predicted value. The average and standard deviation of these residual errors can be computed using the following methods:

$$\overline{\delta} = \frac{\delta_1 + \cdots + \delta_n}{n} \tag{15}$$

$$\sigma = \sqrt{\frac{\sum\limits_{i=1}^{n} (\delta_i - \overline{\delta})^2}{n}} \tag{16}$$

In this context, $\overline{\delta}$ denotes the average of the residual errors, while σ represents the corresponding standard deviation. A lower value of $\overline{\delta}$ and σ indicates a greater accuracy in the curve model fitting.

To evaluate the model's predictive capability, the coefficient of determination (R^2) is calculated. This metric assesses the degree of closeness between the predicted and experimental values, thereby providing insights into the model's effectiveness in estimating CD. The formula for calculating R^2 is as follows:

$$R^2 = 1 - \frac{\sum_{i=1}^{n} (y_e - y_p)^2}{\sum_{i=1}^{n} (y_e - \overline{y}_e)^2} \quad (17)$$

where y_e = experimental value; $\overline{y}_e$ = mean of experimental values; y_p = predicted value, and n = total number of observations

3. Model Verification and Comparison via Concrete Experiment Data

We validated the results of the new evaluation model proposed in this paper by collecting experimental data from references [38–47] (Case 1–10). The fitting results are compared with the existing Fick and exponential models to demonstrate the superiority of the new model.

Case 1: In citation [38], some fly ash concrete specimens measuring 100 mm × 100 mm × 400 mm were cast and subjected to standard curing for 28 days. Initially, the CD of the concrete was measured after 0 days of curing. Subsequently, the fly ash concrete underwent rapid carbonation treatment following the initial measurement. The concentration of carbon dioxide for rapid carbonation is set to 20% + 3%; the temperature is set to 20 °C + 2 °C, and the relative humidity is set to 70% ± 5%. The CD was measured after three days, seven days, 14 days, and 28 days of rapid carbonation treatment. The average value of these measurements for each time point was considered the CD. Table 1 shows the material composition of the fly ash concrete, while Table 2 presents the CD data for each time point.

Table 1. The material composition of the fly ash concrete test block in Case 1 (kg/m^3).

Cement	Sand	Spall	Water	Fly Ash	Water–Binder Ratio
156	695	1135	175	234	0.45

Table 2. CD of concrete specimens at different ages in Case 1 (mm).

Age	0d	3d	7d	14d	28d
CD	1.4	6.2	9.4	11.2	16.6

The obtained data were then fitted into the Fick, exponential, and mixed models for analysis. The resulting fitting curves can be seen in Figure 3. Table 3 displays the models' average fitting error, standard deviation, and coefficient of determination (R^2). The specific fitting equations of the three models are as follows:

$$z(t) = 3.1799\sqrt{t} \quad (18)$$

$$z(t) = 17.9559 - 15.8783 \cdot e^{-t/12.9828} \quad (19)$$

$$z(t) = 2.8194\sqrt{t} + 1.7914 \cdot \frac{\sqrt{t}}{\sqrt{t}+1} \quad (20)$$

Case 2: In citation [39], some standard-sized lime powder concrete test blocks were poured and cured under standard conditions for seven days before being cured under natural conditions. Then, concrete CD at 14, 28, 60, and 90 days of curing was measured. The materials used include the following: Class II fly ash with a fineness of 18.0% and a water requirement of 101%; S105 grade ground granulated blast-furnace slag; limestone powder with 20% residue on a 45 μm sieve and a flowability ratio of 101%; Polycarboxylate superplasticizer; medium sand with a fineness modulus of 2.67 and an apparent density of

2630 kg/m^3; granite gravel with a particle size ranging from 5 to 25 mm and an apparent density of 2670 kg/m^3; and P·O 42.5 grade cement. Table 4 shows the material composition of concrete, while Table 5 displays the CD data at each time point.

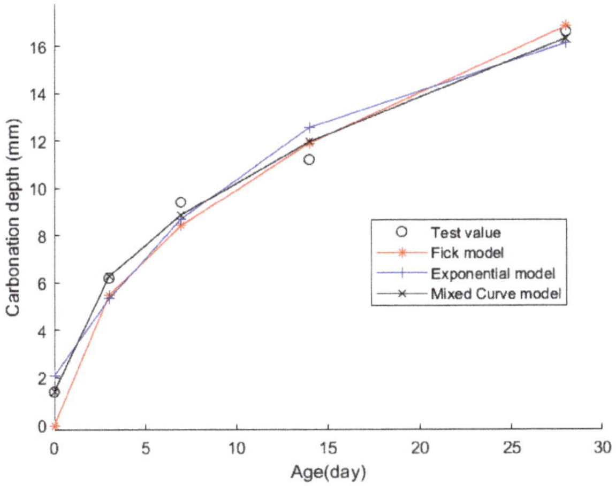

Figure 3. The fitting results of Case 1.

Table 3. The fitting mean errors, standard deviations, and coefficient of determination from the three models in Case 1.

Index	Fick Model	Exponential Model	Mixed-Curve Model
Fitting mean error $\bar{\delta}$	0.8007	0.8130	0.3313
Fitted standard deviation σ	1.4056	1.3674	0.6854
Coefficient of determination (R^2)	0.9692	0.9709	0.9927

Table 4. The material composition of the lime powder concrete test block in Case 2 (kg/m^3).

Cement	Sand	Spall	Water	Limestone Powder	Admixture
306	850	1030	170	34	5.1

Table 5. CD of concrete specimens at different ages in Case 2 (mm).

Age	14d	28d	60d	90d
CD	0	1.1	1.9	2.8

Then, the obtained data were put into the Fick, exponential, and mixed models for analyses. Figure 4 shows the final fitting curve. Table 6 shows the model's average fitting error, standard deviation, and coefficient of determination (R^2). The specific fitting equations for the three models are as follows:

$$z(t) = 0.2453\sqrt{t} \tag{21}$$

$$z(t) = 3.5205 - 4.4714 \cdot e^{-t/52.53} \tag{22}$$

$$z(t) = 0.4617\sqrt{t} - 1.8014 \cdot \frac{\sqrt{t}}{\sqrt{t}+1} \tag{23}$$

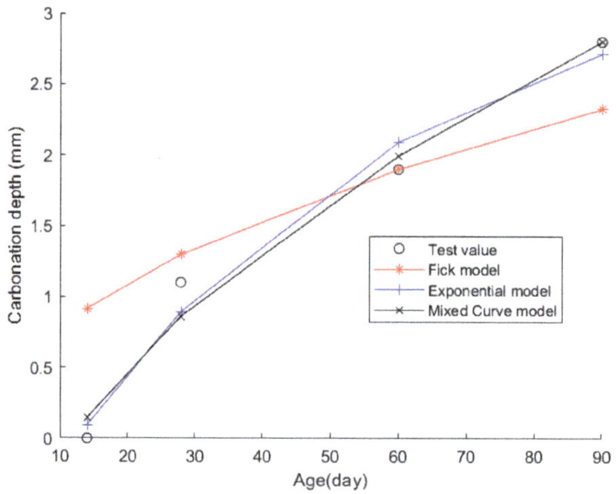

Figure 4. The fitting results of Case 2.

Table 6. The fitting mean errors, standard deviations, and coefficient of determination from the three models in Case 2.

Index	Fick Model	Exponential Model	Mixed-Curve Model
Fitting mean error $\overline{\delta}$	0.3972	0.1445	0.1201
Fitted standard deviation σ	1.0513	0.3086	0.2952
Coefficient of determination (R^2)	0.7399	0.9776	0.9795

Case 3: In citation [40], several C30 concrete cube samples measuring 100 mm on each side were cast and underwent conventional curing procedures. The CD of the concrete was measured after 3, 7, 14, and 28 days of the curing process. This design employs 42.5 grade ordinary Portland cement as the primary binding material. For fine aggregate, high-quality river sand with an apparent density of 2640 kg/m^3, a fineness modulus of 2.9, and a low mud content of only 0.5% is selected, belonging to the Grade II gradation zone with excellent gradation. As for coarse aggregate, two types of continuously graded crushed stones are used: the first is granite crushed stone with an apparent density of 2800 kg/m^3, a crushing value of 6.0%, a mud content of 1.0%, and a particle size range of 2.5 to 20.0 mm; the second is limestone crushed stone with an apparent density of 2740 kg/m^3, a crushing value of 7.9%, a similar mud content of 1.0%, and an extended particle size range of 2.5 to 40.0 mm. Additionally, Grade I fly ash is introduced, featuring a fineness of 8.4, a moisture content of 0.3%, and a water requirement of 95%. The material composition of the concrete is detailed in Table 7, with Table 8 presenting the CD values recorded at the specified time intervals.

Table 7. The material composition of the concrete test block in Case 3 (kg/m^3).

Cement	Sand	Spall	W/C	Fly Ash
300	856	975	0.5	60

Table 8. CD of concrete specimens at different ages in Case 3 (mm).

Age	3d	7d	14d	28d
CD	0.8	2.4	3.2	5.0

Then, the obtained data were put into the Fick, exponential, and mixed models for analyses. Figure 5 shows the final fitting curve. Table 9 shows the model's average fitting error, standard deviation, and coefficient of determination (R^2). The specific fitting equations for the three models are as follows:

$$z(t) = 0.8878\sqrt{t} \tag{24}$$

$$z(t) = 6.0251 - 6.1031 \cdot e^{-t/16.2279} \tag{25}$$

$$z(t) = 1.1291\sqrt{t} - 1.1847 \cdot \frac{\sqrt{t}}{\sqrt{t}+1} \tag{26}$$

Case 4: In citation [41], 12 concrete cube test blocks, each with a side length of 100 mm, were cast. Following standard curing for 26 days, the test blocks underwent drying at a constant temperature of 60 °C. Once they reached a curing age of 28 days, wax was applied to seal the test blocks around all sides except for two opposing ones. Subsequently, they were placed in a standard carbonation box for carbonation testing. The temperature, humidity, and CO_2 concentration within the carbonation chamber all meet the requirements specified in the standard "Test Scheme for Long-term Performance and Durability of Ordinary Concrete" (GB/T 50082-2009) [48]. The CD of the concrete was measured at 3 days, 7 days, 14 days, and 28 days. The material composition of the concrete is outlined in Table 10, and the CD data at each time point are presented in Table 11.

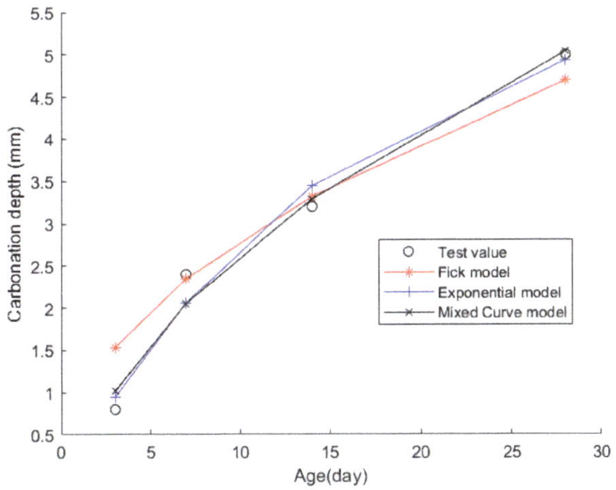

Figure 5. The fitting results of Case 3.

Table 9. The fitting mean errors, standard deviations, and coefficient of determination from the three models in Case 3.

Index	Fick Model	Exponential Model	Mixed-Curve Model
Fitting mean error $\overline{\delta}$	0.3032	0.2008	0.1741
Fitted standard deviation σ	0.8081	0.4523	0.4233
Coefficient of determination (R^2)	0.9286	0.9776	0.9804

Table 10. The material composition of the concrete test block in Case 4 (kg/m^3).

Cement	Sand	Spall	Water Reducing Agent	Fly Ash	Silica Fume	Water	W/C
380	655	983	4.9	40	74	101.1	0.38

Table 11. CD of concrete specimens at different ages in Case 4 (mm).

Age	3d	7d	14d	28d
CD	5.82	6.70	7.47	8.73

Then, the obtained data were put into the Fick, exponential, and mixed models for analyses. Figure 6 shows the final fitting curve. Table 12 shows the model's average fitting error, standard deviation, and coefficient of determination (R^2). The specific fitting equations for the three models are as follows:

$$z(t) = 1.9606\sqrt{t} \tag{27}$$

$$z(t) = 9.897 - 4.6611 \cdot e^{-t/20.437} \tag{28}$$

$$z(t) = 0.8046\sqrt{t} + 5.6756 \cdot \frac{\sqrt{t}}{\sqrt{t}+1} \tag{29}$$

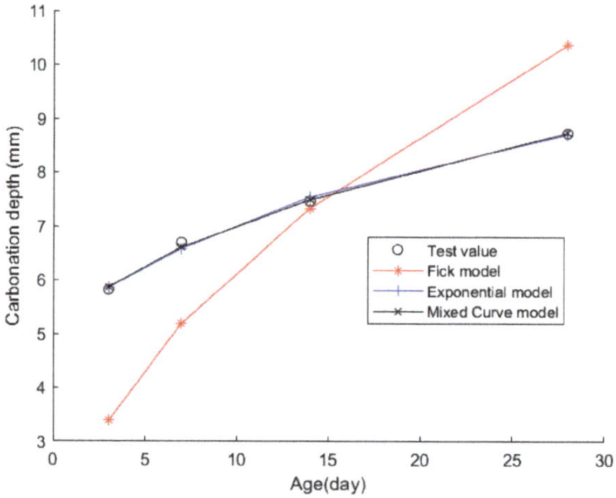

Figure 6. The fitting results of Case 4.

Table 12. The fitting mean errors, standard deviations, and coefficient of determination from the three models in Case 4.

Index	Fick Model	Exponential Model	Mixed-Curve Model
Fitting mean error $\overline{\delta}$	1.4289	0.0648	0.0444
Fitted standard deviation σ	3.2996	0.1471	0.1079
Coefficient of determination (R^2)	−1.3841	0.9953	0.9974

Case 5: In citation [42], C30 cubic specimens measuring 100 × 100 × 100 mm were cast, demolded, and subsequently placed in a standard curing chamber for a period of 28 days. After this curing phase, the specimens were removed and subjected to a 48-hour drying

process in an oven at 60 °C to halt hydration. Following this, the specimens were coated with wax for sealing and then placed in a concrete carbonation chamber for accelerated carbonation testing. The specimens were retrieved at various carbonation ages of 3d, 7d, 14d, and 28d to measure their CD. The experiment employed P·O 42.5 ordinary Portland cement as the primary cementitious material, complemented by a continuously graded limestone gravel ranging from 5 to 25 mm in size (with an apparent density of 2690 kg/m^3, bulk density of 1470 kg/m^3, crushing value of 14%, void ratio of 45%, and mud content of 0.2%). As fine aggregate, granite-based manufactured sand containing black mica was used (featuring an apparent density of 2670 kg/m^3, stone dust content of 12.2%, mica content of 11.8%, water absorption rate of 2.2%, and fineness modulus of 3.1). To further enhance the mixture, a polycarboxylate superplasticizer with a water-reducing rate of 22.3% and a solids content of 20.0% was added, along with an admixture to regulate the workability and compactness of the concrete, boasting a solids content of 10.0%. Table 13 outlines the concrete material composition, while Table 14 displays the CD data recorded at each time point.

Table 13. The material composition of the concrete test block in Case 5 (kg/m^3).

Cement	Sand	Spall	Water Reducing Agent	Regulator	Water
400	818	1038	6.8	40	165

Table 14. CD of concrete specimens at different ages in Case 5 (mm).

Age	3d	7d	14d	28d
CD	3.7	4.8	5.1	7.6

Then, the obtained data were put into the Fick, exponential, and mixed models for analyses. Figure 7 shows the final fitting curve. Table 15 shows the model's average fitting error, standard deviation, and coefficient of determination (R^2). The specific fitting equations for the three models are as follows:

$$z(t) = 1.3347\sqrt{t} \tag{30}$$

$$z(t) = 7.9998 - 5.2683 \cdot e^{-t/23.7058} \tag{31}$$

$$z(t) = 0.8577\sqrt{t} + 2.3417 \cdot \frac{\sqrt{t}}{\sqrt{t}+1} \tag{32}$$

Case 6: In citation [43], C25 prismatic specimens with dimensions of 100 × 100 × 400 mm were cast and subjected to accelerated carbonation testing in a concrete carbonation test chamber in accordance with the standard "Standard Test Methods for Long-term Performance and Durability of Ordinary Concrete" GB/T 50082-2009. The carbonation environment within the chamber was maintained at a temperature of (20 ± 2) °C, relative humidity of (70 ± 5)%, and a CO_2 concentration of (20 ± 3)%. After carbonation ages of 3, 7, 14, 28, and 56 days, the specimens were retrieved to measure their CDs. The concrete mix design employed P·O 42.5 ordinary Portland cement. The fine aggregate was a blend of river sand and iron tailings sand in a mass ratio of 9:1, with an apparent density of 2659 kg/m^3, a bulk density of 1548 kg/m^3, a fineness modulus of 2.94, and a void ratio of 41%. The coarse aggregate was a continuous gradation from 5 mm to 20 mm, achieved by mixing iron tailings stones of 5 mm to 10 mm continuous grain size and 10 mm to 20 mm single-grain size in a mass ratio of 1:4. This coarse aggregate had an apparent density of 2695 kg/m^3, a bulk density of 1576 kg/m^3, a mud content of 2.6%, a void ratio of 39%, and a crushing index of 8.2%. The mineral admixtures included S95 grade slag powder (with an apparent density of 2910 kg/m^3 and a moisture content of 0.44%), Class II fly ash (with an apparent density of 2240 kg/m^3, a fineness of 19%, a water requirement of 104%, and a moisture content of 0.06%), and iron tailings powder (with an apparent density of 2960 kg/m^3 and a specific surface area of 468 m^2/kg). Additionally, polycarboxylate-based superplasticizer was added as an admixture, with a water-reducing rate of 25% and a solid content of 40%. Table 16 presents the material composition of the concrete, while Table 17 displays the CD data at each time point.

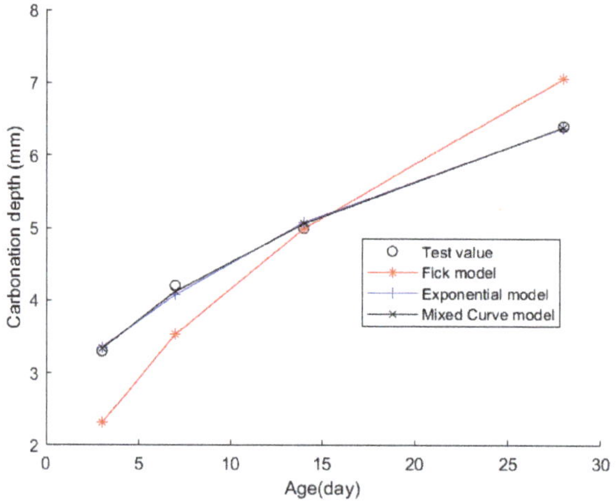

Figure 7. The fitting results of Case 5.

Table 15. The fitting mean errors, standard deviations, and coefficient of determination from the three models in Case 5.

Index	Fick Model	Exponential Model	Mixed-Curve Model
Fitting mean error $\bar{\delta}$	0.5814	0.0694	0.0468
Fitted standard deviation σ	1.3649	0.1580	0.1069
Coefficient of determination (R^2)	0.6409	0.9952	0.9978

Table 16. The material composition of the concrete test block in Case 6 (kg/m^3).

Cement	Water	Sand-Aggregate	Water–Binder Ratio	Iron Tailings Powder	Fly Ash	Slag Powder	Water Reducing Agent	Spall	Sand
141	180	43.5	0.57	31	47	94	1.85	1055	812

Table 17. CD of concrete specimens at different ages in Case 6 (mm).

Age	3d	7d	14d	28d	56d
CD	6.13	8.34	11.33	15.04	21.42

Then, the obtained data were put into the Fick, exponential, and mixed models for analyses. Figure 8 shows the final fitting curve. Table 18 shows the model's average fitting error, standard deviation, and coefficient of determination (R^2). The specific fitting equations for the three models are as follows:

$$z(t) = 2.9162\sqrt{t} \tag{33}$$

$$z(t) = 29.0566 - 24.0750 \cdot e^{-t/49.2683} \tag{34}$$

$$z(t) = 2.6481\sqrt{t} + 1.6646 \cdot \frac{\sqrt{t}}{\sqrt{t}+1} \tag{35}$$

Case 7: In citation [44], prismatic specimens of 100 mm × 100 mm × 300 mm were cast. Following the specifications outlined in the "Standard for Test Methods of Long-Term Performance and Durability of Ordinary Concrete" (GB/T50082-2009), a rapid carbonation method was employed. The test blocks were placed in a concrete carbonation test chamber for accelerated carbonation testing. Within the chamber, the concentration of carbon dioxide was maintained at (20 ± 3)%, humidity at (70 ± 5)%, and temperature at (20 ± 2) °C. The specimens were removed at intervals of 7, 14, 28, and 56 days for CD measurements. This experiment utilized a cementitious material consisting of ordinary Portland cement (with a strength grade of P·O 42.5R) mixed with fly ash having a fineness of 12 and a water requirement ratio of 93%. Natural crushed stone was used, along with fine aggregates composed of river sand (with an apparent surface area of 2689 kg/m^3, a loose bulk density of 1655 kg/m^3, a water absorption rate of 0.5%, and a mud content of 3.1%) and iron tailings sand (with an apparent surface area of 2745 kg/m^3, a loose bulk density of 1824 kg/m^3, a void ratio of 8.7%, a water absorption rate of 2.9%, and a mud content of 19.53%). Additionally, a naphthalene-based superplasticizer with a water-reducing rate of 20% was incorporated. Table 19 details the concrete's material composition, while Table 20 presents the CD data for each time point.

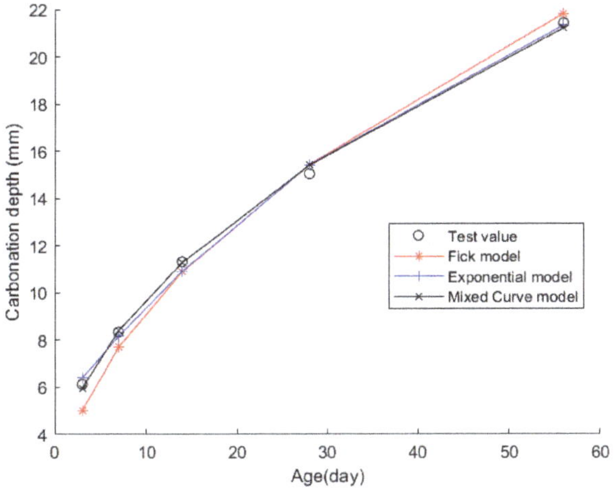

Figure 8. The fitting results of Case 6.

Table 18. The fitting mean errors, standard deviations, and coefficient of determination from the three models in Case 6.

Index	Fick Model	Exponential Model	Mixed-Curve Model
Fitting mean error $\bar{\delta}$	0.5832	0.2609	0.1642
Fitted standard deviation σ	1.0111	0.4527	0.3198
Coefficient of determination (R^2)	0.9859	0.9971	0.9986

Table 19. The material composition of the concrete test block in Case 7 (kg/m^3).

Cement	Water	Sand	Spall	Iron Tailings Sand	Fly Ash	Water Reducing Agent
440	220	459.2	1103	196.8	110	2.2

Table 20. CD of concrete specimens at different ages in Case 7 (mm).

Age	7d	14d	28d	56d
CD	5.42	6.62	8.04	10.40

Then, the obtained data were put into the Fick, exponential, and mixed models for analyses. Figure 9 shows the final fitting curve. Table 21 shows the model's average fitting error, standard deviation, and coefficient of determination (R^2). The specific fitting equations for the three models are as follows:

$$z(t) = 1.5189\sqrt{t} \tag{36}$$

$$z(t) = 13.8148 - 9.4456 \cdot e^{-t/55.2963} \tag{37}$$

$$z(t) = 1.0198\sqrt{t} + 3.2447 \cdot \frac{\sqrt{t}}{\sqrt{t}+1} \tag{38}$$

Case 8: In citation [45], researchers cast standard-sized concrete blocks and put them in a chamber for quick carbonation tests, adhering to the "Standard for Test Methods of Long-Term Performance and Durability of Ordinary Concrete" (GB/T 50082-2009). They took out the blocks on days 3, 7, 14, 28, and 56 after carbonation to precisely gauge their carbonization depth. This study used P·O 42.5 cement, fly ash, natural and artificial sand, crushed stone, and other additives. Concrete composition details are in Table 22, while CDs are logged in Table 23.

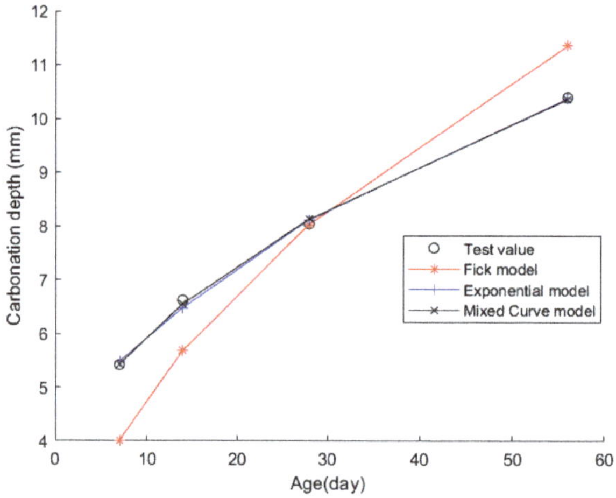

Figure 9. The fitting results of Case 7.

Then, the obtained data were put into the Fick, exponential, and mixed models for analyses. Figure 10 shows the final fitting curve. Table 24 shows the model's average fitting error, standard deviation, and coefficient of determination (R^2). The specific fitting equations for the three models are as follows:

$$z(t) = 0.9651\sqrt{t} \tag{39}$$

$$z(t) = 7.5522 - 5.9846 \cdot e^{-t/28.7005} \tag{40}$$

$$z(t) = 0.8167\sqrt{t} + 0.9210 \cdot \frac{\sqrt{t}}{\sqrt{t}+1} \tag{41}$$

Table 21. The fitting mean errors, standard deviations, and coefficient of determination from the three models in Case 7.

Index	Fick Model	Exponential Model	Mixed-Curve Model
Fitting mean error $\bar{\delta}$	0.8269	0.0771	0.0518
Fitted standard deviation σ	1.9430	0.1769	0.1201
Coefficient of determination (R^2)	0.7253	0.9977	0.9990

Table 22. The material composition of the concrete test block in Case 8 (kg/m^3).

Cement	Water	Artificial Sand	Natural Sand	5–10 Stone	10–20 Stone	Fly Ash	Proportion of Additives (%)
317	175	244	569	250	779	56	2.1

Table 23. CD of concrete specimens at different ages in Case 8 (mm).

Age	3d	7d	14d	28d	56d
CD	1.5	3.7	4.1	4.7	6.9

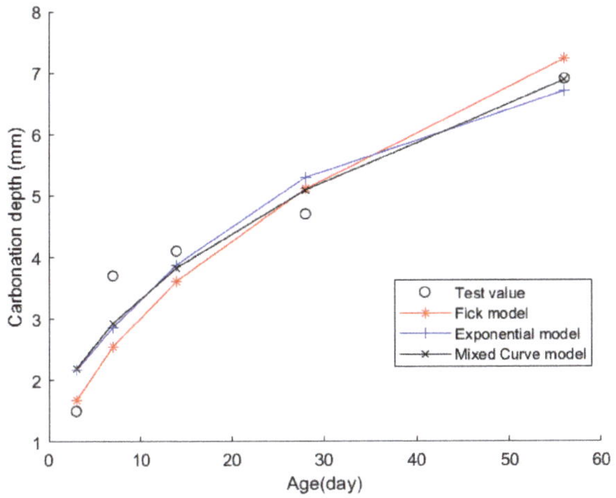

Figure 10. The fitting results of Case 8.

Table 24. The fitting mean errors, standard deviations, and coefficient of determination from the three models in Case 8.

Index	Fick Model	Exponential Model	Mixed-Curve Model
Fitting mean error $\bar{\delta}$	0.5073	0.5031	0.4281
Fitted standard deviation σ	0.9624	0.8896	0.8028
Coefficient of determination (R^2)	0.8772	0.8951	0.9146

Case 9: In citation [46], ten sets of prismatic specimens with dimensions of 100 mm × 100 mm × 400 mm were cast. These standard specimens were cured in a standard curing chamber for 28 days before being placed in a drying oven set at 60 °C for continuous drying for 48 h. Following the drying process, the specimens underwent wax sealing treatment and were then placed in a carbonation chamber where the CO_2 concentration was maintained at (20 ± 3)%; the temperature was controlled at (20 ± 2) °C, and the relative humidity was regulated at (70 ± 5)%. The specimens were retrieved at 3d, 7d, 14d, and 28d intervals during the carbonation process to measure their CD. The cementitious material system employed in this experiment consisted primarily of P.O 42.5 ordinary Portland cement (with a specific surface area of 355 m^2/kg), fly ash, and silica fume. For reinforcement, 18 mm-long basalt fibers were selected, featuring a density of 2.5 g/cm^3 and an elastic modulus ranging from 91 to 110 GPa. To optimize the concrete's performance, a polycarboxylate liquid water-reducing agent was incorporated at a dosage of 0.5% by the total mass of cementitious materials, offering a high water-reducing rate of 37%. Regarding fine aggregate, medium sand with a fineness modulus of 2.6 was chosen. This sand exhibits an apparent density of 2.610 g/cm^3, a bulk density of 1.630 g/cm^3, a moisture content controlled below 0.8%, and a silt content of less than 1.9%. As for coarse aggregate, crushed stone with a particle size ranging from 5 to 25 mm was utilized, possessing an apparent density of 2.753 g/cm^3, a bulk density of 1.634 g/cm^3, a moisture content of 0.15%, and a silt content of 0.4%. Tables 25 and 26, respectively, showcase the detailed material composition of the concrete and the recorded CD data at each specified time point.

Table 25. The material composition of the concrete test block in Case 9 (kg/m^3).

Cement	Water	Fine Aggregate	Coarse Aggregate	Water Reducing Agent (%)
441	150	511.68	1314.18	0.5%

Table 26. CD of concrete specimens at different ages in Case 9 (mm).

Age	3d	7d	14d	28d
CD	5.1	7.5	10.1	13.7

Then, the obtained data were put into the Fick, exponential, and mixed models for analyses. Figure 11 shows the final fitting curve. Table 27 shows the model's average fitting error, standard deviation, and coefficient of determination (R^2). The specific fitting equations for the three models are as follows:

$$z(t) = 2.6723\sqrt{t} \tag{42}$$

$$z(t) = 17.0893 - 13.8396 \cdot e^{-t/20.0009} \tag{43}$$

$$z(t) = 2.4048\sqrt{t} + 1.3136 \cdot \frac{\sqrt{t}}{\sqrt{t}+1} \tag{44}$$

Case 10: In citation [47], a carbonation test was conducted in accordance with the "Standard Test Methods for Long-Term Performance and Durability of Ordinary Concrete". The test specimens measured 100 mm × 100 mm × 100 mm and were subjected to carbonation testing after being cured for 7 days. The specimens were retrieved, and their CDs were measured at 3 days, 7 days, 14 days, 28 days, and 56 days of carbonation. The raw materials used in this project comprised P.O42.5 grade cement with 3-day and 28-day compressive strengths reaching 31.5 MPa and 48.0 MPa respectively; Class F, Grade I fly ash (FA) featuring a water requirement ratio of 93% and a fineness of 7.3% (sieve residue on a 45 μm square hole sieve); S95 grade ground granulated blast furnace slag (KF) with a specific surface area of 425 m^2/kg, maintaining a fluidity ratio of 100%, and an excellent 28-day mortar activity index of 104%; fine aggregate in the form of river sand with a fineness modulus of 2.81 and strict control of mud content below 1.6%; coarse aggregate composed of crushed stone with a crushing value of 15.1% and a flaky and elongated particle content of 4.9%, employing a two-stage grading of 5~20 mm, with a blending ratio of 4.75~9.5 mm to 9.5~19.0 mm particles at 2:8; and polycarboxylate superplasticizer as an admixture, boasting a water-reducing efficiency of 27%. Table 28 presents the concrete material composition, while Table 29 displays the CD data at each time point.

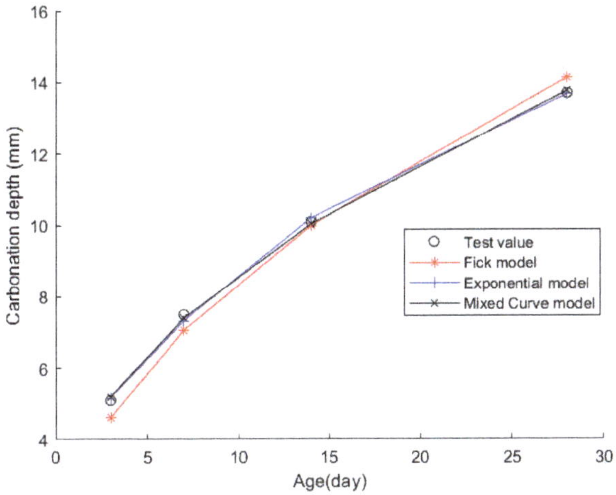

Figure 11. The fitting results of Case 9.

Table 27. The fitting mean errors, standard deviations, and coefficient of determination from the three models in Case 9.

Index	Fick Model	Exponential Model	Mixed-Curve Model
Fitting mean error $\bar{\delta}$	0.3607	0.0954	0.0824
Fitted standard deviation σ	0.7818	0.2164	0.1692
Coefficient of determination (R^2)	0.9850	0.9989	0.9993

Table 28. The material composition of the concrete test block in Case 10 (kg/m^3).

Cement	Water	Fly Ash	Slag Powder	Sand	Stone	Water–Binder Ratio	Water Reducing Agent Dosage
329	146	94	47	743	1114	0.31	1.25%

Table 29. CD of concrete specimens at different ages in Case 10 (mm).

Age	3d	7d	14d	28d	56d
CD	0.2	1.0	2.5	4.0	6.6

Then, the obtained data were put into the Fick, exponential, and mixed models for analyses. Figure 12 shows the final fitting curve. Table 30 shows the model's average fitting error, standard deviation, and coefficient of determination (R^2). The specific fitting equations for the three models are as follows:

$$z(t) = 0.7676\sqrt{t} \tag{45}$$

$$z(t) = 9.1571 - 9.5449 \cdot e^{-t/43.0316} \quad (46)$$

$$z(t) = 1.1189\sqrt{t} - 2.1803 \cdot \frac{\sqrt{t}}{\sqrt{t}+1} \quad (47)$$

Through a comprehensive analysis of the errors and coefficient of determination across the ten case studies, it is discernible that the novel model outperforms both the Fick and exponential models, exhibiting a diminished mean fitting error and a reduced standard deviation in its assessments. Additionally, the determination coefficients of the new model surpass those of the Fick and exponential models, underscoring the superior fitting quality of the new model. For example, compared to the second-best model, the new model exhibits a reduction of 59% and 50% in average fitting error and standard deviation in Case 1, 31% and 27% in Case 4, 33% and 32% in Case 5, and 37% and 29% in Case 6. These results demonstrate that this new model significantly enhances fitting accuracy. Upon meticulous examination of Figures 3–12, the fitting curve of the novel model is observed to be more closely aligned with the experimental data points, indicating the model's heightened precision in replicating the empirical observations. To sum up, the new model demonstrates superior data fitting capabilities relative to the Fick and exponential models, rendering it more precise and dependable in assessing concrete CD.

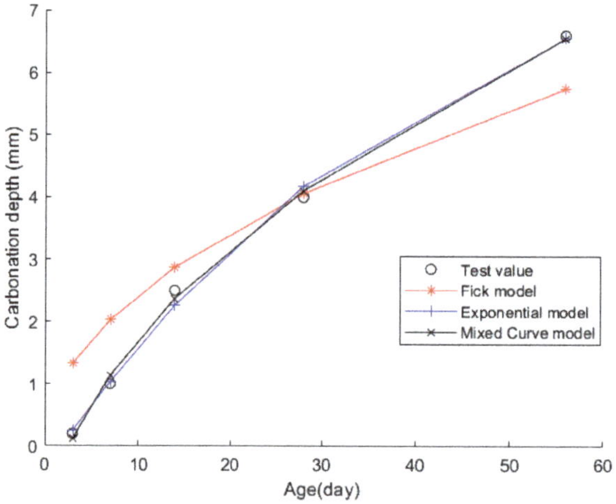

Figure 12. The fitting results of Case 10.

Table 30. The fitting mean errors, standard deviations, and coefficient of determination from the three models in Case 10.

Index	Fick Model	Exponential Model	Mixed-Curve Model
Fitting mean error $\overline{\delta}$	0.6900	0.1111	0.0998
Fitted standard deviation σ	1.2675	0.2173	0.1680
Coefficient of determination (R^2)	0.8762	0.9964	0.9978

4. Conclusions

This article introduces a novel mathematical curve model for assessing the CD of concrete. By fitting mathematical curves using limited initial experimental data, this model offers a more accurate evaluation of concrete CD. The validity of this new model was confirmed through validation with various experimental data sets. The calculated results lead to the following key conclusions:

- The new model introduced in this article combines elements of the Fick and hyperbolic models. Unlike the index model, which may prematurely reach its limit value, the new model offers a more precise fit for the later stages of concrete carbonation;
- Compared to the Fick and exponential models, the new model introduced in this paper shows reduced fitting errors and higher R^2, indicating superior fitting precision. Unlike the exponential model, which necessitates fitting three unknown parameters, this new model only needs the fitting of two unknown parameters. This reduced parameter requirement makes this new model more efficient for practical engineering applications, requiring less data for fitting and easier implementation.

This article primarily focuses on the relationship between the CD of concrete and time without extensively addressing other factors that may influence this depth. The newly proposed model does not simultaneously account for multiple influencing variables about concrete CD. However, in contrast to multi-factor models, this new model requires only limited experimental data to derive the mathematical curve equation for evaluation. In scenarios where experimental data are scarce, this model can be integrated with machine learning techniques to simulate and generate additional data for training purposes, thereby enhancing the generalization capabilities of the machine learning models. The validation of the model presented in this article is based on a subset of the available data, indicating a need for further research to confirm that the model is fitting, evaluate capabilities for concrete, and optimize the fitting conditions. Additionally, this model may be relevant for the property parameters of other materials, and future experiments could be conducted to assess the model's applicability in those contexts.

Author Contributions: Conceptualization, X.W., Q.Y., and H.C.; methodology, X.W. and Q.Y.; investigation, X.W., H.C., and F.Q.; writing—original draft preparation, X.W., Q.Y., H.C., and F.Q.; writing—review and editing, X.W., Q.Y., H.C., and F.Q. All authors have read and agreed to the published version of the manuscript.

Funding: This work is supported by the Zhejiang Public Welfare Technology Application Research Project (LGF22E080021, LY24E080010), Chongqing Transportation Science and Technology Project (Grant No. 2022-01), and Natural Science Foundation of China (52192663, 52008215).

Institutional Review Board Statement: Not applicable.

Informed Consent Statement: Not applicable.

Data Availability Statement: The original contributions presented in the study are included in the article, further inquiries can be directed to the corresponding authors.

Conflicts of Interest: The authors declare that there are no conflicts of interest. The funders had no role in the collection, analyses, or interpretation of data or in the decision to publish the results.

References

1. Liu, L.L. Study on the Absorption of CO_2 by Fresh Cement Slurry and Its Reverse Mechanism. Ph.D. Thesis, China University of Mining and Technology, Beijing, China, 2021. [CrossRef]
2. Wang, X.; Yang, Q.; Peng, X.; Qin, F. A Review of Concrete Carbonation Depth Evaluation Models. *Coatings* **2024**, *14*, 386. [CrossRef]
3. Felix, E.F.; Carrazedo, R.; Possan, E. Carbonation model for fly ash concrete based on artificial neural network: Development and parametric analysis. *Constr. Build. Mater.* **2021**, *266*, 121050. [CrossRef]
4. Ventura, A.; Ta, V.L.; Kiessé, T.S.; Bonnet, S. Design of concrete: Setting a new basis for improving both durability and environmental performance. *J. Ind. Ecol.* **2021**, *25*, 233–247. [CrossRef]
5. Medvedev, V.; Pustovgar, A. A Review of Concrete Carbonation and Approaches to Its Research under Irradiation. *Buildings* **2023**, *13*, 1998. [CrossRef]
6. Chen, G.; Lv, Y.; Zhang, Y.; Yang, M. Carbonation depth predictions in concrete structures under changing climate condition in China. *Eng. Fail. Anal.* **2021**, *119*, 104990. [CrossRef]
7. Peng, X.; Shi, F.; Yang, J.; Yang, Q.; Wang, H.; Zhang, J. Modification of construction waste derived recycled aggregate via CO_2 curing to enhance corrosive freeze-thaw durability of concrete. *J. Clean. Prod.* **2023**, *405*, 137016. [CrossRef]
8. Malami, S.I.; Akpinar, P.; Lawan, M.M. Preliminary investigation of carbonation problem progress in concrete buildings of north Cyprus. In Proceedings of the MATEC Web of Conferences, Kuala Lumpur, Malaysia, 13–14 August 2018; p. 20306007.

9. Jiang, C.; Gu, X.; Huang, Q.; Zhang, W. Carbonation depth predictions in concrete bridges under changing climate conditions and increasing traffic loads. *Cem. Concr. Compos.* **2018**, *93*, 140–154. [CrossRef]
10. Forsdyke, J.C.; Lees, J.M. Model fitting to concrete carbonation data with non-zero initial carbonation depth. *Mater. Struct.* **2023**, *56*, 22. [CrossRef]
11. Fuhaid, A.F.A.; Niaz, A. Carbonation and corrosion problems in reinforced concrete structures. *Buildings* **2022**, *12*, 586. [CrossRef]
12. Jeong, H.; Jung, B.J.; Kim, J.H.; Seo, S.Y.; Kim, K.S. Development and assessment of Nile blue-immobilized pH sensor to monitor the early stage of concrete carbonation. *J. Build. Eng.* **2022**, *62*, 105319. [CrossRef]
13. Majlesi, A.; Koodiani, H.K.; de Rincon, O.T.; Montoya, A.; Millano, V.; Torres-Acosta, A.A.; Troconis, B.C.R. Artificial neural network model to estimate the long-term carbonation depth of concrete exposed to natural environments. *J. Build. Eng.* **2023**, *74*, 106545. [CrossRef]
14. Tran, V.Q. Using Artificial neural network containing two hidden layers for predicting carbonation depth of concrete. In *CIGOS 2021, Emerging Technologies and Applications for Green Infrastructure: Proceedings of the 6th International Conference on Geotechnics, Civil Engineering and Structures*; Springer: Berlin/Heidelberg, Germany, 2022; pp. 1945–1952.
15. Concha, N.C. A robust carbonation depth model in recycled aggregate concrete (RAC) using neural network. Expert Systems with Applications. *Expert Syst. Appl.* **2024**, *237*, 121650. [CrossRef]
16. Akpinar, P.; Uwanuakwa, I. Investigation of the parameters influencing progress of concrete carbonation depth by using artificial neural networks. *Mater. Construcción* **2020**, *70*, e209. [CrossRef]
17. Zhang, K.; Zhang, K.; Bao, R.; Liu, X. A framework for predicting the carbonation depth of concrete incorporating fly ash based on a least squares support vector machine and metaheuristic algorithms. *J. Build. Eng.* **2023**, *65*, 105772. [CrossRef]
18. Huang, S.; Wang, L.; Li, Z.; Su, D.; Luo, Q. Machine learning-based prediction model for CO_2-induced corrosion on oil well cement under high-pressure and high-temperature condition. *Constr. Build. Mater.* **2024**, *414*, 134999. [CrossRef]
19. Çevik, A.; Kurtoğlu, A.E.; Bilgehan, M.; Gülşan, M.E.; Albegmprli, H.M. Support vector machines in structural engineering: A review. *J. Civ. Eng. Manag.* **2015**, *21*, 261–281. [CrossRef]
20. Li, Z.; He, H.; Zhao, S. Research on support vector machine's prediction of concrete carbonization. In Proceedings of the 2008 International Seminar on Business and Information Management, Wuhan, China, 19 December 2008; Volume 1, pp. 319–322.
21. Xiang, R. Prediction of concrete carbonation depth based on support vector regression. In Proceedings of the 2009 Third International Symposium on Intelligent Information Technology Application, Nanchang, China, 21–22 November 2009; Volume 3, pp. 172–175.
22. Londhe, S.; Kulkarni, P.; Dixit, P.; Silva, A.; Neves, R.; De Brito, J. Tree based approaches for predicting concrete carbonation coefficient. *Appl. Sci.* **2022**, *12*, 3874. [CrossRef]
23. Wang, J.; Zhang, Z.Q.; Liu, X.M.; Shao, Y.; Liu, X.Y.; Wang, H.M. Prediction and interpretation of concrete corrosion induced by carbon dioxide using machine learning. *Corros. Sci.* **2024**, *233*, 112100. [CrossRef]
24. Moghaddas, S.A.; Nekoei, M.; Golafshani, E.M.; Nehdi, M.; Arashpour, M. Modeling carbonation depth of recycled aggregate concrete using novel automatic regression technique. *J. Clean. Prod.* **2022**, *371*, 133522. [CrossRef]
25. Lee, H.; Lee, H.; Suraneni, P. Evaluation of carbonation progress using AIJ model, FEM analysis, and machine learning algorithms. *Constr. Build. Mater.* **2020**, *259*, 119703. [CrossRef]
26. Kumar, A.; Arora, H.C.; Kapoor, N.R.; Kontoni, D.P.N.; Kumar, K.; Jahangir, H.; Bhushan, B. Practical applicable model for estimating the carbonation depth in fly-ash based concrete structures by utilizing adaptive neuro-fuzzy inference system. *Comput. Concr.* **2023**, *32*, 119–138.
27. Koichi, K. *Durability of Iron Reinforced Concrete*; Hiroshima Institute of Construction Technology Press: Hiroshima, Japan, 1963.
28. Zhu, A.M. Carbonization of Concrete and Durability of Reinforced Concrete. *Concrete* **1992**, *6*, 18–22.
29. Liu, Z.Y. Research on Durability Testing and Life Prediction Method of Marine Concrete Based on Environment. Ph.D. Thesis, Southeast University, Nanjing, China, 2006.
30. Niu, D.T. *Durability and Life Prediction of Concrete Structures*; Science Press: Beijing, China, 2003.
31. Xu, L.P.; Huang, S.Y. Mathematical Model for Predicting Carbonation Depth in Concrete. *J. Shanghai Build. Mater. Inst.* **1991**, *4*, 347–357.
32. Gong, L.S.; Su, M.Q.; Wang, H.L. Multi coefficient carbonation equation of concrete and its application. *Concr. Reinf. Concr.* **1985**, *6*, 10–16.
33. Papadakis, V.G.; Vayenas, C.G.; Fardis, M.N. Fundamental modeling and experimental investigation of concrete carbonation. *Mater. J.* **1991**, *88*, 363–373.
34. Possan, E.; Andrade, J.; Dal Molin, D.; Ribeiro, J. Model to estimate concrete carbonation depth and service life prediction. In *Hygrothermal Behaviour and Building Pathologies*; Springer: Cham, Switzerland, 2021; pp. 67–97.
35. Ekolu, S.O. Model for practical prediction of natural carbonation in reinforced concrete: Part 1-formulation. *Cem. Concr. Compos.* **2018**, *86*, 40–56. [CrossRef]
36. Liang, M.T.; Lin, S.M. Mathematical modeling and applications for concrete carbonation. *J. Mar. Sci. Technol.* **2003**, *11*, 3. [CrossRef]
37. Si, X.Y.; Hu, W.H.; Pan, H.M. Anti carbonation performance and prediction model of mineral admixture concrete. *Concrete* **2024**, *1*, 17–20+27.
38. Deng, S.F. Research on predicting the full life of concrete based on carbonation durability. *Constr. Mach.* **2023**, *11*, 119–125.

39. Zhang, J.Y.; Gu, S.Y.; Li, Q.F.; Huang, H.L. The influence of mineral admixtures on the carbonation performance of concrete. *Jiangxi Build. Mater.* **2022**, *11*, 40–42.
40. Zhang, C.; Song, C.Q.; Yao, J.; Song, X.Y.; Shi, P.C. The influence and mechanism analysis of Basf early strength agent on the durability performance of concrete. *Technol. Innov. Appl.* **2023**, *13*, 74–77.
41. Ding, Y.H.; Guo, S.Q.; Zhang, X.G.; Xu, P.; Wu, J.; Zhang, M.X. The influence of basalt fiber on the carbonation resistance of recycled concrete. *J. Compos. Mater.* **2022**, *39*, 1228–1238.
42. Qin, M.T. Preparation and Performance Study of Granite Mechanized Sand Concrete. Master's Thesis, Guilin University of Technology, Guilin, China, 2023.
43. Liang, X.G. *Carbonization and Mix Proportion Optimization of C20~C30 Concrete Mixed with Iron Tailings Powder*; Hebei Agricultural University: Baoding, China, 2022. [CrossRef]
44. Zheng, W.B. *Experimental Study on Carbonation Performance of Hybrid Fiber-Reinforced Iron Tailings Sand Concrete*; Xijing University: Xi'an, China, 2022. [CrossRef]
45. Xue, L.Y.; Hao, Q.J.; Zhang, W.K.; Sun, Q. Experimental study on the influence of different factors on the carbonation resistance of concrete. *Cem. Eng.* **2021**, *5*, 9–12. [CrossRef]
46. Wang, S.L. *Research on the Mechanical and Carbonization Properties of Mineral Admixtures and Basalt Fiber Reinforced Concrete*; Zhongyuan Institute of Technology: Zhengzhou, China, 2021. [CrossRef]
47. Xu, X.H.; Xie, J.F.; Xia, W.H.; Li, B.X. The influence of mineral admixtures on the performance and microstructure of bridge tower concrete. *China Harb. Constr.* **2021**, *41*, 40–45.
48. *GB/T 50082-2009*; Standard for Test Methods of Long-Term Performance and Durability of Ordinary Concrete. China National Standardization Administration Committee: Beijing, China, 2009.

materials

MDPI

Article

Experimental Determination of the Equivalent Moment of Inertia and Stresses of Aluminium Profiles with Thermal Breaks

Dawid Rusin [1,2,*], Janusz Juraszek [3] and Piotr Woźniczka [4]

[1] CUT Doctoral School, Cracow University of Technology, Warszawska 24, 31-155 Cracow, Poland
[2] Aluprof S.A., Warszawska 153, 43-300 Bielsko-Biała, Poland
[3] Faculty of Materials, Civil and Environmental Engineering, University of Bielsko-Biala, Willowa 2, 43-309 Bielsko-Biała, Poland; jjuraszek@ubb.edu.pl
[4] Faculty of Civil Engineering, Cracow University of Technology, Warszawska 24, 31-155 Cracow, Poland; piotr.wozniczka@pk.edu.pl
* Correspondence: dawrusin@gmail.com

Abstract: This paper presents the results of experimental tests and computer simulations on the stiffness of composite aluminium mullions used in unitised façades. The elements analysed were subjected to bending in order to simulate the actual operating conditions of aluminium façades subjected to significant wind pressure or suction loads. The basic mechanical and physical properties of the materials from which the analysed type of aluminium façade is made (Aluminium EN AW-6060 in the T66 temper and polyamide PA66 25GF), the test method, and the results obtained are described. As a result of the tests, equivalent moments of inertia of the composite profile (aluminium profile with the thermal break) were determined, which are strongly dependent on the strength of the connection between the individual elements, the asymmetry of the cross-section, and the properties of the thermal break. Strain measurements carried out using FBG (Fiber Bragg Grating) strain sensors installed in the profiles under tests allowed for determining the actual stress values of the aluminium profiles under consideration. The results obtained were compared to theoretical (numerical) values, indicating discrepancies at higher load values. The methodology presented in this article is to be used to monitor the deformation of the aluminium façade mullions of HRB (High-Rise Buildings).

Keywords: aluminium façade; FBG sensors; moment of inertia; stress; wind loads; structural monitoring

1. Introduction

Façades made of aluminium are popular in the construction industry because they offer a range of advantages [1,2]. Modern design, thermal and light comfort for users and no need for maintenance (f.e. corrosion resistance [3,4]) are the most valuable of these advantages. Moreover, automated production in workshops and fast installation are important for manufacturers [5]. According to estimates (2017), 12 billion dollars were spent per year on aluminium unitized facades alone [6], which were used twice as often as stick facades (2012) [7]. In city centres in the second half of the 20th century, the number of new buildings with aluminium facades exceeded the number of buildings made of traditional masonry walls [8]. Aluminium profiles made from EN AW-6060 alloy (EN 573-3 [9]) in the T66 temper (EN 755-2 [10]) are lightweight and are characterised by high tensile strength [11]. The basic mechanical and physical properties of the alloy used are given in Table 1 [9], and its chemical composition is in Table 2.

Table 1. Physical and mechanical properties of EN AW-6060 alloy for the 0–3 mm wall thickness range.

Property	Value	Unit
Alloy–Condition	AW-6060	-
Density	2.7	g/cm^3
Elasticity modulus E	69,500	MPa
Shear modulus G	26,100	MPa
Poisson's ratio	0.33	-
Solidification temperature	610	°C
Melting temperature	655	°C
Specific heat at 20 °C	898	J/kgK
Thermal expansion coefficient at 20 °C	23.4	µm/mK
Specific resistance	32	nWm
Thermal conductivity	209	W/mK
Electrical conductivity	54	[%IACS]
Tensile strength Rm min.	215	MPa
Tensile strength Rm max.	-	MPa
Yield point Rp02 min.	160	MPa
Yield point Rp02 max.	-	MPa
Elongation min. A50 mm	8	%
Elongation min. A	6	%

Table 2. Chemical composition AW-6060 alloy (data provided by the manufacturer).

Alloy Element	Amount [%]
Mg	0.35–0.60
Mn	≤0.10
Fe	0.10–0.30
Si	0.30–0.60
Si + Fe	-
Cu	≤0.10
Zn	≤0.15
Cr	≤0.05
Mn + Cr	-
Ti	≤0.10
Bi	-
Ni	-
Pb	-
Sn	-
Zr	-
Ga	-
V	-
Zr + Ti	-
Other	≤0.05
Other − total	≤0.15
Al	residual

The basic types of aluminium façades are mullion-and-transom façades and unitised façades. The first ones are characterized by the fact that most of the assembly work—such as screwing together aluminium structural elements and installing glass—takes place on the construction site. Unitised façades [12,13], on the other hand, are characterised by the possibility of transferring the burden of fabrication of the aluminium frames and their glazing to the manufacturing plant and compensating for significant expansion movements [7,11]. This has a huge impact on speeding up the work and reducing the possibility of execution errors.

An important feature of façades made of aluminium is that they can be recycled [13], making them environmentally friendly construction elements. Ease of machining and high tolerance of their manufacture allows a wide variety of shapes and structures to be created [14,15], while large glazing provides better access to natural light and contributes to thermal comfort inside the building.

Due to the fact that as much as 80% of energy in buildings is currently used for heating purposes [16], each façade should be characterized by the lowest possible thermal transmittance coefficient. This is currently 0.9 W/m^{2}K [N] for window and door structures and façades [17].

In order to reduce the heat transfer coefficient and improve the thermal properties of the façade, the external and inner aluminium profiles are separated by thermal breaks, usually made of polyamide. Polyamides (PA) are a group of semi-crystalline thermoplastic polymers characterized by high chemical, thermal and mechanical resistance. PA has a low density (approx. 1.14 g/cm^3) and a relatively low melting point (approx. 250 °C), which makes it easier to process. An important advantage of PA is its stability of shape at high temperatures, which is important in the context of sections with a thermal break due to their exposure to high temperatures during powder coating and during the actual use of the façade. To improve mechanical and thermal properties (Table 3), polyamides are often modified by the addition of glass fibre, which leads to increased stiffness, creep resistance, and dimensional stability. In construction applications, specially in façade systems, polyamides reinforced with 25% glass fibre (PA 66 GF25) are commonly used. The use of such an addition leads to a thermal expansion coefficient close to that of aluminium, which prevents the profiles from being damaged by changes in their length caused by large temperature fluctuations. The main function of PA profiles is to reduce the heat transfer coefficient between the exterior and interior of the building, which is achieved through their insulating properties.

Table 3. Physical and mechanical properties of the PA66 GF25 (data provided by the manufacturer).

Property	Standard	Unit	Dry Material
Melting temperature	ISO 3146 [18]	C	min. 250
Density	ISO 1183-1- or -3 [19,20]	g/cm^3	1.3 +/− 0.05
Glass fibre content	ISO 1172 [21]	%	25 +/− 2.5
Shore D	ISO 868 [22]		82 +/− 4
Impact strength	ISO 179-1 [23]	kJ/m^2	min. 30
Tensile strength	527-2 and -4 [24,25]	N/mm^2	min. 80
Young's modulus	527-2 and -4 [24,25]	N/mm^2	4500
Elongation at rupture	527-2 and -4 [24,25]	%	min. 3

Despite the advanced technical solutions and the aforementioned advantages of aluminium façades, it should be noted that there are some challenges in designing this type of structure. The force resulting from the action of the wind on the glass panels of the façade is transferred to the aluminium frame (made of mullions and transoms), which transmits it to the building structure via hooks and consoles [3,4]. Aluminium walls are characterised by lower rigidity compared to traditional masonry or reinforced concrete walls. Therefore, appropriate structural solutions are required. Aluminium profiles yield when exposed to wind [1,3,4,7], so sections with appropriate wall thicknesses must be used. In thin-walled aluminium profiles, tangential stresses may occur, causing excessive deformation and even loss of stability [14]. When designing aluminium façades, the requirements of the PN-EN 13830 [26] and PN-EN 1999-1-1 [27] standards should be taken into account.

In this context, this paper presents the results of our own experimental research and computer simulations concerning composite aluminium profiles. The aim of this research was to determine the actual deflections and, consequently, the actual moments of inertia of the profiles with the thermal break in the direction of wind pressure, taking into account the interaction between the individual components, i.e., the aluminium profiles and the thermal break. The obtained results were used to calibrate the FEM model intended for further parametric analyses. Deformation measurements of the aluminium profile were also carried out using FBG (Fibre Bragg Grating) sensors, which allowed the calculation of the actual stress values. The developed innovative sensor system is to be ultimately used to monitor deformations of aluminium façade mullions of HRB (High-Rise Building).

2. Experimental Research

2.1. Methodology

The tests consisted of three-point bending of aluminium profiles (with and without a thermal break). The load from the cross-beam of the testing machine was transferred to the profiles using steel blocks (made of S235 steel, with dimensions of 18 × 20 × 20 mm for profiles without a thermal break and 18 × 20 × 85.7 mm for profiles with a thermal break) placed in the glazing gasket seat zone in order to reproduce the positive wind load (pressure) as accurately as possible [Figure 1].

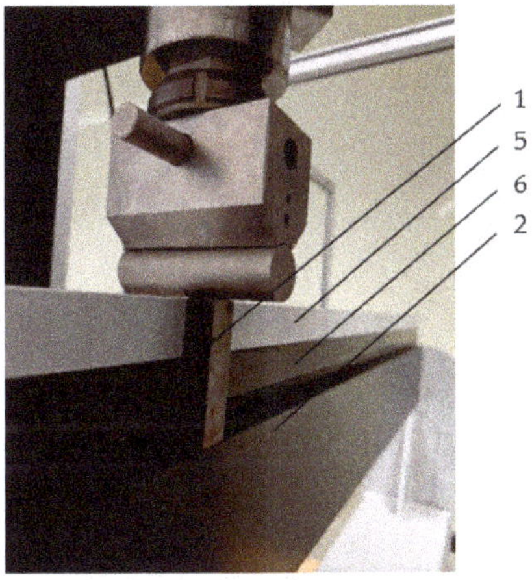

(**a**)

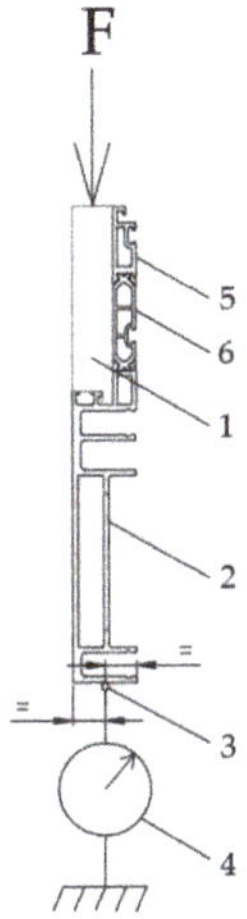

(**b**)

Figure 1. (**a**) Method of transferring the load to the profile; (**b**) diagram of the loading of the profiles by the steel cube and placement of the measuring instruments; 1—steel block, 2—inner half of the profile with a thermal break (aluminium), 3—FBG sensor, 4—mechanical displacement sensor, 5—external half of the profile with a thermal break (aluminium), 6—the thermal break profile (PA GF25).

The experiments were conducted on a Galdabini Quasar 50 testing machine. The machine's load-cell reading resolution was over three million divisions (24-bit A/D converter). A load program was developed that ensures that the applied force, in steps of 1 kN, increases from 0 to 10 kN and decreases from 10 kN to 0. Displacements in the middle and two opposite ends of the tested profile (sensors P1, P2 and P3) [Figure 2], as well as deformations using the FBG sensor, were recorded. The supporting structure consisted of a steel tube (S235 JR) with a 200 × 200 × 6 mm cross-section, to which two flat bars with cut-outs (saddles) were welded, in which the tested profiles were placed. The saddles ensured vertical and horizontal stabilization of the tested profile. The spacing between supports was 1910 mm [Figure 2]. The points of measurement (P1, P2, P3) of the displacement were introduced to eliminate the effect of deformation of the supporting structure and cancel out any play in the saddles.

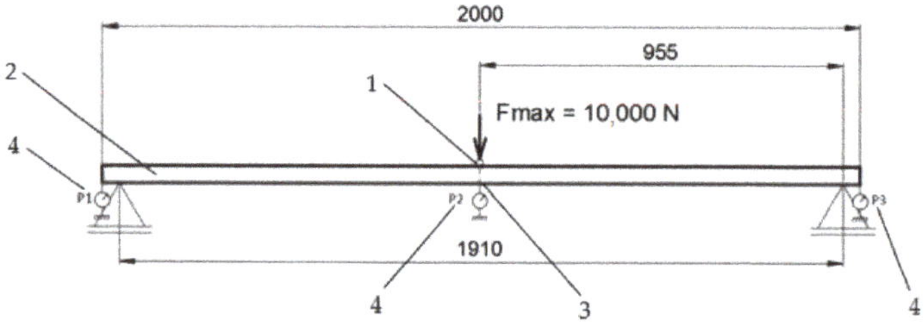

Figure 2. Test bench diagram; 1—steel block, 2—profile (with or without a thermal break), 3—FBG sensor, 4—mechanical displacement sensor.

Movements were measured using mechanical dial sensors with digital displays. The resolution of the measurement was 0.01 mm, and its accuracy was 0.02 mm [Figure 3].

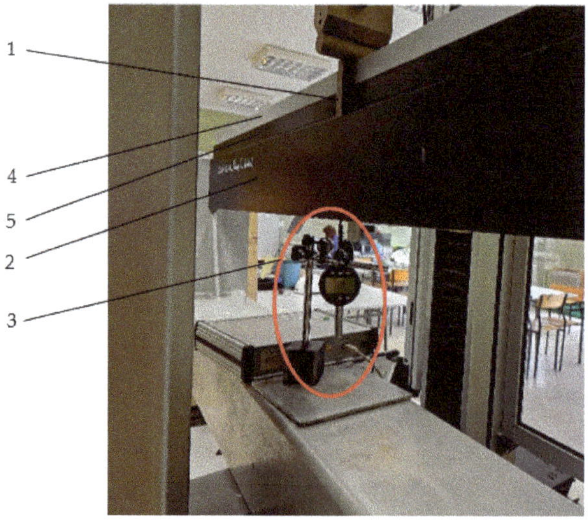

Figure 3. View of the mechanical displacement sensor; 1—steel block, 2—inner half of the profile with a thermal break (aluminium), 3—mechanical displacement sensor, 4—external half of the profile with a thermal break (aluminium), 5—the thermal break profile (PA GF25).

FBG sensors with wavelengths from 1530–1560 nm and a maximum measurement range of 3000 microstrain (με) were used to record the strain. These were glued to the rear inner walls of the column profile halves [Figures 1b and 4] using a specialized 3M DP490 Scotch-Weld epoxy adhesive.

The system was powered by light using a two-channel FOS&S FBG-scan 800 interrogator capable of operating at wavelengths of 1515–1590 nm and at a scanning frequency of 2 kHz. The ILLumiSense Strainnumer software (version v2.1.3) was used to record the strain results.

Figure 4. View of FBG sensor glued to aluminium; 1—inner half of the profile with a thermal break (aluminium), 2—FBG sensor, 3—mechanical displacement sensor.

2.2. Subject of Tests

The mullion profiles of unitised aluminium façade with a length of 2000 mm were tested. These sections are characterized by high slenderness and cross-section asymmetry. They mainly transfer bending forces resulting from wind action. The tensile forces resulting from the weight of the glazing transferred to the mullions through the angle connectors are small, such that they can be neglected. Torsion [7] resulting from section asymmetry and the eccentricity of the load application point was neglected due to the relatively high torsional stiffness for a given profile length. The characteristic dimension of these mullions is their depth calculated from the sealing seat of the inner profiles to their rear wall. In the case of the tested pieces, it was 150 mm [Figure 5].

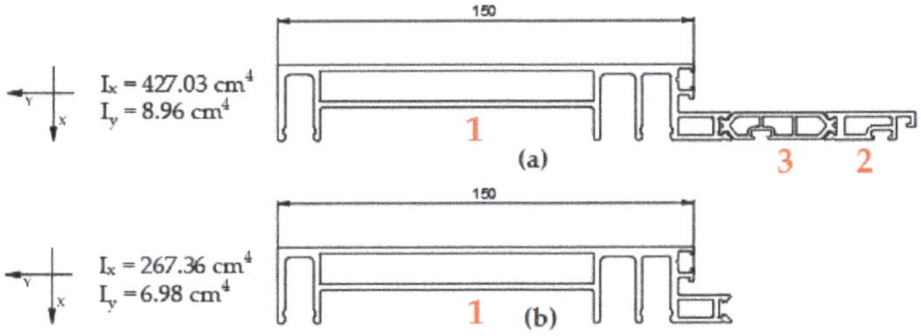

Figure 5. (**a**) Aluminium profile with the thermal break; (**b**) Inner half of the profile with the thermal break; 1—inner half of the profile with the thermal break (aluminium), 2—external half of the profile with the thermal break (aluminium), 3—the thermal break (PA GF25).

The tested profiles with the thermal break can be manufactured using one of two different technologies. The first of these (the so-called "single-colour")—hereinafter referred to as technology A—involves mechanically connecting (crimping) the two profiles together using a plastic spacer and then subjecting them to a varnishing process, during which both sections are subjected to a temperature of approx. 200 °C for approx. 20 min. The profiles then go through the processes of degreasing, etching, washing, and drying [28]. The second technology (so-called "two-colour")—hereinafter referred to as technology B—involves separating the painting of outer and inner aluminium profiles and then crimping them together without any painting processes, which could affect the plastic layer itself [28]. The first method of manufacturing is more advantageous because of logistical and economic aspects.

In addition to the profiles with the thermal break, the aluminium inner sections of these profiles were also tested [Figure 5b] for comparison purposes. Six series of tests were carried out—three for the profiles with the thermal break and three for the inner halves of the profile.

The main sources of measurement errors for this study are inaccuracies of the FBG sensors and mechanical displacement sensors and testing machine load precision. The values of these inaccuracies have been previously described.

3. Laboratory Results and Discussion

3.1. Inner Halves of Profiles with the Thermal Break

Figures 6–9 show the results obtained for the inner halves of the profile. In each of these, the results obtained are compared with theoretical values. The cross points in the graph are the minimum and maximum values obtained during experiments on the discussed samples.

Figure 8 shows the strain values determined based on physical relationships at the mid-length of the tested profile based on the strain measurement.

The deflection, deformation, and strain results, irrespective of the applied force, coincide with the theory, especially in the load build-up phase. A slight decrease in the mechanical parameters can be observed in the stress-relieving phase of the tested piece. The dependence of the moment of inertia on the force, especially at lower values, differs slightly from the theoretical values.

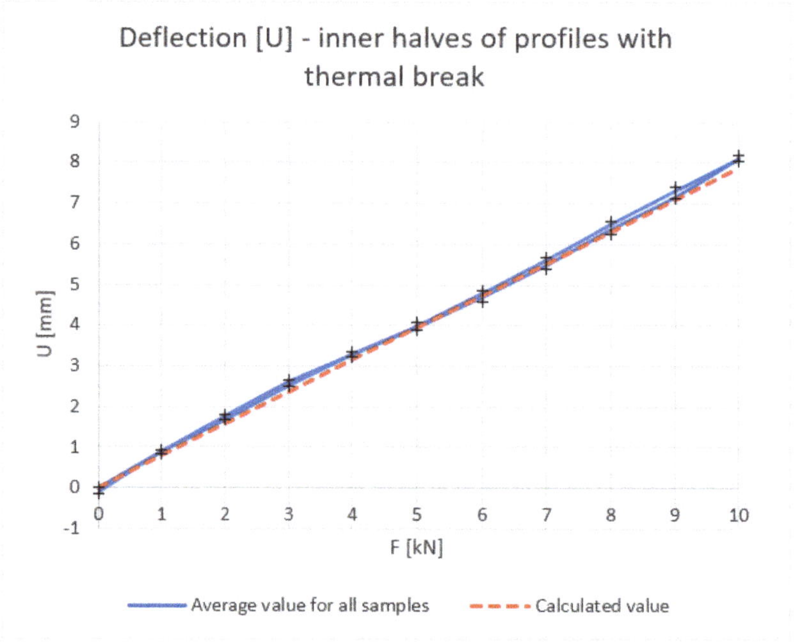

Figure 6. Deflection U [mm] of inner halves of profiles with the thermal break (average value for loading and unloading of the beam). Calculated value—according to the theoretical calculation.

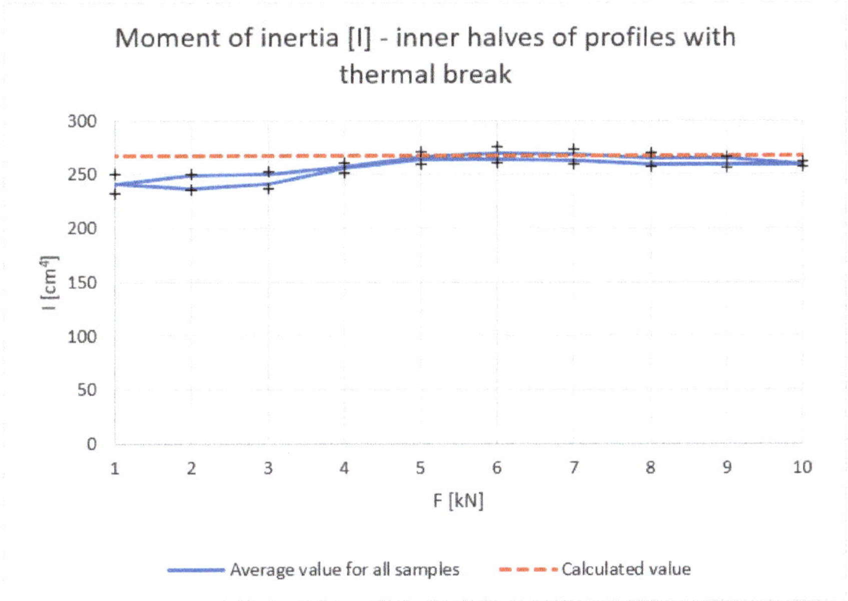

Figure 7. Moment of inertia I (cm^4) of inner halves of profiles with the thermal break (average value for loading and unloading of the beam). Calculated value—according to the theoretical calculation.

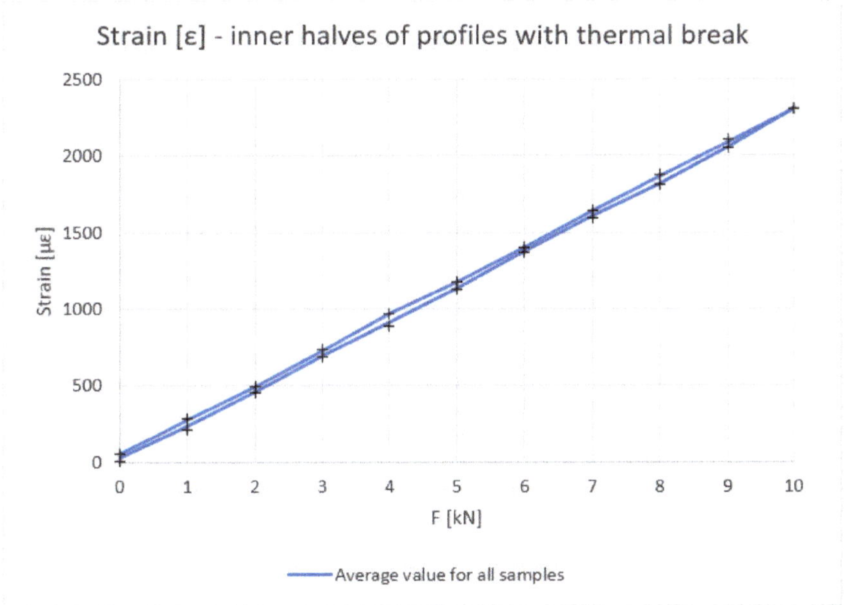

Figure 8. Strain ε (με) of inner halves of profiles with the thermal break (average value for loading and unloading of the beam).

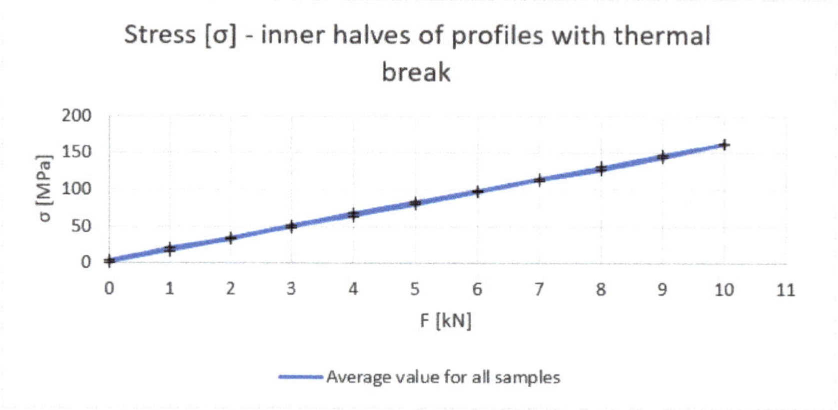

Figure 9. Stress σ (MPa) of inner halves of profiles with the thermal break (average value for loading and unloading of the beam).

3.2. *Profiles with the Thermal Break*

Figures 10–13 show the results obtained for the profiles with thermal breaks. In the case of components manufactured according to single-colour technology (A), the results were presented as an average of two samples. The theoretical moment of inertia for profiles with the thermal break was calculated following the methodology given in PN-EN 14024:2004 Annex C [29]. The presented graphs show that both the deflection and deformation of the profile do not return to zero points after unloading.

Figure 12 shows the strain values determined for the profiles with the thermal break based on physical relationships at the mid-length of the profile and the deformation measurement. Values up to 8 kN bending force can be considered correct. Above this value, non-linear effects occur, probably related to the operation of the thermal break.

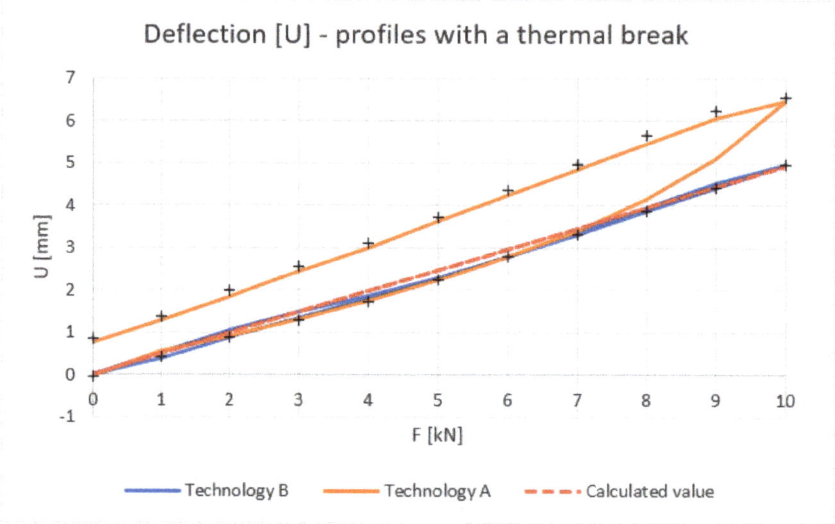

Figure 10. Deflection U [mm] of profiles with a thermal break (average value for loading and unloading of the beam). Calculated value—according to [29].

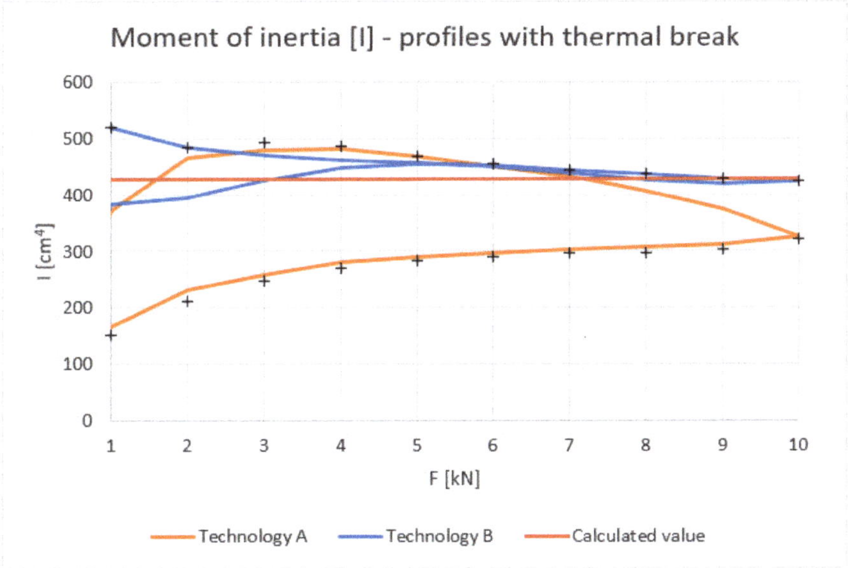

Figure 11. Moment of inertia I (cm^4) of profiles with the thermal break (average value for loading and unloading of the beam). Calculated value—according to [29].

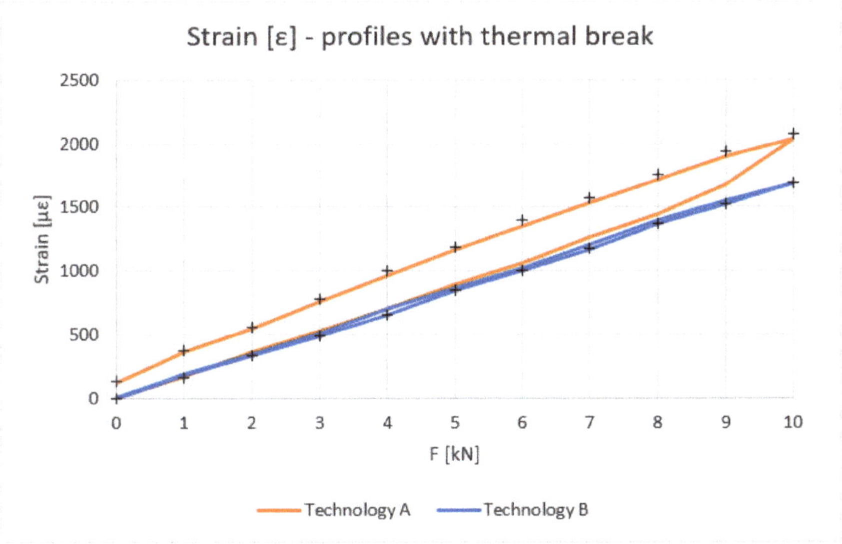

Figure 12. Strain ε (με) of profiles with the thermal break (average value for loading and unloading of the beam).

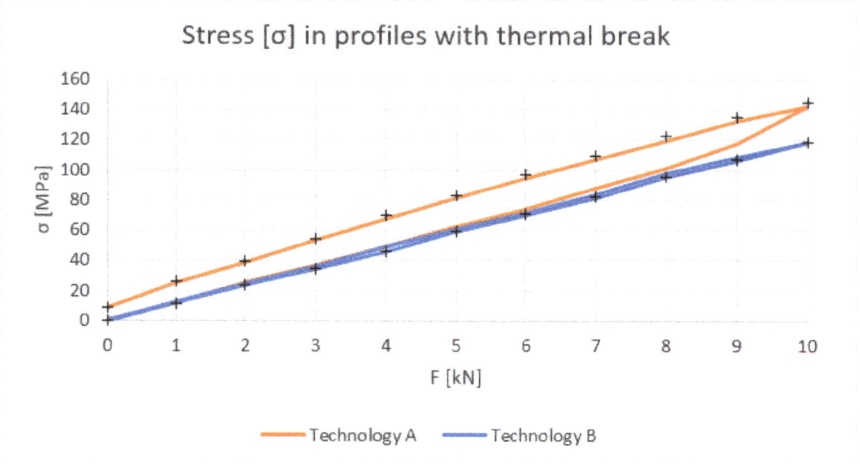

Figure 13. Stress σ (MPa) in profiles with the thermal break (average value for loading and unloading of the beam).

3.3. *Evaluation of the Test Results*

The tests showed that the results for the inner halves of the façade mullion profiles are highly consistent with theory. The average deviation of the experimental values from the theoretical values at a load of 10 kN was approximately 3%. When the load was relieved to 1 kN, the deviation averaged 9%. Profiles with the thermal break manufactured using the single-colour technology (A) at a load value of 10 kN showed a discrepancy of about 25% in relation to the theoretical values, whereas after relieving the load to a value of 1 kN, this discrepancy was, on average, 60%. The profiles manufactured in two-colour technology (B) at a maximum load of 10 kN showed a 1% discrepancy with the theory, and after their unloading to 1 kN, a deviation of about 22% was observed. The high convergence of the results with the theory in the case of the inner profiles of the profile halves is caused by the lack of additional plastic elements used as a thermal break. The beam material is homogeneous and behaves linearly in elastic range. These profiles are never used as stand-alone elements in frame structures and are always a component of profiles with thermal breaks manufactured by either A or B technology. The results of these tests confirmed that the aluminium profiles of the mullions operate within their elastic range under a given load and demonstrated the consistency of the calculated results with the obtained experimental ones. Profiles with thermal breaks that include a plastic thermal break are characterized by non-linearity. Their mechanical properties depend not only on their length, described in [29], but also on the value of the applied load, which can be seen in the result graphs. In the case of profiles manufactured using two-colour technology (A), the deviation values are the highest. As already described, their length was 2000 mm, but the distance between the points of support (saddles) was 1910 mm. According to standards [26,30], the maximum permissible deflection of curtain wall profiles between support points should not exceed $d \leq L/200$ or 15 mm, whichever is smaller. For the profiles tested, it was, therefore, 9.5 mm. This value was not exceeded in any of the cases tested, i.e., for both profiles with the thermal break manufactured using single-colour technology (A) (maximum deflection of 6.5 mm) and two-colour technology (B) (maximum deflection of 4 mm). However, when analysing the nature of the decrease in mechanical parameters and the high hysteresis of profiles manufactured using technology A, it should be assumed that despite the lack of visual signs of their damage, these profiles may not guarantee their structural performances. This behaviour of the profiles can be attributed to the effect of the coating temperature on the thermal break, which is approximately 200 °C, relatively close to the melting point of its material (PA 66 GF25) of 250 °C. Additionally, the previously mentioned technological processes (degreasing, etching, rinsing, and drying [28]) performed during the painting process, if we take into account the geometry of the connection of the aluminium profiles with the thermal break, may also result in the deterioration of their mechanical properties.

The primary sources of measurement error may include inaccuracies in the load application by the Quasar 50 testing machine and imprecision in the strain measurement using mechanical displacement sensors.

All results are summarized in the Figures 14–17.

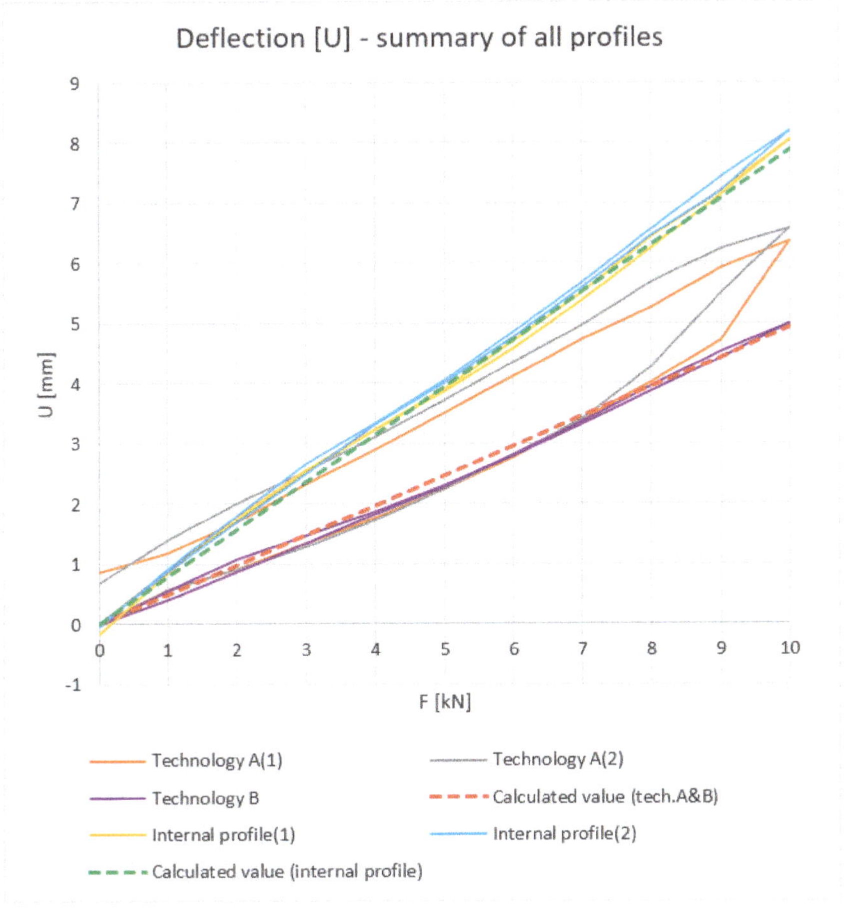

Figure 14. Load [kN]–Deflection U (mm): summary values of inner halves of profiles and profiles with the thermal break made in single-colour (A) and two-colour technology (B).

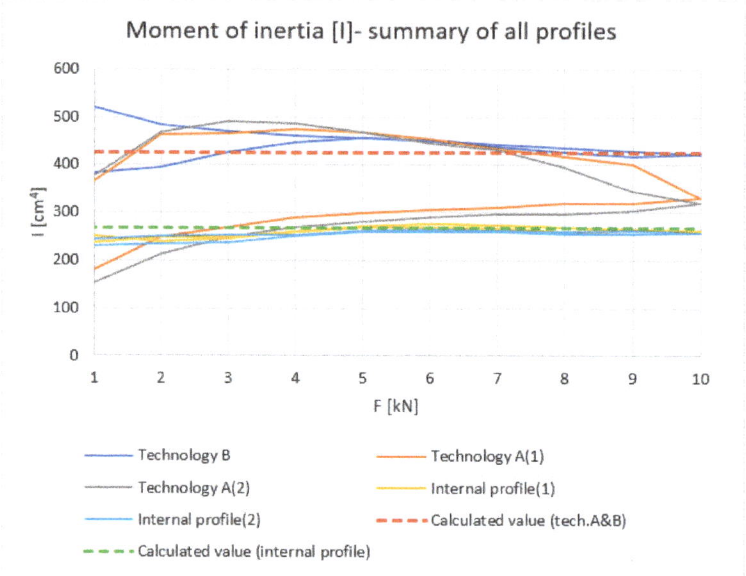

Figure 15. Load [kN]–Moment of inertia I (cm^4): summary values of inner halves of profiles and profiles with the thermal break made in single-colour (A) and two-colour technology (B).

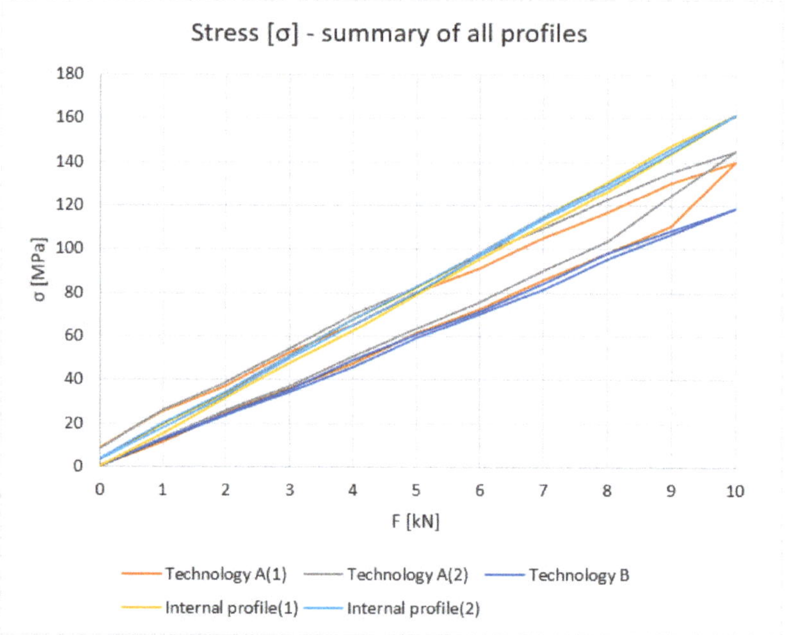

Figure 16. Load [kN]–Stress σ (MPa): summary values of inner halves of profiles and profiles with the thermal break made in single-colour (A) and two-colour technology (B).

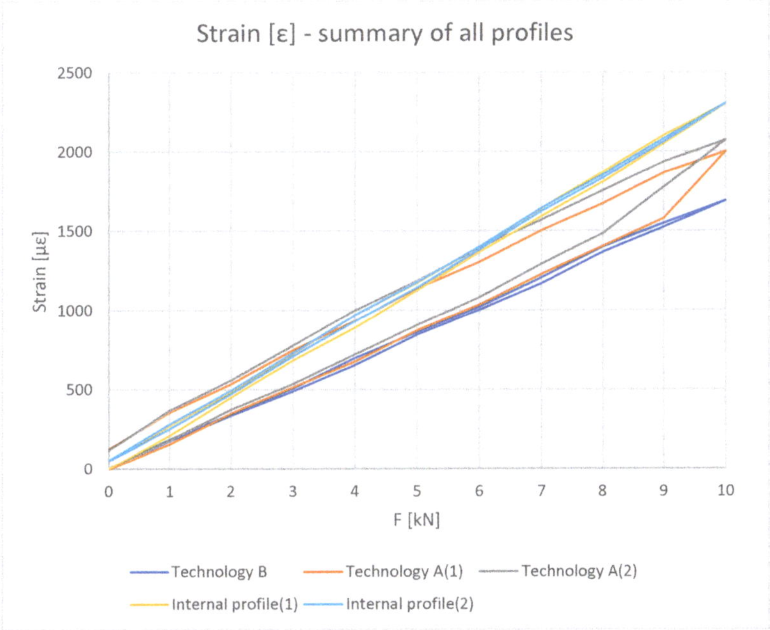

Figure 17. Load [kN]–Strain ε (με): summary values of inner halves of profiles and profiles with the thermal break made in single-colour (A) and two-colour technology (B).

3.4. Evaluation of the Use of FBG Sensors

During the tests, it was observed that FBG sensors perform well as devices measuring the deformation of aluminium frame profiles of unitised façades and can be implemented as part of a continuous monitoring system in HRBs. This is supported by their maintenance-free design, durability, compact dimensions, and high measurement accuracy. Their use allows the calculation of the stress occurring in the loaded profile. A major advantage is that the sensor can be calibrated to also measure profile deflections, which was verified during the tests. This process consisted of a simultaneous reading of the displacement measured using traditional displacement sensors and the deflection measured with an FBG sensor. The relationship between these values was linear; hence, by reading the deformation, the exact value of the displacement is known. The measurement method is described in more detail in the paper [31]. This makes it possible to eliminate the use of a mechanical displacement measurement system, which consists of at least three sensors (for each profile) mounted on an independent substructure, and replace it with a single optical fibre bonded to the tested element. Regarding the HRB's continuous façade monitoring system, it is important to limit the size of the measuring and recording equipment due to the inconvenience to normal building use. In addition, FBG sensors can also record other parameters, such as the intrinsic vibrations of the profiles or their temperature. A review of the scientific literature [31,32] also confirms the validity of their selection for these purposes.

The strain measurement could be influenced by the compliance of the epoxy adhesive used to bond the FBG sensor to the profile, although this effect is difficult to quantify and is likely negligible.

4. Numerical Analysis and Discussion

4.1. Computer Model

The previously described experimental studies were used to validate the computer model developed using the Abaqus program [33]. Aluminium profiles and gaskets were modelled using general-purpose linear four-sided shell elements (S4R). The transition zones between the individual profiles, the gaskets, and the steel load element were, in turn, modelled using C3D8R solid elements (general-purpose eight-node linear brick element, with reduced integration and hourglass control). Similar approaches have previously been successfully used by the authors for computer modelling of full-scale experimental studies of a wind-loaded aluminium curtain wall [1] and also for other complex analyses of thin-walled shell structures [34,35].

The aluminium sections and gaskets were, in each case, divided into rectangular finite elements with dimensions equal to 5 × 5 mm (locally 2 to 4 mm). For the steel loading bar, a finite element mesh with dimensions of 4 × 5 × 5 mm was used. Further refinement of the basic mesh (for aluminium members, steel bar and gaskets) to a side length of 2.5 mm did not significantly affect the results. The calculated deflection values and stresses changed up to 2.5%. Taking into account computational cost, the further mesh size reduction was omitted. However, in the central part of the cross-section where the stress values were read, the mesh was densified until the convergence was reached. Finally, a local dimension of 1 × 1 mm was adopted. Simulations have been conducted on a Fujitsu Celsius M740 unit (Xeon E5-2630v3 processor; 2 × 16 GB DDR4-2133 rg ECC). Depending on the case and mesh density, the required time for a single simulation varied from just a few minutes to nearly one hour.

Computer simulations were run as geometrically and materially non-linear (GMNA) simulations. The computational approach used, which does not take into account the influence of self-stress and geometric imperfections, can be justified by the significant slenderness of the component under test while, at the same time, the measured values of geometric deviations are negligible. For all elements, the elastic-plastic material model was used. Regarding thermal breaks made of PA 66 GF25, this approach should be considered simplistic [36] but can be justified by the expected work of the component in the elastic range resulting from the relatively low stress values. Full integration of the individual components in the computer model was assumed. The contact problem was only considered for the contact surface between the aluminium profile and the steel load bar. In the case of the aforementioned zone, a two-surface contact problem was applied. Normal ('hard contact') and tangential behaviour (friction) were considered. The beam load was modelled as a uniformly distributed load applied to the upper surface of the steel element. The load value was changed linearly following the assumptions made during the experimental studies. The model was supported in a fork-like manner. Additionally, a horizontal support for the steel loading element was introduced to the model. This ensured a better representation of the conditions prevailing during the tests and eliminated the numerical instabilities of the model. Views of the developed model, together with the adopted boundary conditions, are shown in Figure 18.

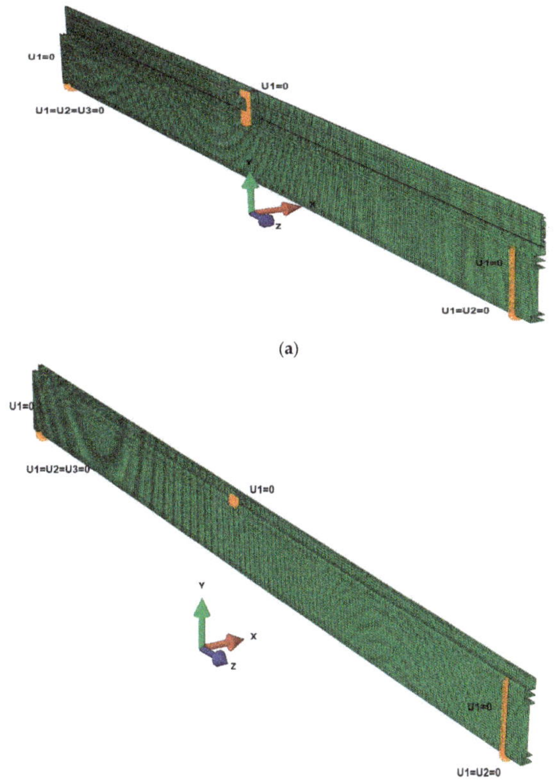

(**a**)

(**b**)

Figure 18. Boundary conditions of the (**a**) profile with the thermal break; (**b**) inner half of profiles with the thermal break.

4.2. Results of the Computer Modelling

Figure 19 compares the deflections obtained during the experimental tests with the results of the computer simulations. Table 4 presents the maximum deflection values determined for individual cases. Finally, Figure 20 shows the deformation of the composite profile subjected to a maximum force of 10 kN.

As can be seen from the presented data, the convergence of the results of computer modelling and experimental studies should be considered satisfactory. Maximum differences of up to approx. 20% between the FEM results and the experimental results were recorded for beams manufactured using single-colour technology A. It should be noted that in the case of models developed for these beams, the influence of heat treatment on the mechanical properties of the thermal break was not considered. However, the aforementioned treatment can lead to slight changes in the mechanical properties of aluminium, amounting to approximately 10% regarding the conventional yield strength and about 1.5% regarding the elastic modulus [37]. For the other elements, the achieved convergence of results can be considered highly satisfactory. In the case of a composite beam manufactured using the B-type ("two-colour") technology, computer modelling led to a slight overestimation of the deflection value (maximum 8%). In contrast, for profiles without a thermal break, the modelling results led to a minimal overestimation of deflection (2% to 4%, cf. Table 4). In the case of the maximum stress values (Table 5), the differences between the computer simulation results and the experimental results were (for single-colour profiles and aluminium profiles) up to a maximum of 10%. Comparison of the stress results (Table 6) determined in the "z" direction (see Figure 18b) with the readings taken using optical fiber indicate greater discrepancies (from 12 to max. 18%). The observed differences may be caused by both the necessary simplifications of the complex geometry of the profiles and the way the seal itself works, which starts to transfer loads with a certain delay. Additionally, as already mentioned, the computer model does not take into account residual stresses, the value of which, in the case of similar aluminium alloys, can be up to 20 MPa [38,39]. Adding the mentioned value of residual stresses to the results of computer modelling leads to a safe estimation of the stress values in relation to the results of measurements performed during experimental tests. In summary, the results obtained should be considered satisfactory.

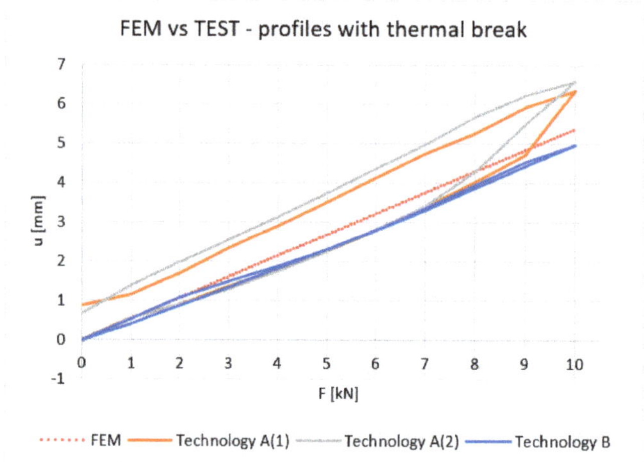

(**a**)

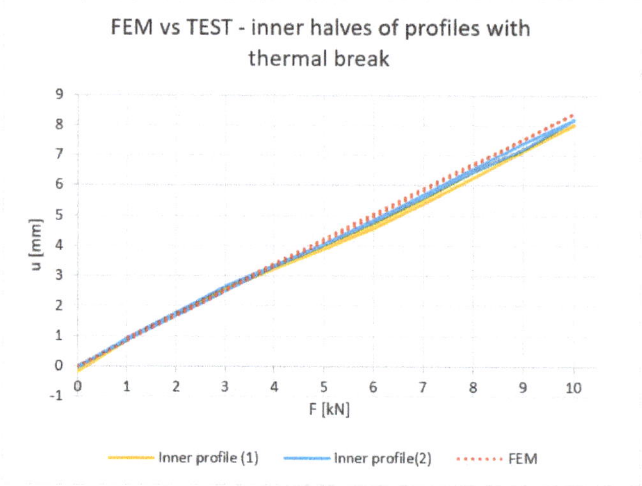

(**b**)

Figure 19. (**a**) Comparison of FEM deflection and tests—profiles with the thermal break, (**b**) Comparison of FEM deflection and tests—inner halves of profiles with the thermal break.

Table 4. Values of deflection calculated by FEM and determined during experimental tests.

	Tech. B	Tech. A (1)	Tech. A (2)	Inner Prof. (1)	Inner Prof. (2)
$U_{max.TEST}$ [mm]	4.98	6.36	6.58	8.04	8.19
$U_{max.FEM}$ [mm]		5.39		8.39	
FEM/TEST	1.08	0.85	0.82	1.04	1.02

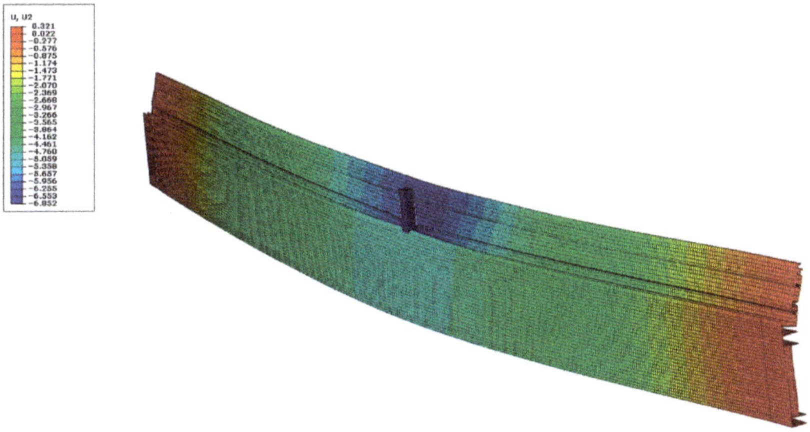

Figure 20. Deflection [mm] of profile with the thermal break for F = 10 kN.

Table 5. Maximum stress values calculated by FEM and obtained during experimental tests. RS—residual stresses of assumed value of 20 MPa [38,39].

	Tech. B	Tech. A (1)	Tech. A (2)	Inner Prof. (1)	Inner Prof. (2)
$\sigma_{max.TEST}$ [MPa]	118.37	140.07	145.04	161.35	161.56
$\sigma_{max.FEM}$ [MPa]		106.4		151.5	
$\sigma_{max.FEM}$ + RS [MPa]		126.4		171.5	
FEM/TEST	0.90	0.76	0.73	0.94	0.94
FEM + RS/TEST	1.07	0.90	0.87	1.06	1.06

Table 6. Stress values in the lower strip axis calculated using FEM and determined during experimental tests. RS—residual stresses of assumed value of 20 MPa [38,39].

	Tech. B	Tech. A (1)	Tech. A (2)	Inner Prof. (1)	Inner Prof. (2)
$\sigma_{max.TEST}$ [MPa]	118.37	140.07	145.04	161.35	161.56
$\sigma_{z.FEM}$ [MPa]		97.0		142.2	
$\sigma_{z.FEM}$ + RS [MPa]		117		162.2	
FEM/TEST	0.82	0.69	0.67	0.88	0.88
FEM + RS/TEST	0.99	0.84	0.81	1.00	1.00

5. Conclusions

1. Experimental tests have confirmed the linear relationship between deformation and load for a uniform inner profile. Satisfactory convergence between theoretical, experimental, and computer simulation results has been demonstrated. For the deflections, the difference between FEM results and experimental results was up to 8% (two-colour and aluminium profiles). In relation to stresses, the final results of FEM and experimental investigations were nearly identical (difference of up to 1%).

2. Significant differences were observed in the values of deflections (up to 32%) and strains (up to 23%) for profiles with thermal breaks made using different single-colour (A) and two-colour (B) technologies.
3. The deformation hysteresis is lower in the case of profiles made using two-colour technology (B) (Figure 10), and the load–deflection curve is nearly the same for both loading and unloading cases.
4. Single-colour technology (A) is characterized by higher strain-deflection hysteresis occurring between the loading and unloading processes. It is as high as 25%. This alters the value of the equivalent moment of inertia.
5. Two-colour technology (B), in comparison to single-colour technology (A), is characterized by higher deformation stability in loading processes and a stable value of the equivalent moment of inertia.
6. Satisfactory convergence was achieved between the computer modelling results and the experimental results for the half sections and beams manufactured using two-colour technology (B). The maximum difference between FEM and experimental results was up to 8%. This should be considered acceptable. Thus, a developed computer model can be used for further parametric tests of aluminium façade elements. In the case of technology A beams, further material tests of the gaskets are necessary to determine their parameters resulting from the thermal treatment.
7. This research has shown that FBG sensors can be successfully used as continuous monitoring elements in the study of the deformation of aluminium façades.

Author Contributions: Conceptualization, J.J.; methodology, D.R.; software, J.J. and P.W.; validation, P.W.; formal analysis, D.R.; investigation, D.R.; resources, J.J. and P.W.; data curation, D.R.; writing—original draft preparation, D.R.; writing—review and editing, D.R. and P.W.; visualization, D.R.; supervision, J.J.; project administration, J.J. and D.R.; funding acquisition, J.J. All authors have read and agreed to the published version of the manuscript.

Funding: Research was funded by the implementation doctorate grant No. DWD/6/0524/2022.

Data Availability Statement: The original contributions presented in this study are included in the article. Further inquiries can be directed to the corresponding author.

Conflicts of Interest: The authors declare no conflicts of interest.

References

1. Juraszek, J.; Woźniczka, P.; Rusin, D. Badania eksperymentalne i symulacje komputerowe ściany osłonowej MB-SR50N HI obciążonej wiatrem. *Mater. Bud.* **2022**, *9*, 53–55. [CrossRef]
2. Górka, M.; Leśniak, A. Systemy fasad aluminiowo–szklanych i ich ocena wielokryterialna. *Zesz. Nauk. Politech. Częstochowskiej* **2018**, *24*, 102–107. [CrossRef]
3. Masoumeh, A.; Shanmuganathan, G.; Nima, T.; Benoit, P.G.; Hong, G. Numerical study on bearing behaviour and design of aluminium sub-heads in façade systems. *Thin-Walled Struct.* **2021**, *168*, 108140. [CrossRef]
4. Masoumeh, A.; Shanmuganathan, G.; Benoit, P.G.; Hong, G. Experimental study on bearing behaviour and design of aluminium sub-heads in façade systems. *Thin-Walled Struct.* **2020**, *156*, 106867. [CrossRef]
5. Coelho, G.B.A.; Ostapska, K.; Kraniotis, D.; Brzozovsky, J.; Loli, A. Development of a hybrid timber and aluminium based unitized façade system resilient to the future weather conditions in Europe via monitoring campaigns and computational models. *Procedia Struct. Integr.* **2024**, *55*, 39–45. [CrossRef]
6. Lee, A.; Shepherd, P.; Evernden, M. Optimizing the cross-sectional shapes of extruded aluminium structural members for unitized curtain wall facades. *Structures* **2017**, *10*, 147–156. [CrossRef]
7. Leea, A.D.; Alimanza, J.A.; Shepherd, P.; Evernden, M.C. Axial Rotation and Lateral Torsional Buckling of Extruded Aluminium Mullions in Curtain Wall Facades. *Structures* **2019**, *20*, 658–675. [CrossRef]
8. Leea, A.D.; Shepherd, P.; Evernden, M.C.; Metcalfe, D. Optimizing the architectural layouts and technical specifications of curtain walls to minimize use of aluminium. *Structures* **2018**, *13*, 8–25. [CrossRef]

9. *PN-EN 573-3*; Aluminium i Stopy Aluminium-Skład Chemiczny i Rodzaje Wyrobów Przerobionych Plastycznie-Część 3: Skład Chemiczny i Rodzaje Wyrobów. Polski Komitet Normalizacyjny: Warszawa, Poland, 2014.
10. *PN-EN 755-2*; Aluminium i Stopy Aluminium-Pręty, Rury i Kształtowniki Wyciskane-Część 2: Własności Mechaniczne. Polski Komitet Normalizacyjny: Warszawa, Poland, 2016.
11. Kozicka, E.; Kukawska, E.; Rabczak, K. Fasady Aluminiowe w Eurokodzie 9. *Builder* **2019**, *263*, 106–111. [CrossRef]
12. Pulling, L.L.; Muessel, D.C.; Wille, H.S. Wall Construction. US Patent 3,037,591, 5 June 1962. Available online: http://www.google.com/patents/US3037591 (accessed on 18 November 2024).
13. Al-Alimi, S.; Yusuf, N.K.; Ghaleb, A.M.; Lajis, M.A.; Shamsudin, S.; Zhou, W.; Altharan, Y.M.; Abdulwahab, H.S.; Saif, Y.; Didane, D.H.; et al. Recycling aluminium for sustainable development: A review of different processing technologies in green manufacturing. *Results Eng.* **2024**, *23*, 102566. [CrossRef]
14. Djuric Mijovic, D.; Cilic, A.; Kostic, D. Lateral Torsional Buckling of Split Aluminium Mullion. *Sci. J. Civ. Eng.* **2019**, *8*, 37–42. [CrossRef]
15. Naqash, M.T.; Formisano, A.; De Matteis, G. Aluminium framing members in Facades. *Key Eng. Mater.* **2016**, *710*, 327–332. [CrossRef]
16. Yılmaz, E.; Aykanat, B.; Çomak, B. Environmental life cycle assessment of rockwool filled aluminium sandwich facade panels in Turkey. *J. Build. Eng.* **2022**, *50*, 104234. [CrossRef]
17. *PN-EN 14351-1*; Okna i Drzwi–Norma Wyrobu, Właściwości Eksploatacyjne–Część 1: Okna i Drzwi Zewnętrzne. Polski Komitet Normalizacyjny: Warszawa, Poland, 2006.
18. *ISO 3146:2022*; Plastics—Determination of Melting Behaviour (Melting Temperature or Melting Range) of Semi-Crystalline Polymers by Capillary Tube and Polarizing-Microscope Methods. BSI: London, UK, 2022.
19. *ISO 1183-1:2019*; Plastics—Methods for Determining the Density of Non-Cellular Plastics, Part 1: Immersion Method, Liquid Pycnometer Method and Titration Method. BSI: London, UK, 2022.
20. *ISO 1183-3:1999*; Plastics—Methods for Determining the Density of Non-Cellular Plastics, Part 3: Gas Pyknometer Method. BSI: London, UK, 2022.
21. *ISO 1172:2023*; Textile-Glass-Reinforced Plastics—Prepregs, Moulding Compounds and Laminates—Determination of the Textile-Glass and Mineral-Filler Content Using Calcination Methods. BSI: London, UK, 2022.
22. *ISO 868:2003*; Plastics and Ebonite—Determination of Indentation Hardness by Means of a Durometer (Shore Hardness). BSI: London, UK, 2022.
23. *ISO 179-1:2023*; Plastics—Determination of Charpy Impact Properties, Part 1: Non-Instrumented Impact Test. BSI: London, UK, 2022.
24. *ISO 527-2:2012*; Plastics—Determination of Tensile Properties, Part 2: Test Conditions for Moulding and Extrusion Plastics. BSI: London, UK, 2022.
25. *ISO 527-4:2023*; Plastics—Determination of Tensile Properties, Part 4: Test Conditions for Isotropic and Orthotropic Fibre-Reinforced Plastic Composites. BSI: London, UK, 2022.
26. *PN-EN 13830*; Ściany Osłonowe. Norma Wyrobu. Polski Komitet Normalizacyjny: Warszawa, Poland, 2015.
27. *PN-EN 1999-1*; Eurokod 9: Projektowanie Konstrukcji Aluminiowych. Cz. 1-1, Reguły Ogólne. Polski Komitet Normalizacyjny: Warszawa, Poland, 2004.
28. *Lakierowanie i Anodowanie–Company Presentation*; Aluprof SA: Bielsko-Biała, Poland, 2017.
29. *PN-EN 14024:2024*; Kształtowniki Metalowe z Przekładką Termiczną. Właściwości Mechaniczne. Wymagania, Sprawdzenia i Badania do Oceny. Polski Komitet Normalizacyjny: Warszawa, Poland, 2024.
30. Sun, L.; Li, C.; Zhang, C.; Liang, T.; Zhao, Z. The strain transfer mechanism of fiber bragg grating sensor for extra large strain monitoring. *Sensors* **2019**, *19*, 1851. [CrossRef]
31. Donghua, X.; Yongxiang, W.; Jingfeng, X. Monitoring and Analysis of Building Curtain Wall Deformation Based on Optical Fiber Sensing Technology. *Trans. Civ. Eng.* **2022**, *46*, 3081–3091. [CrossRef]
32. Rusin, D.; Juraszek, J. Innowacyjna metoda ciągłego pomiaru odkształceń ściany osłonowej aluminiowo szklanej oraz opracowanie doświadczalno-numerycznej metody analizy stanu naprężeń w konstrukcji. *Przegląd Bud.* **2024**, *3*, 30–33. [CrossRef]
33. Dassault Systemes Simulia, I. *ABAQUS: Analysis User's Manual*; Version 6.10; Dassault Systemes Simulia, Inc.: Waltham, MA, USA, 2010.
34. Woźniczka, P. Critical temperature of laterally unrestrained steel plate girders with slender section. *Eng. Struct.* **2023**, *291*, 116242. [CrossRef]
35. Woźniczka, P. Fire resistance and stability of non-uniformly heated steel plate girders with slender section. *Thin-Walled Struct.* **2024**, *203*, 112234. [CrossRef]
36. Woźniczka, P. Computer modelling of point supported laminated glass panes. *Arch. Civ. Eng.* **2022**, *68*, 3. [CrossRef]

37. Sun, Y.; Zhang, K.; Gong, G. Material properties of structural aluminium alloys after exposure to fire. *Structures* **2023**, *55*, 2105–2111. [CrossRef]
38. Zhao, Y.; Zhai, X.; Wang, J. Buckling behaviors and ultimate strengths of 6082-T6 aluminium alloy columns under eccentric compression—Part I: Experiments and finite element modeling. *Thin-Walled Struct.* **2019**, *143*, 106207. [CrossRef]
39. Li, B.; He, P.; Mo, S.; Wang, J.; Wang, Y. Stability and design of high-strength aluminium alloy RHS members under eccentric compression. *Thin-Walled Struct.* **2024**, *205*, 112325. [CrossRef]

materials

MDPI

Article

3D Scanning of Wood–Plastic Composite Decking After Cyclic Thermal Action

Artur Piekarczuk *, Ewa Szewczak , Ewelina Kozikowska and Łukasz Gołębiowski

Instytut Techniki Budowlanej, 00-611 Warsaw, Poland; e.szewczak@itb.pl (E.S.); e.kozikowska@itb.pl (E.K.); l.golebiowski@itb.pl (Ł.G.)

* Correspondence: a.piekarczuk@itb.pl

Abstract: Wood–plastic composites (WPC) combine the properties of polymers and wood, providing an attractive alternative to traditional materials, particularly for terrace flooring. When exposed to various environmental conditions, WPCs are affected by factors, such as water and ultraviolet (UV) radiation. Although most test methods for assessing the durability of these products have focused on changes in mechanical properties and linear dimensions, out-of-plane deformations (concavity and convexity) are often overlooked. This study focusses on evaluating the usefulness of the test method that allows for precise determination of these deformations after ageing. The test procedure involves exposure to classic weathering for decking boards, including moisture, UV radiation, and water spray, followed by three-dimensional (3D) scanning to track deformation after different exposure times. Analysis of variance was used to assess whether the sensitivity of this method is sufficient to detect minor deformations. Additionally, scanning electron microstructural images of the aged samples were examined to determine whether there was a relationship between the deformation and the microstructural changes. This study demonstrated the potential to use scanning methods for assessing the aspects of ageing resistance of this type of composite product in the context of deformation.

Keywords: construction products; wood–plastic composites; 3D scanning; SEM test; test method uncertainty; plane deformations

1. Introduction

Environmental concerns and the recycling of construction materials have led to great interest in polymeric materials with natural fillers and their properties, and the literature in this field is quite rich, for example, [1–4]. According to the bcc Research report [5], the global market for biobased composite materials is estimated to be worth USD 5.9 billion in 2023 and will grow to USD 9.0 billion by 2029. Wood–polymer composites (WPC) [6–8] are an excellent solution for those looking for a material that has properties similar to wood but has better water resistance, can be economically processed into a final product, and has satisfactory properties stability in many environments. One application of such composites is in the form of prefabricated planks to cover floors on outdoor terraces [9]. Prefabricated products, that is, floorboards, are exposed to a wide range of environmental conditions, and their durability is determined by their service life and intensity of exposure to the elements. Floorboards are usually in the form of a chambered profile several tens of millimetres wide with air voids inside the shape of the chambers. This is a relatively simple shape, which should provide sufficient rigidity and at the same time have a relatively low dead weight. The structural structure of the pre-cast is considerably more varied. In a WPC composite, the matrix is a polymer (usually polyethene, polypropylene, or polyvinyl chloride [4]), and the filler is wood in various forms, such as shavings, sawdust, splinters, wood chips, or, most commonly, wood flour. Matrix selection depends on the intended use. Civil engineering applications are dominated by WPC with polyvinyl chloride (PVC) or high-density polyethylene (HDPE) matrix [4]. They are used in solid or cellular profiles intended for outdoor floors, such as terraces and swimming pools. Because of the unique combination of polymer and wood properties as well as their attractive appearance, WPC composites have become an interesting alternative to polymers and products in the wood industry. Wood flour is a material that is typically produced by mechanically grinding wood in several stages in rotary hammer-type shredders. It has lipo- and hydrophilic properties and is insoluble in water. Depending on the amount of wood flour introduced into the polymer, the adsorption properties of the composite produced can be modified [10]. The composition of wood moss depends on the species of the tree, while its purity depends on the condition and region of origin, as well as the process from which the waste is produced [11].

The WPC composite profiles were manufactured by extrusion or coextrusion. In this technology, profiles are extruded from thermoplastic materials that change their aggregation state from solid to flexible under the influence of heat and pressure. Extrusion is a profile-forming process that uses a single mixture of raw materials. The mixture for the WPC composite, consisting of wood flour and polymers, was pushed through a moulding die, which gave the desired shape to the profile. Thus, the final product consisted entirely of a single raw material. Coextrusion, on the other hand, is a method of multilayer manufacturing of composite profiles, in which two blends of raw materials are used, which are combined during a single production process, allowing better strength, aesthetics, or functional performance.

The tests to which terrace decking profiles made of wood–polymer composites are subjected before they are placed on the market derive primarily from the EN 15534-4 standard [12] on decking profiles and tiles for external use, made from wood–polymer composites or natural fibre composites (NFC), and the European Assessment Document (EAD) 190005-00-0402 [13] on terrace decking kits. The two documents, which are partially interrelated, provide for tests of the mechanical and physical properties of products immediately after manufacture, as well as tests of resistance to external factors, including weathering. According to EN 15534-4, the resistance to weathering is determined by changes in colour and resistance to moisture by changes in flexural strength and water absorption. EAD also recommends testing for UV radiation, taking as an indicator change in Charpy impact strength. Linear deformation of the boards is determined only by swelling when exposed to moisture. Out-of-plane deformation changes (concavity and convexity) are not included as a durability parameter. On the basis of the authors' own experience in this publication, this type of deformation can deteriorate the functional characteristics of floors. Furthermore, out-of-plane deformations can be indicative of structural degradation of the composite material itself, with major consequences for its durability.

Traditional measurement methods (e.g., using point sensors), especially for small out-of-plane deformations, are time-consuming and technically difficult to implement. The most effective method appears to be vision-based observation and, in particular, image acquisition using 3D scanning technology and image processing using inspection engineering methods [14,15]. Three-dimensional scanning technology has been developed for more than two decades and has been applied in various areas of technology. It is developing particularly rapidly in the precision industry for quality control of retail production, especially in precision quality control measurements. In general, scanning is the optical capture of information about the geometry of physical objects. Currently, the most commonly used scanning techniques use different types of projection units. These include laser, structured light, and infrared light scanners. Regardless of the technique used, each scanning process produced a numerical image that represented the topography of the object. The 3D images obtained can have different representation accuracies depending on the device used. For precision measurements, a measurement accuracy of ±0.01 mm is achieved. After the object, the information was digitally stored as a point cloud. The point cloud was then converted to STL surface meshes (standard triangle language) using appropriate graphics tools [16]. The numerical representation of the topography of the object completes the model acquisition stage and can be used in various survey scenarios. The simplest is that the scan is used to reconstruct the object documentation, that is, to build a computer-aided design (CAD) model. Reverse engineering has been used for this purpose [17,18]. The result of the reverse engineering implementation is the acquisition of a digital model, that is, a representation of the object's geometry with features similar to the original. This model can be further processed as a batch file for numerical simulations using the finite element method, or it can form wider computer structures, such as a digital twin [19]. Other applications of models are control engineering [20] and inspection engineering [14]. The scanned model has all the geometrical features of the object, including its defects. Juxtaposing the scan with a 3D CAD design or another scan creates an inspection pair. Objects juxtaposed in this manner can be compared with each other by determining the shape deviations. It is precisely this application of scanning and inspection techniques that is included in the example analysed.

The 3D scanning process and inspection analysis were the only tools used to obtain test measurement results. Typical testing of building products involves many steps, including measurements, exposure to environmental impacts, and visual assessments. The suitability of a measurement or test method for a given application can be considered from two perspectives: the relationship to the conditions of use of the product and the metrological properties of the method, such as its accuracy. The aim of this study is related to the metrological aspect of test methods and is to assess whether the test method based on standard exposure of terrace boards to atmospheric factors and then measurement of their plane deformations (concavity and convexity) can be used, along with other methods, to assess the resistance of composite terrace boards to atmospheric factors.

The suitability of a test method should be based on its ability to reveal differences in the phenomena that occur or their effects. To some extent, the uncertainty value of the result contains information on the accuracy of the test method. This includes the resolution of the measurement itself and many other factors that influence the dispersion of the results. There are many discrepancies in the uncertainty estimation of the test results ([21–23]), including those related to its dependence on the material under study. Therefore, the value of the uncertainty estimated according to commonly used rules [24] alone may not give a clear answer to the usefulness of the method. In this paper, a variance analysis was used to assess the suitability of the 3D scanning method for observing the deformations that occur in building products after ageing. Composite planks with relatively small deformations during ageing were chosen for this study to determine whether, with small deformations, the method can capture differences in the deformation of individual points on the samples, differences due to the effect of exposure time, or differences between individual samples.

To confirm the sense of the out-of-plane deformation test as an indicator of durability (weather resistance), microstructure SEM (scanning electron microscope) testing was conducted to determine whether there was a relationship between the deformation and microstructural changes.

2. Materials and Methods

2.1. Test Sample and Climatic Impacts

The tests were carried out on the chamber profiles (Figure 1) for floor coverings on outdoor terraces. Five samples (AE) were taken for testing. The width and thickness of the profiles were 180 and 22 mm, respectively. The length of the samples was 300 mm. Profiles were made from NFPC (natural fibre-reinforced polymer composite) with a PVC matrix and a filler of fine lignocellulosic fibres in the form of the so-called wood flour, recycled from wood industry waste, with the addition of plasticisers and modifiers, including UV absorbers. The composition of the composite is a company secret that has not yet been disclosed. Profiles were produced using an extrusion process.

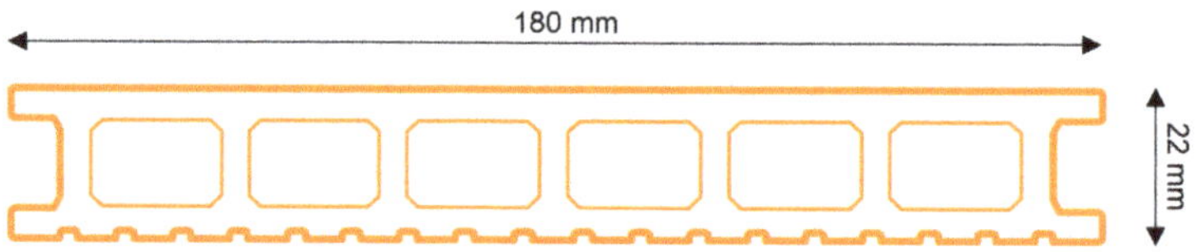

Figure 1. Sample cross-section.

The profiles had two wear surfaces: fluted and plane (Figure 2). Typically, both surfaces are mechanically brushed with a metal brush to give them a wood-like appearance.

(**a**) (**b**)

Figure 2. Useful surfaces of profiles (**a**) fluted (**b**) plane (brushed).

The ageing procedure was carried out using a UV test apparatus (Atlas, Linsengericht, Germany) equipped with type 1A fluorescent lamps (UVA-340), according to ISO 16474-3 [25], which emit light with a wavelength between 300 and 400 nm, with an emission maximum at 343 nm (Table 1). The exposure pattern was in accordance with EN 927-6 [26]. The samples were cycled, consisting of a long condensation phase followed by exposure to UV lamps at an irradiance of 0.89 W/m^2 measured at 349 nm, alternating with wetting by water spray. Specimens cut from the flat surfaces of usable profiles from a brushed surface were exposed. At the time of exposure, the samples were placed at an angle of approximately 80 ° to allow the water to drain freely. Exposures lasting a total of 3024 h were applied. The course of exposure for each cycle is shown in Table 1.

Table 1. Ageing exposure.

Total Duration of Exposure	Total Number of Cycles	Exposure Pattern Over the Cycle
3024	18	24 h of humidification by condensation at temperature 45 ± 3 °C 168 h of alternating irradiation and sprinkling, in the sequence: (a) 2.5 h of irradiation with UVA-340 lamps, irradiance 0.89 watt per square metre (340 nm), temperature (Black Standard Thermometer) 60 ± 3 °C, (b) 0.5 h of water sprinkling, no UV, sprinkling rate 6–7 litre per minute.

2.2. 3D Scanning Method and Inspection Analysis

Three-dimensional imaging was performed using a 3D handheld scanner FreeScan Trio 3D (SHINING 3D, Hangzhou, China (HQ)). Sample scanning was performed using a handheld laser scanner, FreeScan Trio 3D [27] (Figure 3). Basic scanner parameters: light source blue laser 26 cross lines; scanning accuracy to 0.02 mm; volumetric accuracy in photogrammetry mode: 0.02 mm + 0.015 mm/m; scanning speed 3,010,000 pct/s.

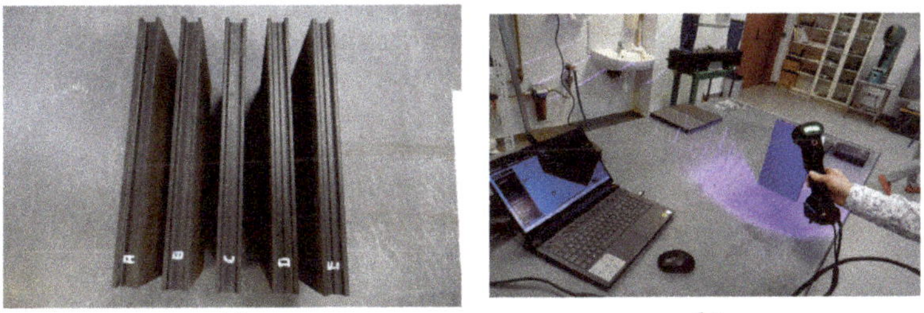

(a) (b)

Figure 3. Scanning of test samples, (**a**) test samples, (**b**) scanning process, A–E: sample symbols.

The scans were performed using several different sequences. The first was the reference scan, that is, before the climatic impact. The following scan sequences were operational scans, that is, after climatic impact. The reference and operational scans were compared with each other in the Geomagic Control X inspection programme (v. 2024.1). The comparison resulted in a quantitative and qualitative assessment of the deformation. The algorithm used to determine the deviations consists of calculating the deformation as a scalar quantity determined according to Relation (1).

$$D = \sqrt{GV_x^2 + GV_x^2 + GV_z^2}, \quad (1)$$

The GV vector is defined as the positional difference between a point in the STL mesh of the deformed operational model and its corresponding point on the undeformed reference model. The calculation of these vectors was based on an established reference system, specifically the normal to the reference surface. Reference points, in general, are determined as follows.

$$P_o = \langle x_0, y_0, z_0 \rangle ;\; P_r = \langle x_r, y_r, z_r \rangle \quad (2)$$

where x, y, and z are the coordinates of the node indexes: o, operational model, r, reference model.

Vector:

$$GV = \langle x_o - x_r, y_o - y_r, z_o - z_r \rangle, \tag{3}$$

The results of the inspection can be presented in two primary formats: quantitatively, as discrete points with precise deformation values, or qualitatively, as a deformation map with an indicative deformation scale. In the quantitative approach, the deformation is calculated according to Relation (1), where the result represents an estimate of the expected value in the immediate vicinity of the measurement point. It should be emphasised that this result is inherently associated with a standard uncertainty that arises from the data acquisition methodology.

When a result is determined for a specific point, it is calculated as the mean value of the STL mesh nodes located within a defined neighbourhood of the selected point. In the case of a deformation map, the standard deviation is determined based on points distributed throughout the observation area, allowing for the assessment of deformation patterns, including the geometric distribution and intensity of the extrema. For sectional analyses, the standard deviation of displacements is calculated specifically for nodes situated along the section line. This enables a detailed evaluation of the deformation within a given region of interest.

2.3. Statistical Analysis

The deformation of the samples was tested by measuring the deviations from the baseline dimensions in five samples at ten measurement points on each sample. Five measurement points were located on the plane surface and five on the fluted surface. To determine the feasibility of drawing conclusions about the behaviour of the samples based on the measurements, analysis of variance (ANOVA) tests were adopted as the primary method for analysing the statistical significance of differences between the results. ANOVA was preceded by the normality test. Although with small sets of results, there are no normality tests that provide fully reliable results, it was decided to conduct the Shapiro–Wilk normality test (with a 5% probability of error). This test was chosen because it is generally considered to have relatively high power (e.g., [28–30]).

Shapiro–Wilk tests were performed on sets of results, where the variable was:

(P) Point on the sample. Sets of results were obtained for testing, each for the same exposure time and the same sample. The sample points on a plane surface ($1 < 5$) and a fluted surface ($6 < 10$) were tested separately. A total of 42 sets of four to five results were tested.

(S) Sample. Sets of results for different samples were obtained for testing, but each for the same point and exposure times. Fifty sets of four to five results were tested.

(E) Exposure times. Sets of results for different exposure times were obtained for the test, but each for the same point and the same sample. A total of fifty sets of four to five results were tested.

A two-way analysis of variance was used to assess the significance of differences between sets of results to initially assess the influence of the following factors on the measured deformation:

(a) Effect of sample and exposure time for each point on the sample separately (ten sets of results, each at a different point on the sample).

(b) Effect of the point on the sample and exposure time for each sample separately. (ten sets of results: five for five samples for points 15 and five for points $6 \div 10$).

(c) Effect of sample and point for each exposure time separately (ten sets of results: five for five exposure times for points 15 and five for points $6 \div 10$).

One-way analysis of variance was performed for pairs of sets of results obtained for individual points on the samples, treating the points on the plane surface and the points on the fluted surface separately. The same analysis was performed for pairs of sets of results for two different exposure times to assess the influence of exposure time on the deformation.

2.4. Microstructure SEM Analysis

The microstructure of the coated surface was investigated using a Sigma 500 VP scanning electron microscope with cold-field emission (Carl Zeiss Microscopy GmbH, Köln, Germany).

The microstructure surfaces of the coatings were examined at an accelerating voltage of 5 keV excitation electron beam using an SE detector on samples sprayed with a coat of gold. SEM microscopic observations of the chamber profiles were made on the brushed surface in the initial state and after ageing for 18 cycles (3024 h). The samples for testing were cut from the corners and central area of the profiles. All observations were made at 200× and 1000× magnification.

3. Results

3.1. Measurement Inspection

The measurement inspection of the WPC composite profiles was performed using Gemoagic Control X (version: 2024.1.0. OQTON, INC, San Francisco CA 94104, USA)The results are presented as an example of a selected sample after two inspection steps, that is, after four and eight cycles of ageing in the form of a strain map on which the measuring points were additionally plotted (Figures 4 and 5). However, point measurements of deformation after five consecutive inspections are summarised in Figure 6 for the brushed surface and in Figure 7 for the fluted surface analysed sample.

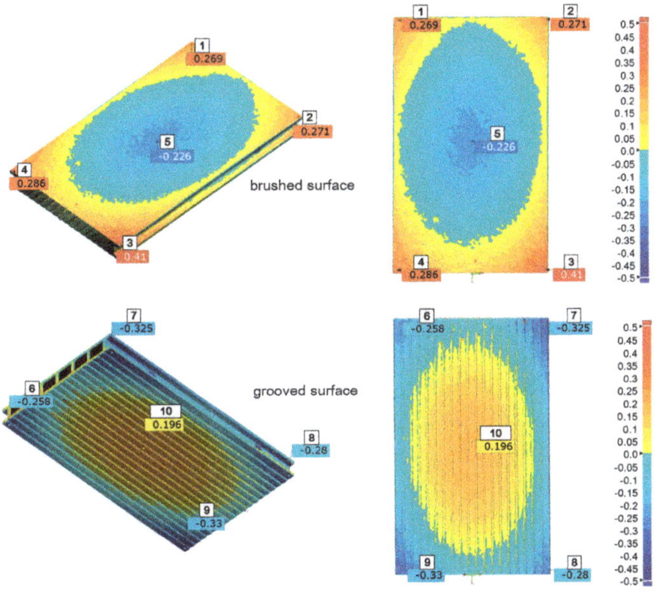

Figure 4. Deformation map of the specimen deformation after four ageing cycles with marked measurement points on the brushed and fluted surfaces. Deformations are expressed in micrometres, 1–10: measuring points.

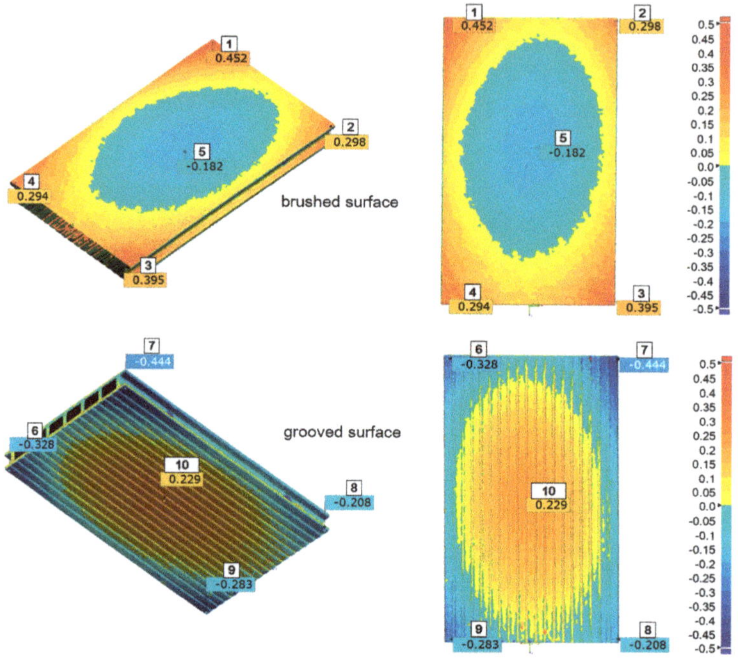

Figure 5. Deformation map of the specimen after eight ageing cycles with marked measurement points on the brushed and fluted surfaces. Deformations are expressed in micrometres, 1–10: measuring points.

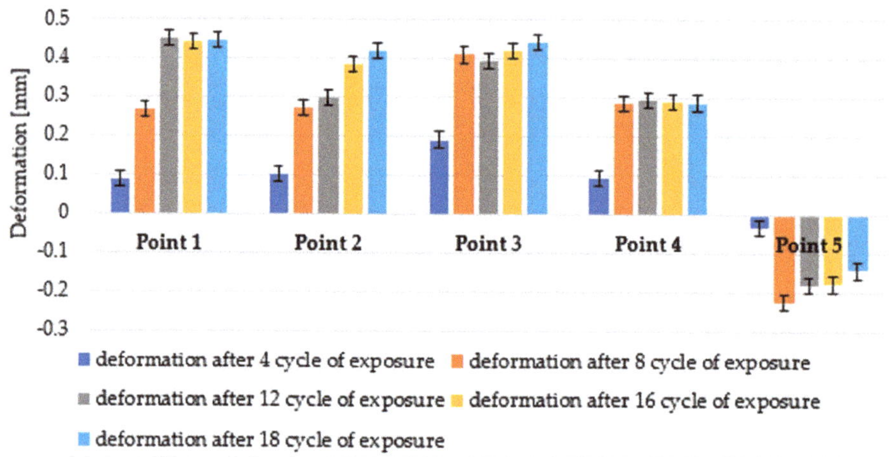

Figure 6. Deformation at measurement points on the brushed surface in successive ageing stages. The error bars show the expanded measurement uncertainty estimated as described in Section 3.4.

Detailed analysis of the statistical significance of differences between deformations is presented at point 3.3.

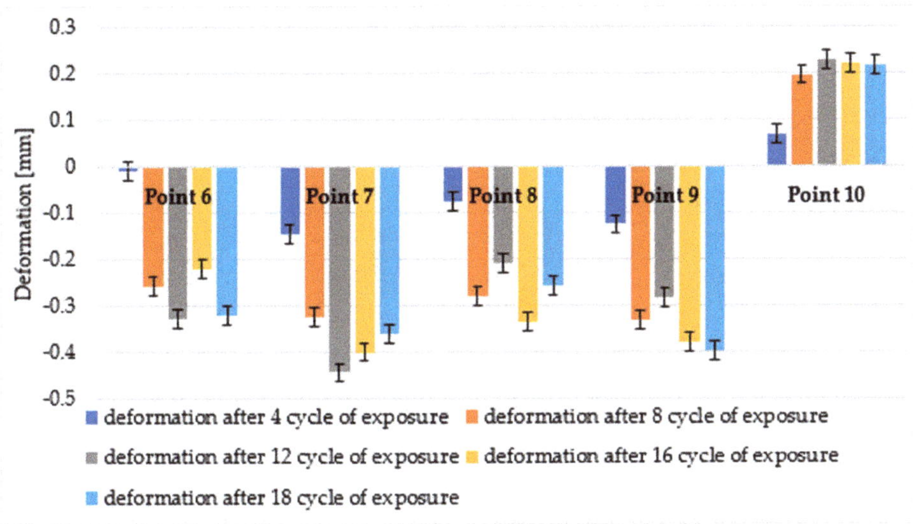

Figure 7. Deformation at the measurement points on the fluted surface during successive ageing stages. The error bars show the expanded measurement uncertainty estimated as described in Section 3.4.

3.2. Microstructure Analysis

The morphological analysis consisted of a qualitative assessment of the surface and verification of damage to the surface layer of the profiles. A comparison was made between the surface taken from the corner of the profile and the fragment taken from the middle part of the profile in the initial state and after ageing. The microstructure of the surface of the WPC wood flour composite samples in the initial state indicated that the fibres were homogeneously coated with polymer, with no visible filler agglomerates or exposed wood fibres in the images. The surfaces of the profiles showed characteristically orientated streaks and scratches, indicative of the surface topography of the composite resulting from the brushing process (Figures 8 and 9). In comparison, subjecting the composite profiles to an ageing process clearly changed the surface morphology. Microscopic images showed melting of the outer layer of the polymer, revealing surfaces of nonmoistened filler in the form of large lamellae and agglomerates of wood fibres, which were observed in composites taken from the centre of the profile and from the corners of the WPC profile. Changes in the morphology of the filler were also observed, that is, wood mace, where when the surface layer of the material was removed, the filler became visible in the form of interconnected fibres that formed characteristic clusters (Figures 10a and 11a). In the composites taken from the central part of the profile, additional surface cracks were visible in the form of fissures, which deepened and widened in various places, forming a network of cracks and leading to degradation of the surface layer (Figure 11b). The aged samples from the central part of the profile also showed additional filler removal, which is visible in the images as oval hollow areas (Figure 11b).

(**a**) (**b**)

Figure 8. Microstructure of the surface of WPC profiles in the initial state (fragment taken from the corner of the profile), at magnification: (**a**) 200×, (**b**) 1000×.

(**a**) (**b**)

Figure 9. Microstructure of the surface of WPC profiles in the initial state (fragment taken from the centre of the profile), at magnification: (**a**) 200×, (**b**) 1000×.

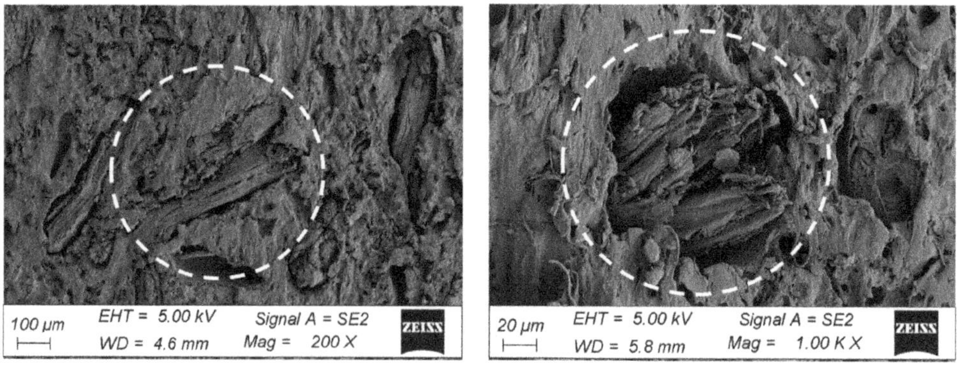

(**a**) (**b**)

Figure 10. Microstructure of the surface of WPC profiles after ageing (fragment taken from the corner of the profile), at magnification: (**a**) 200×, (**b**) 1000×.

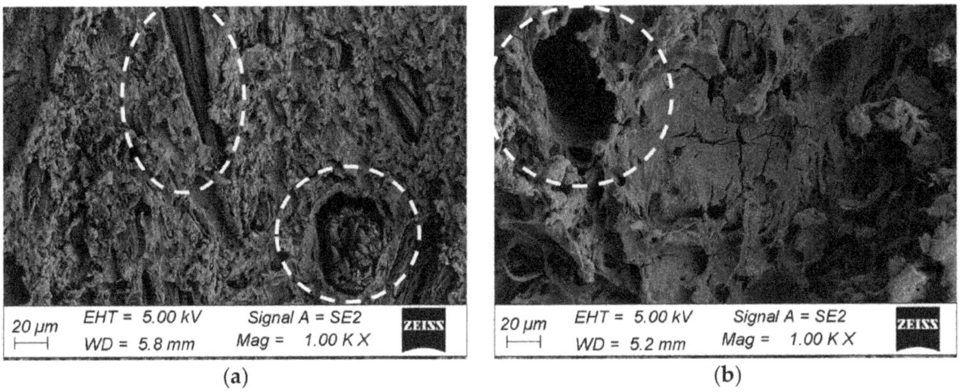

Figure 11. Microstructure of the surface of WPC profiles after ageing (fragment taken from the centre of the profile), at magnification: (**a**) 200×, (**b**) 1000×.

3.3. Statistical Results

Despite the small number of samples in the result sets, the large number of sets subjected to Shapiro–Wilk tests allowed us to draw conclusions about the distributions of the results.

(P) In most sets (60%) in which the variable was a point in the sample, the null hypothesis that the set came from a normally distributed population was not confirmed.

(S) The sets in which the variable was a sample, the Shapiro–Wilk test confirmed the hypothesis that the sets came from a normally distributed population in 96% of the sets in which the variable was a sample.

(E) The Shapiro–Wilk test confirmed the hypothesis that the sets came from a normally distributed population in 90% of the sets, in which exposure time was a variable.

Based on the above results, an ANOVA analysis should be performed only for the sets in which the variable was a sample or exposure time because the normal distribution was not confirmed in the sets in which the variable was a point on the sample. However, to obtain a general picture of the situation, a two-way analysis of variance (ANOVA) was performed for all variables. The results of the analysis are presented in Table 2.

Table 2. Results of the two-way ANOVA analysis determining the effect of the sample, the sample point and exposure time on sample deformation. (a) Effect of the sample and exposure time tested for each point separately, (b) effect of the sample point and exposure time tested for each sample separately, and (c) effect of the sample and sample point tested for each exposure time separately. SSD: statistically significant differences (influence of the factor is significant).

(a)					
	Sample Point Number				
Influences	1	2	3	4	5
Effect of the sample	SSD	SSD			
Effect of the time of exposure	SSD	SSD	SSD		SSD
	Sample Point Number				
Influences	6	7	8	9	10
Effect of the sample					
Effect of the time of exposure		SSD		SSD	

Table 2. *Cont.*

(b)					
	Sample				
Influences	A	B	C	D	E
Effect of the point on the sample (points 1–5)	SSD	SSD	SSD	SSD	SSD
Effect of exposure time of exposure (points 1–5)			SSD		
Effect of the point on the sample (points 6–10)	SSD	SSD	SSD	SSD	SSD
Effect of exposure time of exposure (points 6–10)	SSD			SSD	SSD
(c)					
	Time of Exposure-Number of Cycles				
Influences	**4 Cycles**	**8 Cycles**	**12 Cycles**	**16 Cycles**	**18 Cycles**
Effect of the Sample (points 1–5)			SSD		
Effect of the point on the sample (points 1–5)	SSD	SSD	SSD	SSD	SSD
Effect of the sample (points 6–10)					
Effect of the point on the sample (points 6–10)	SSD	SSD	SSD	SSD	SSD

Considering that the influence of the sample on the results was statistically insignificant in most cases (Table 2a,b,c), the results obtained for all samples were averaged. Factors on which the deformation results may depend are assumed to be the point of the sample and the number of stages of ageing. Table 3 presents the differences between the deformations measured in the individual age stages separately for each point of the samples (1, 2, . . ., 10 markings according to Figure 4). The notation "x y cycles" means the difference between the average deformation for five samples at a given point after the y ageing cycles and after the x cycles. The differences for which statistical significance was confirmed are shown in the grey background.

Table 3. Differences between mean deformations (five samples) in subsequent ageing stages at individual sample points.

The Point on the Sample	**Mean Deformation After 4 Cycles**	**4÷ 8 Cycles**	**8 ÷ 12 Cycles**	**12÷ 16 Cycles**	**16÷ 18 Cycles**	**8 ÷ 18 Cycles**	**Total Mean Deformation After 18 Cycles**
	Deformation Differences, mm						
Point 1	0.146	0.177	0.035	−0.042	0.104	0.097	**0.42**
Point 2	0.179	0.145	0.007	0.023	0.058	0.088	**0.41**
Point 3	0.251	0.142	0.059	−0.078	0.045	0.026	**0.42**
Point 4	0.223	0.167	0.006	−0.074	0.063	−0.005	**0.38**
Point 5	−0.082	−0.166	−0.009	0.015	0.020	0.027	**−0.22**
Point 6	−0.175	−0.208	0.002	−0.031	0.040	0.011	**−0.37**
Point 7	−0.178	−0.189	−0.088	0.111	−0.051	−0.028	**−0.39**
Point 8	−0.102	−0.278	0.022	−0.015	0.089	0.097	**−0.28**
Point 9	−0.159	−0.208	−0.026	0.041	−0.030	−0.015	**−0.38**
Point 10	0.139	0.0780	0.033	−0.051	−0.049	−0.066	**0.15**

3.4. Measurement Uncertainty

The measurement uncertainty was estimated from ten measurements taken at the same point (each time, the entire measurement procedure was repeated).

For each measurement, a result was obtained together with the standard deviation s_1 assigned to the individual result estimated by the inspection software. This standard deviation results from the distribution of the surface point cloud of the compared area representing the measurement point. Because the images are three-dimensional, the measurement point is represented by a sphere with a corresponding diameter. In this study, a sphere with a diameter of 3 mm was used. All points inside this sphere were subjected to statistical analysis, for which the standard deviation was determined.

Based on the ten deviations s_{1i}, $i = 1, \ldots, 10$, the mean standard deviation for the point, s_1, was obtained. $s_1 = 0.0078$ mm.

The standard deviation s_2 of the ten results obtained was taken as the standard deviation of repeatability. $s_2 = 0.0030$ mm.

The standard uncertainty of the measurement of um was assumed to be:

$$u_m = \sqrt{s_1^2 + s_2^2} \quad (4)$$

The value of the standard uncertainty of the measurement obtained in this way is $u_m = 0.008$ mm. Considering the coverage interval of 95%, the final result is an extended uncertainty of measurement of $U_m = 0.02$ mm.

Table 4 shows the differences in deformation between the individual points of the samples after the subsequent stages of exposure. The numbers in the table indicate the difference in the sample deformation at the point x-listed horizontally in the header and y-listed vertically, after the subsequent stages of ageing. The differences between the points for which statistical significance was confirmed are shown in the grey background.

Table 4. Differences between deformations of individual points in the samples.

Exposure		Point 1	Point 2	Point 3	Point 4
		Deformation Differences, mm			
4 cycles	Point 2	−0.032	-	-	-
	Point 3	−0.105	−0.073	-	-
	Point 4	−0.076	−0.044	0.029	-
	Point 5	−0.229	0.261	0.334	0.305
8 cycles	Point 2	−0.001	-	-	-
	Point 3	−0.070	−0.069	-	-
	Point 4	−0.066	−0.065	0.004	-
	Point 5	0.571	0.572	0.642	0.638
12 cycles	Point 2	0.028	-	-	-
	Point 3	−0.094	−0.121	-	-
	Point 4	−0.037	−0.065	0.057	-
	Point 5	0.616	0.588	0.709	0.652
16 cycles	Point 2	−0.038	-	-	-
	Point 3	−0.058	−0.020	-	-
	Point 4	−0.005	0.033	0.053	-
	Point 5	0.558	0.596	0.616	0.563
18 cycles	Point 2	0.009	-	-	-
	Point 3	0.001	−0.008	-	-
	Point 4	0.036	0.028	0.036	-
	Point 5	0.642	0.634	0.642	0.606

Table 4. *Cont.*

Exposure		Point 6	Point 7	Point 8	Point 9
		Difference, mm			
4 cycles	Point 7	0.002	-	-	-
	Point 8	−0.073	−0.075	-	-
	Point 9	−0.017	−0.019	0.056	-
	Point 10	−0.315	−0.317	−0.241	−0.298
8 cycles	Point 7	0.017	-	-	-
	Point 8	0.003	−0.014	-	-
	Point 9	0.017	0.000	0.014	-
	Point 10	0.601	0.584	0.598	0.584
12 cycles	Point 7	−0.073	-	-	-
	Point 8	0.023	0.096	-	-
	Point 9	−0.011	0.062	−0.034	-
	Point 10	0.632	0.705	0.609	0.643
16 cycles	Point 7	0.069	-	-	-
	Point 8	0.039	−0.030	-	-
	Point 9	0.061	−0.008	0.022	-
	Point 10	0.612	0.543	0.573	0.551
18 cycles	Point 7	−0.022	-	-	-
	Point 8	0.089	0.111	-	-
	Point 9	−0.009	0.014	−0.097	-
	Point 10	0.523	0.546	0.435	0.532

4. Discussion

Three-dimensional scanning enables the acquisition of detailed images of the test models. The accuracy of the obtained results depended on the scanning technique used. In this study, a metrology-grade laser scanner was used. The small size of the models and stable observation conditions allow for a highly accurate three-dimensional mapping of the surface of the test model. At least two scans were required to quantitatively and qualitatively assess changes in geometry. The first is a reference scan before the geometry changes, and the second is a model scan after the model deformation. A comparison of these scans was performed using dedicated inspection software. This task was performed in this study. The use of a high-accuracy measurement method does not guarantee that the results obtained in a test procedure involving this measurement will have high accuracy. The standard uncertainty of the measurement for the presented test method was estimated to be 0.008 mm, which, assuming a normal distribution and 95% coverage interval, gives an approximate expanded uncertainty value of 0.016 mm. Therefore, presenting the results with an accuracy of hundreds of millimetres is metrologically justified.

However, an essential feature of a test method is that after completion of the full test procedure, the results for internally different products or phenomena should differ from each other, which can be called the "resolution of the test method". The test method used, which consists of ageing the composite profiles and then assessing their deformation using 3D scanning, was analysed to determine their "resolution".

The results of the two-way ANOVA were not entirely conclusive (Table 1). However, the authors assumed that although the sets of results were few, the number of sets was relatively large. Based on the number of sets in which statistically significant differences were confirmed, it can be summarised as follows:

- Because in most cases (nine out of ten when the second influencing factor was the point in the sample and eight out of ten when the second influencing factor was exposure time) the differences between the results for individual samples were found to be statistically insignificant, it can be concluded that the deformation differences for individual samples are statistically insignificant. This means that the effect of the sample is insignificant; therefore, the set of results obtained from several samples can be regarded as a set in which the dispersion of results is random and the mean of the results for the individual samples can be regarded as the authoritative result. Confirmation of such conclusions can also be found in the Shapiro–Wilk tests, which, in sets where the sample was variable, confirmed the hypothesis of a normal distribution in 96% of cases.
- In all sets of results, a two-way analysis of variance (ANOVA) showed statistically significant differences in deformation at different points on the samples, both when the other influencing factors were the sample and the exposure time.
- For deformation differences according to the exposure time, the situation is not clear. Only half of the ANOVA results indicated that these differences were significant.

Although the two-way ANOVA was not fully conclusive, the one-way analysis of variance, comparing pairs of single sets of results, provides more detailed information (Tables 2 and 3).

At all sample points, statistically significant changes in deformation occurred up to cycles four or eight. For the differences between the deformations found in successive ageing stages: after 12 cycles, after 16 cycles, and after 18 cycles, one-way analysis of variance did not confirm statistical significance at any point in the samples. Thus, using the measurement technique in question, it is possible to assess at what stage of ageing the most significant changes occur.

For all exposure times, the differences between the deformations of the points placed in the corners of the specimens: 1-2, 1-3, 1-4, 2-3, 2-4, and 3-4 (as indicated in Figures 4 and 5) are insignificant, while the differences between the individual corner points 1, 2, 3, 4, and point 5 in the centre of the specimen are statistically significant. Similarly, on the fluted side of the sample, the deformation of the corner points (6, 7, 8, and 9) was not statistically significantly different, while the deformation of the centre point (10) was significantly different from that of the corner points. This can also be visually observed in the strain maps shown in Figures 4 and 5. Therefore, statistical analysis confirms that the deformation measurement technique is suitable for determining the nature of deformation. In this case, it was the formation of a convexity in the central part of the samples. The nature of the deformation was the same for all samples.

Depending on the sets of results analysed, the smallest differences between deformation values of approximately 0.08 mm were assessed as statistically significant. The largest values assessed as nonsignificant did not exceed 0.17 mm. This provides information on the resolution of the method, which can be approximated at 0.2 mm.

As mentioned in Chapter 1 of this paper, the durability in relation to the geometry of composite decking is limited to swelling tests after exposure to moisture. The requirements for the change in linear dimensions under the influence of moisture (28 days) presented in the EAD for individual results are as follows:

- 5% in thickness, which for a thickness of 25 mm is <1.25 mm
- 1.2% in width, which for a width of 180 mm is 2.16 mm.

The actual swelling results in the thickness of composite boards with a PVC matrix and a filler of fine lignocellulosic fibres being generally in the range of 0.2 to 0.5 mm.

In the case in question, the out-of-plane deformations under ageing were 0.15 to 0.42 (Table 3) at the individual points, and the average differences in the deformations between the points reached 0.7 mm (Table 4). Therefore, out-of-plane deformation may be noteworthy in terms of the durability of decking boards and is comparable to the changes caused by swelling under the influence of moisture.

3D scanning, unlike other methods of measuring deviation from flatness (e.g., using a feeler gauge or bow gauge for measuring cupping shown in EN 15534-1 [31], allows for capturing the nature of the sample deformation. Statistical analysis showed that the measurement method provided the ability to determine the effect of various ageing stages and to determine the nature of the deformation based on the deformation at various points in the sample.

The subjecting of the composite profiles to an ageing process clearly changes the surface morphology. Microscopic images highlighted the melting of the outer layer of the polymer, revealing the surfaces of the unweathered filler in the form of large lamellae and agglomerates of wood fibres, which can be observed in the composites after ageing for fragments coming from both the corner and middle areas of the profile (Figures 10a and 11a). Changes in filler were also observed, swelling, and visible in the form of clusters [32], which is particularly evident for fragments taken from the corner of the profile (Figure 10b). Microstructural analysis showed a change in the nature of the failure of the surface layer of the profile, depending on the location of the material used for testing. The sample taken from the corner of the profile was less degraded compared to the fragments coming from the centre of the profile. In the case of the corners, the filler is exposed, detached from the matrix, and swells (Figure 10a,b). In the ageing specimens taken from the central part of the profile, additional filler salvage occurred, which can be seen in the images as oval hollow areas. Furthermore, the polymer matrix itself changes with cracking of the surface layer [33] of polyvinyl chloride (Figure 11a,b), which is particularly evident at higher magnifications (Figure 11b).

Microstructural analysis of the samples after ageing showed that exposure of the profiles to light emitted by fluorescent lamps, alternating with humidification, led to significant surface degradation, depending on the nature of the damage, depending on the place the material was taken for testing. The greatest changes in the surface of the profile were observed in the sample obtained from the central part of the profile. In the morphology analysis, melting of the top layer of the polymer matrix [34], exposure of the wood fibre surface [35], fracturing of the top layer of polyvinyl chloride, and loss of filler from the polymer matrix became apparent [36]. In the 3D scan inspection tests, relatively large deformations were also observed in the central part of the sample, which means that this area was subjected to strong geometrical changes during environmental impacts.

The results of the SEM observations confirmed the differences that could also be observed in the deformation measurements. The largest deformations detected by the inspection programme indicate, at the same time, areas of marked changes in the structure of the material.

5. Conclusions

Both the estimated measurement uncertainty and the statistical evaluation of the resolution of the test method show that it is reasonable to conclude that the 3D scanning method used allows the differences between the deformations of the composite samples to be assessed at a level not worse than 0.2 mm. On the basis of the comparison with the requirements for swelling (change in linear dimensions) of composite decking boards under the influence of moisture, it was concluded that with a similar level of requirements for out-of-plane deformation under ageing, the method would be metrologically suitable for its evaluation. The method allows conclusions to be drawn about the deformation occurring in the samples under the influence of ageing and the differences in deformation at individual locations and after successive stages of ageing. The 3D scanning method, in combination with inspection analysis, has proven to be effective in detecting micro-deformations in zones that have not yet been considered during standard measurements. This is particularly important in combination with changes in the microstructure of the material. This is because it turns out that the largest deformations detected by the inspection programme indicate areas of marked changes in the structure of the material at the same time. Thus, including an out-of-plane deformation assessment after environmental impacts may be important in assessing the durability of composite decking for outdoor applications. The 3D scanning method is a suitable and convenient method for this purpose.

In the future, the research team plans to measure deformations in larger elements, such as a fragment of a wall covered with WPC composite panels. Large-area elements will be exposed to humidity and thermal factors. This research will contribute to the method developed for the diagnostics of composite facade cladding using 3D scanning.

Author Contributions: Conceptualisation, A.P. and E.S.; methodology, A.P., E.K., E.S. and Ł.G.; validation, E.S. and A.P.; formal analysis, E.S.; investigation, E.K. and Ł.G.; resources, A.P. and E.K.; data curation, A.P.; writing—original draft preparation, A.P., E.K., E.S. and Ł.G.; writing—review and editing, A.P.; visualisation, A.P., E.K., E.S. and Ł.G.; supervision, A.P.; project administration, E.S.; funding acquisition, E.S. and E.K. All authors have read and agreed to the published version of the manuscript.

Funding: This research was funded by "Ministerstwo Nauki i Szkolnictwa Wyższego" Poland, as part of the project number ITB L-2/2024 and NZK-080/2024.

Institutional Review Board Statement: Not applicable.

Informed Consent Statement: Not applicable.

Data Availability Statement: The data presented in this study are available on request from the corresponding author (for privacy reasons).

Conflicts of Interest: The authors declare that they have no conflicts of interest. The sponsors had no role in the study's design; in the collection, analyses, or interpretation of data; in the writing of the manuscript; or in the decision to publish the results.

References

1. Morreale, M.; Liga, A.; Mistretta, M.C.; Ascione, L.; La Mantia, F.P. Mechanical, Thermomechanical and Reprocessing Behavior of Green Composites from Biodegradable Polymer and Wood Flour. *Materials* **2015**, *8*, 7536–7548. [CrossRef] [PubMed]
2. Hejna, A.; Barczewski, M.; Kosmela, P.; Mysiukiewicz, O.; Piasecki, A.; Tercjak, A. Compatibility of Sustainable Mater-Bi/Poly(ε-Caprolactone)/Cellulose Biocomposites as a Function of Filler Modification. *Materials* **2023**, *16*, 6814. [CrossRef] [PubMed]
3. Peças, P.; Carvalho, H.; Salman, H.; Leite, M. Natural Fibre Composites and Their Applications: A Review. *J. Compos. Sci.* **2018**, *2*, 66. [CrossRef]
4. Sudoł, E.; Kozikowska, E.; Szewczak, E. Artificial Weathering Resistance Test Methods for Building Performance Assessment of Profiles Made of Natural Fibre-Reinforced Polymer Composites. *Materials* **2022**, *15*, 296. [CrossRef]
5. Publishing, B. Global Bio-Based Composite Material Market Analysis Report (Report Code: AVM232B). Available online: https://www.bccresearch.com/market-research/advanced-materials/bio-composites-market.html?srsltid=AfmBOook68u3WozEAvDQjmizH0gqvRtITw4vqCpYTSEP2mVEUyTy5JPa (accessed on 21 November 2024).
6. Mazzanti, V.; Mollica, F. A Review of Wood Polymer Composites Rheology and Its Implications for Processing. *Polymers* **2020**, *12*, 2304. [CrossRef] [PubMed]
7. Ndiaye, D.; Fanton, E.; Morlat-Therias, S.; Vidal, L.; Tidjani, A.; Gardette, J.L. Durability of Wood Polymer Composites: Part 1. Influence of Wood on the Photochemical Properties. *Compos. Sci. Technol.* **2008**, *68*, 2779–2784. [CrossRef]

8. Rowell, R.M. Advances and Challenges of Wood Polymer Composites—Fpl_2006_rowell001.Pdf. In Proceedings of the Pacific Rim Bio-Based Composites Symposium, Kuala Lumpur, Malaysia, 20–23 November 2006; pp. 2–12.
9. Dale Ellis, W. Wood-Polymer Composites: Review of Processes and Properties. In *Proceedings of the Molecular Crystals and Liquid Crystals Science and Technology, Section A: Molecular Crystals and Liquid Crystals*; Gordon and Breach Science Publishers Inc.: London, UK, 2000; Volume 353, pp. 75–84.
10. Marcuello, C.; Foulon, L.; Chabbert, B.; Aguié-Béghin, V.; Molinari, M. Atomic Force Microscopy Reveals How Relative Humidity Impacts the Young's Modulus of Lignocellulosic Polymers and Their Adhesion with Cellulose Nanocrystals at the Nanoscale. *Int. J. Biol. Macromol.* **2020**, *147*, 1064–1075. [CrossRef] [PubMed]
11. Nukala, S.G.; Kong, I.; Kakarla, A.B.; Tshai, K.Y.; Kong, W. Preparation and Characterisation of Wood Polymer Composites Using Sustainable Raw Materials. *Polymers* **2022**, *14*, 3183. [CrossRef] [PubMed]
12. *EN 15534-4:2014*; Composites Made from Cellulose-Based Materials and Thermoplastics (Usually Called Wood-Polymer Composites (WPC) or Natural Fibre Composites (NFC)—2014, Part 4: Specifications for Decking Profiles and Tiles. European Committee for Standardization (CEN): Brussels, Belgium, 2014.
13. European Organisation for Technical Assessment (EOTA). *European Assessment Document—EAD 190005-00-0402, Terrace Decking Kit*; European Organisation for Technical Assessment (EOTA): Brussels, Belgium, 2017.
14. Yilmaz, B.; Marques, V.R.; Donmez, M.B.; Cuellar, A.R.; Lu, W.E.; Abou-Ayash, S.; Çakmak, G. Influence of 3D Analysis Software on Measured Deviations of CAD-CAM Resin Crowns from Virtual Design File: An In-Vitro Study. *J. Dent.* **2022**, *118*, 103933. [CrossRef]
15. Taraben, J.; Morgenthal, G. Automated Linking of 3D Inspection Data for Damage Analysis. In *Proceedings of the Bridge Maintenance, Safety, Management, Life-Cycle Sustainability and Innovations—Proceedings of the 10th International Conference on Bridge Maintenance, Safety and Management, IABMAS 2020*; CRC Press/Balkema: Leiden, The Netherlands, 2021; pp. 3714–3720.
16. Szilvási-Nagy, M.; Mátyási, G. Analysis of STL Files. *Math. Comput. Model.* **2003**, *38*, 945–960. [CrossRef]
17. Botín-Sanabria, D.M.; Mihaita, S.; Peimbert-García, R.E.; Ramírez-Moreno, M.A.; Ramírez-Mendoza, R.A.; Lozoya-Santos, J.D.J. Digital Twin Technology Challenges and Applications: A Comprehensive Review. *Remote Sens.* **2022**, *14*, 1335. [CrossRef]
18. Buonamici, F.; Carfagni, M.; Furferi, R.; Governi, L.; Lapini, A.; Volpe, Y. Reverse Engineering Modeling Methods and Tools: A Survey. *Comput. Aided. Des. Appl.* **2018**, *15*, 443–464. [CrossRef]
19. Lim, K.Y.H.; Zheng, P.; Chen, C.H. A State-of-the-Art Survey of Digital Twin: Techniques, Engineering Product Lifecycle Management and Business Innovation Perspectives. *J. Intell. Manuf.* **2020**, *31*, 1313–1337. [CrossRef]
20. Stałowska, P.; Suchocki, C.; Rutkowska, M. Crack Detection in Building Walls Based on Geometric and Radiometric Point Cloud Information. *Autom. Constr.* **2022**, *134*, 104065. [CrossRef]
21. Szewczak, E.; Winkler-Skalna, A.; Czarnecki, L. Sustainable Test Methods for Construction Materials and Elements. *Materials* **2020**, *13*, 606. [CrossRef] [PubMed]
22. Szewczak, E.; Piekarczuk, A. Performance Evaluation of the Construction Products as a Research Challenge. Small Error—Big Difference in Assessment? *Bull. Polish Acad. Sci. Tech. Sci.* **2016**, *64*, 675–686. [CrossRef]
23. Szewczak, E. Does Standardisation Ensure a Reliable Assessment of the Performance of Construction Products? *Standards* **2022**, *2*, 260–275. [CrossRef]
24. Working Group 1 of the Joint Committee for Guides in Metrology (JCGM/WG 1). *Evaluation of Measurement Data—Guide to the Expression of Uncertainty in Measurement*, 1st ed.; Committee for Guides in Metrology: Sevres, France, 2008; Volume 50, ISBN 9267101889.
25. *ISO 16474-3:2021*; Paints and Varnishes. Methodsof Exposure to Laboratory Light Sources. Part 3: Fluorescent UV Lamps. International Organization for Standardization: Geneva, Switzerland, 2021.
26. *EN 927-6:2018*; Paints and Varnishes—Coating Materials and Coating Systems for Exterior Wood—Part 6: Exposure of Wood Coatings to Artificial Weathering Using Fluorescent UV Lamps and Water. European Committee for Standardisation: Brussels, Belgium, 2018.
27. SHINING 3D Tech Co., Ltd. *FreeScan Trio User Manual*; SHINING 3D Tech Co., Ltd.: Hangzhou, China, 2024.
28. Kot, S.M.; Jakubowski, J.; Sokołowski, A. *Statystyka*; Wyd. 2, po.; Difin: Warsaw, Poland, 2011; ISBN 9788376413495.
29. Keskin, S. Comparison of Several Univariate Normality Tests Regarding Type I Error Rate and Power of the Test in Simulation Based Small Samples. *J. Appl. Sci. Res.* **2006**, *2*, 296–300.
30. Razali, N.M.; Wah, Y.B. Power Comparisons of Shapiro-Wilk, Kolmogorov-Smirnov, Lilliefors and Anderson-Darling Tests. *J. Stat. Model. Anal.* **2011**, *2*, 21–33.
31. *EN 15534-1:2014*; Composites made from cellulose-based materials and thermoplastics (usually called wood-polymer composites (WPC) or natural fibre composites (NFC)) - Part 1: Test methods for characterisation of compounds and products. European Committee for Standardisation: Brussels, Belgium, 2014.

32. Gao, X.; Li, Q.; Cheng, W.; Han, G.; Xuan, L. Optimization of High Temperature and Pressurized Steam Modified Wood Fibers for High-Density Polyethylene Matrix Composites Using the Orthogonal Design Method. *Materials* **2016**, *9*, 847. [CrossRef]
33. Wang, X.; Song, K.; Ou, R. Effects of Carbon Black and Titanium Dioxide on Ultraviolet Weathering of Wood Flour-HDPE/Using Multi-Phase Co-Extrusion Technology. *BioResources* **2017**, *12*, 6173–6186. [CrossRef]
34. Gunwant, D. Moisture Resistance Treatments of Natural Fiber-Reinforced Composites: A Review. *Compos. Interfaces* **2024**, *31*, 979–1047. [CrossRef]
35. Turku, I.; Kärki, T.; Puurtinen, A. Durability of Wood Plastic Composites Manufactured from Recycled Plastic. *Heliyon* **2018**, *4*, e00559. [CrossRef] [PubMed]
36. Andrady, A.L.; Heikkilä, A.M.; Pandey, K.K.; Bruckman, L.S.; White, C.C.; Zhu, M.; Zhu, L. Effects of UV Radiation on Natural and Synthetic Materials. *Photochem. Photobiol. Sci.* **2023**, *22*, 1177–1202. [CrossRef] [PubMed]

materials

MDPI

Article

Full-Locked Coil Ropes with HDPE Sheath: Studies of Mechanical Behavior of HDPE Under Accelerated Aging

Benjamin Schaaf [1], Björn Abeln [1,*], Markus Feldmann [1], Elisabeth Stammen [2] and Klaus Dilger [2]

1 Institute of Steel Construction, RWTH Aachen University, 52074 Aachen, Germany
2 Institute of Joining and Welding, Technical University of Braunschweig, 38106 Braunschweig, Germany
* Correspondence: abeln@stb.rwth-aachen.de; Tel.: +49-241-80-25825

Abstract: In accordance with German guideline ZTV-ING Part 4, full-locked coil ropes are provided with a three-layer corrosion protection coating based on epoxy resin and polyurethane, which must be renewed regularly. An alternative method is to use a coating of high-density polyethylene (HDPE), which is extruded onto the rope. In this article, the mechanical behavior of the thermoplastic material is studied, taking into account various accelerated aging processes, which are derived from the climatic boundary conditions of a real bridge structure and implemented in tests. In addition to the quasi-static material behavior, which is described using the uniaxial tensile test, the cyclic conditioning, relaxation, type of production and oxidation stability are also investigated. Finally, the results obtained are evaluated with regard to the applicability of the material as corrosion protection for full-locked coil ropes.

Keywords: HPDE; bridge cables; corrosion protection; accelerated aging; mechanical characterization

1. Introduction

If spans of more than 150 m have to be overcome for traffic routes, cable-span bridges are an economical construction method in bridge building. Since the beginning of the 2000s, the construction of this type of bridge has increased significantly. A distinction can be made between suspension bridges, cable-stayed bridges, arch bridges and other bridge types. Cable-stayed bridges in particular are frequently used in Germany due to their design advantages, such as the absence of costly abutments [1]. For example, the new replacement Rhine bridges near Leverkusen and Duisburg-Neuenkamp are being built as cable-stayed bridges. Regardless of the type of bridge, cables play a central role as high-strength tension members for the load-bearing capacity of the bridge structure. The requirements for corrosion protection of the bridge cables are therefore correspondingly high. In accordance with the state of the art as per German guideline ZTV-ING Part 4 [2], three-layer corrosion protection systems based on epoxy resin and polyurethane are used, which are applied after the ropes have been installed. A new alternative is the use of HDPE sheathing, which is extruded onto the cable by the factory. This makes it possible to dispense with the need for subsequent applications of the corrosion protection coating in the structure and maintenance coatings during the service life. HDPE-coated fully locked cables were used for the first time on the Schlunzig Mulde bridge (Figure 1a).

(a)

(b)

Figure 1. (**a**) Mulde Bridge Schlunzig (photo: Schulze + Rank Ingenieurgesellschaft); (**b**) Fehmarnsund Bridge (photo: dpa).

The investigations presented here show that the HDPE coating is a competitive alternative to conventional corrosion protection coatings for full-locked coil ropes (FLCR) due to its resistance to climatic influences.

Experimental studies are discussed, considering various accelerated aging methods derived from the climatic requirements of the network arch bridge over the Fehmarnsund (Figure 1b).

1.1. Bridge Cables

1.1.1. Conventional Bridge Cables

The most common type of bridge rope used in Germany are FLCRs. They consist of an inner core of round wires several layers thick and at least two outer layers of shaped wires (Z-wires), Figure 2. The first layer contains a core wire with six round wires laid in a helix. The other wire layers are stranded spirally in alternating directions [3]. Under tensile load, the interlocking Z-wires of the outer layers form a closed surface, which prevents the intrusion of corrosive media [1].

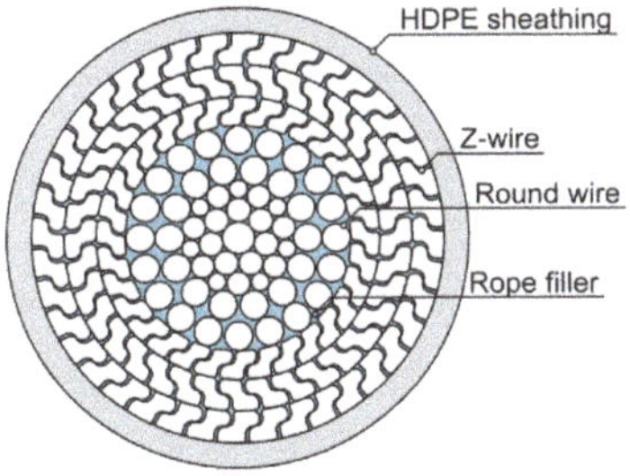

Figure 2. Technical illustration of full-locked coil rope (FLCR).

FLCRs have a total of three corrosion protection barriers. The first anti-corrosion barrier is the outer coating of the rope. The current construction method involves a multi-layer coating system with an epoxy resin-based primer. Several polyurethane-based functional layers are applied on top of this.

With the exception of the primer coat, the coating system must only be applied after installation in accordance with ZTV-ING [2], which requires the use of scaffolding or appropriate coating robots [4]. A rope filler (lube) is inserted into the rope to prevent internal friction and as additional corrosion protection (the second corrosion protection barrier). Only hot-dip galvanized wires may be used as an internal corrosion protection barrier.

In Germany, bridge cables are designed in accordance with DIN EN 1993 Part 1-11 [5] and the associated national annex. Technical test specifications and delivery conditions for fully locked coil ropes are regulated within the TL/TP-VVS [6].

1.1.2. FLCR Sheathed with HDPE

The use of an FLCR with an HDPE sheath offers a new and economical alternative to conventional FLCR with an anti-corrosion coating. A major advantage is the complete production of the outer corrosion protection barrier in the factory by means of extrusion, which eliminates the need to apply the corrosion protection after the bridge cables have been installed, as is the case with conventional FLCRs. The HDPE sheath is extruded onto the bridge cable at approx. 160 °C after stranding. The thickness of the HDPE sheath is usually 4 to 7 mm. A wide range of colors can also be selected using appropriate additives. Figure 3 shows a sample of the new bridge cables of the Fehmarnsund Bridge in the characteristic green color of the original cables.

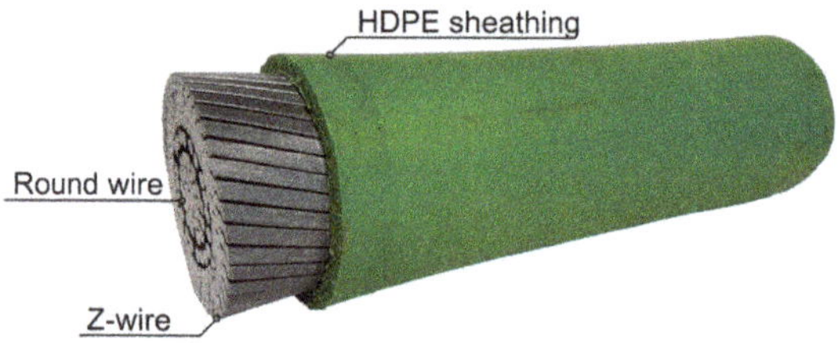

Figure 3. Sample of the new cables of the Fehmarnsund Bridge with green HDPE sheathing.

The specifications for HDPE-coated bridge cables can be found in fib Bulletin 30 [7].

On the one hand, requirements are made for the material composition, such as density, carbon black content or thermal stability. On the other hand, minimum requirements are specified with regard to mechanical characteristics such as modulus of elasticity, strength and fracture strain. These are differentiated into requirements relating to the raw material and the extruded pipe. All mechanical tests are carried out using uniaxial tensile tests on the dog bone specimen in accordance with ISO 527-2 [8]. In Germany, the use of HDPE-coated FLCR requires approval from building authorities [1].

To anchor the FLCR, the rope ends are formed into a uniform rope broom and placed in a tapered rope sleeve. If FLCRs with HDPE sheathing are used, the anchorage of the cable broom is not to be made with a metal filling as usual, but with a synthetic resin filling (Figure 4a).

According to TL/TP-VVS [6], the rope sleeves must be cast with the alloy "Zamak Z 610", which consists mainly of zinc, whereby the casting takes place at approx. 450 °C and the cable head is heated to 350 °C. However, the melting point of HDPE ranges from approx. 120 °C to 160 °C, meaning that the Zamak casting would melt or thermally decompose the HDPE, which would be equivalent to destroying the corrosion protection. In addition, HDPE softens considerably in temperature ranges below the melting point, which would lead to inhomogeneities in the HDPE sheath in cooler areas away from the cable head.

For this reason, socketing with a synthetic resin filling is used, Figure 4. These are usually polyester or epoxy resins. WIRELOCK®, a two-component polyester resin with a quartz powder as filler with approx. 49% by mass, is an established product on the market [9]. The socketing compound hardens within a short time, developing heat up to approx. 120 °C. The heat can be dissipated quickly via the potting sleeve and the rope. Requirements for synthetic resin potting for steel wire ropes can be found in DIN EN 13411-4 [10].

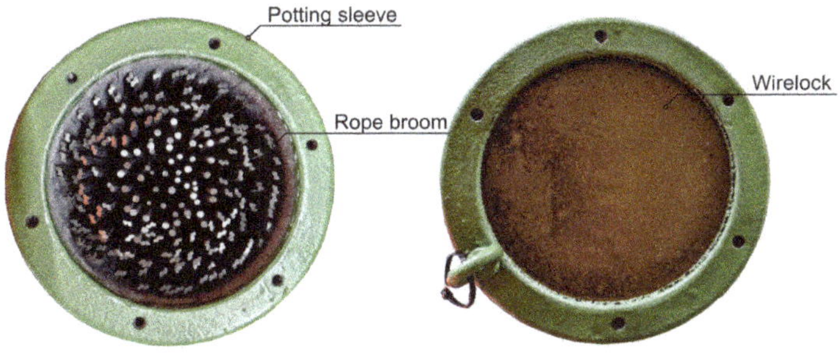

(a) (b)

Figure 4. (**a**) Cable sleeve prepared for socketing with fanned-out cable broom; (**b**) cable head cast with WIRELOCK®.

The use of HDPE-coated FLCRs has hardly been widespread in Germany to date. This type of FLCR was used for the first time on the Schlunzig Mulde Bridge (Figure 1b). In the course of exchanging the cables of the Fehmarnsund Bridge, FLCRs with HDPE sheathing and synthetic resin grouting were used. This type of bridge cable is already more established internationally, for example on the Strömsund Bridge in Sweden and the BØkfjord Brigde in Norway, as well as some bridges in the Far East and the USA.

1.2. High-Density Polyethylene

1.2.1. Basic Properties

High-density polyethylene (HDPE) is a thermoplastic polymer material produced by the polymerization of ethylene on certain catalysts in a medium to low pressure range. The high density of HDPE (more than 0.94 g/cm^3) results from a low degree of visualization of its molecular structure and the resulting increased crystallinity, which leads to improved strength and rigidity. These properties make HDPE particularly suitable for applications where structural integrity and durability are required [11].

HDPE is also known for its resistance to a wide range of aggressive chemicals, acids and bases. UV stabilizers and antioxidants can be added during production to increase UV and oxidation stability [12]. The material is particularly attractive for use in chemically explosive environments.

The crystallinity of HDPE is typically in the range of 60–80%. With increasing crystallinity, the modulus of elasticity, tensile strength, hardness, density and melting temperature also increase, while elongation decreases. Extruded HDPE has a modulus of elasticity of approx. 800–1000 MPa in the linear-elastic range. The strength is usually in the range of 20–25 MPa. HDPE generally has a pronounced yield potential, with a correspondingly high fracture strain in the range of 300–600%. Depending on the manufacturing process, however, the yield potential can be considerably reduced. As for most plastics, a dependence of the mechanical parameters on the loading rate and temperature can be observed [13]. Their influence on the crystallinity, density, tensile strength and modulus of elasticity, among other things, is determined by the superstructure of the semi-crystalline HDPE. In addition to the amorphous tangle structures of the polymer chains, the crystallites form first orders lamellae and later spherulites. With slow cooling in the manufacturing process, fewer spherulites are formed, but these can become comparatively large, whereas with rapid cooling only very small spherulites are formed [14]. The manufacturing process can therefore influence the material properties and must therefore be taken into account when testing the material properties. As described by Chaoui et al. in [15], the extrusion process for HDPE pipes is responsible for a multitude of morphological and intrinsic heterogeneities through the pipe wall. The fast quenching of the outer wall and the slow cooling at the inner surface leads to a gradient of stresses, (semi-)crystallinity and content of antioxidants.

In the project, HDPE material was provided by the manufacturer, which was taken directly from the rope sheaths, and that represents a complete cross-section through the extruded mass. The following material characteristics were taken from the supplied quality assurance protocols:

- The impact resistances are 14.2 kJ/m^2 from the unnotched Charpy test according to ISO 179-1 [16], with an elongation at break of 136% at -20 ± 2 °C and 50 mm/min (ISO 6259-1 and -3 [17,18]);
- The bulk density is 0.958 g/cm^3 in accordance with ASTM D1505 [19]. The melt flow index according to ASTM D1238 [20] is 0.051 gm/10 min.

Due to the small number of samples provided, these values could not be tested and were adopted.

1.2.2. Resistance of HDPE

HDPE has excellent resistance to a wide range of influences. For use as a corrosion protection and sealing barrier for the FLCRs of bridge structures, it is exposed to weathering. The relevant climatic influences are thus temperature changes, water and moisture, in a maritime environment or near roads, saline solutions and UV radiation. In addition, there are dynamic loads due to traffic loads and wind effects.

Albozahid et al. [21] studied the influence of moisture and UV radiation. Moisture leads to a slight degradation of the modulus of elasticity and strength. Sunlight or UV radiation from artificial sources lead to chain defects and thus also to a reduction in mechanical properties. The aging of HDPE bottles in sea water for up to 72 days at ambient and high temperature (80 °C) by Elkori et al. [22] showed similar results.

Hsueh et al. [23] and Carrasco et al. [24] observed an embrittlement of the material under UV irradiation (an increase in modulus, reduction in fracture strain). The higher the temperature prevailing during irradiation, the greater the embrittlement. Zhao et al. [25] found in thermal oxidation tests that the tensile strength increases with the aging time, while the elongation first increases (by specific cross-linking reactions) and then decreases.

There have been extensive studies on the aging behavior of HDPE, particularly with regard to pipelines or geomembranes. Mueller and Jakob [26] showed changes in the mechanical properties and oxidative induction times (OITs) of HDPE geomembranes during aging behavior in hot air and hot water, where oxidation is essentially dependent on the slow loss of antioxidants. The oxidation starts only after the depletion of antioxidants and then leads quickly to the brittleness of the sample, but no complete oxidative deterioration has been observed. Ewais and Rowe [27,28] have studied HDPE geomembranes in leachate at different temperatures, and morphological changes were observed before the antioxidants were depleted. Guermazi et al. [29] have investigated the accelerated aging of HDPE coatings in marine applications under the influence of seawater at different temperatures. A change in the mechanical properties was found, which depends on the aging temperature and the aging cycle.

In bridge construction, HDPE has only been used occasionally to protect bridge cables against corrosion.

Friedrich [4] compared the advantages and disadvantages of fully sealed spiral cables and strand bundle cables for cable-stayed bridges as a decision-making aid for future planning. Only the classic corrosion protection and butyl cheek bands were considered, not HDPE sheathing. Aging tests were not used for the comparison.

Saul and Nützel [30] investigated automatic wrapping with butyl bonding tapes as an alternative to the conventional coating of fully sealed bridge cables for the refurbishment of a suspension bridge in Norway. The focus was on the development of the process; no aging tests were carried out and no comparison to HDPE.

Yu et al. [31] evaluated the corrosion of steel cables using accelerated corrosion tests on semi-parallel cables with different HDPE sheath breaks. The focus here was on the corrosion of the steel, not the aging of the HDPE.

Huang et al. [32] examined the fatigue behavior of HDPE sheathing with regard to mechanical behavior (friction losses, slip, wear resistance).

Liu et al. [33] analyzed HDPE jackets using small part samples that were subjected to up to 5000 h of accelerated aging with a xenon arc lamp. The focus here was on the change in mechanical behavior. A critical time was determined, after which the mechanical characteristics decrease and the service life is reduced. No extrapolation was made to a specific service life, and the aging was carried out solely under xenon illumination.

Mkacher et al. [34] have presented a general methodology for assessing the durability of HDPE outer jackets. Their investigations are based on a series of experiments to clarify aging mechanisms by means of thermal oxidation tests on stabilized and pure PE, the determination of the chemical consumption of antioxidants and the formation of oxidation products.

Lin and Zhang [35] examined cable sheaths made of HDPE in a dry environment and carried out an artificial aging test in daylight and a tensile test under different load levels. The results showed that as the HDPE jacket aged, damage gradually occurred and the mechanical properties of the HDPE jacket also decreased.

Dynamic permanent loads influence the yield elongation and stress as well as the modulus of elasticity of HDPE [32,36,37]. However, the strain amplitudes required for this must be relatively large in order to achieve significant changes. In addition, HDPE exhibits time-dependent (viscoelastic) material behavior. Lai und Bakker [38] studied the creep behavior of HDPE and observed non-linear creep. With the help of a time–tension superposition, accelerated statements can be made about long-term creep. In addition, retardation spectra are shifted towards longer periods of time due to physical aging. A rate dependence of the material was experimentally proven by Elleuch and Taktak [12] for compressive and tensile loads.

The previous investigations show the influence of different aging effects on the material HDPE by (UV) light, oxygen and temperature and loads. For the application of FLCRs, the accelerated aging processes described in Section 2.2. are derived using the Fehmarnsund Bridge, which is very exposed to the climate in a maritime environment. Compatibility with the various substances with which the HDPE pipe comes into contact during its service life is also analyzed. The influences of hydrocarbon compounds like crude oil or motor oils and fuels on HDPE were tested by Chaoui et al. [15] and Rezakalla and Gennadyevech [39]. Within the first weeks, a decrease in modulus and strength became obvious. Due to this major influence, engine oil is taken into account within our analyses.

2. Materials and Methods

2.1. High-Density Polyethylene

The influence of artificial aging, as well as the compatibility with other materials and temperature dependence, are primarily studied on the basis of the change in mechanical behavior under tensile load. Based on fib Bulletin 30 [7], tensile tests are carried out in accordance with DIN EN ISO 527-2 [8] on test specimen type 1B (Figure 5).

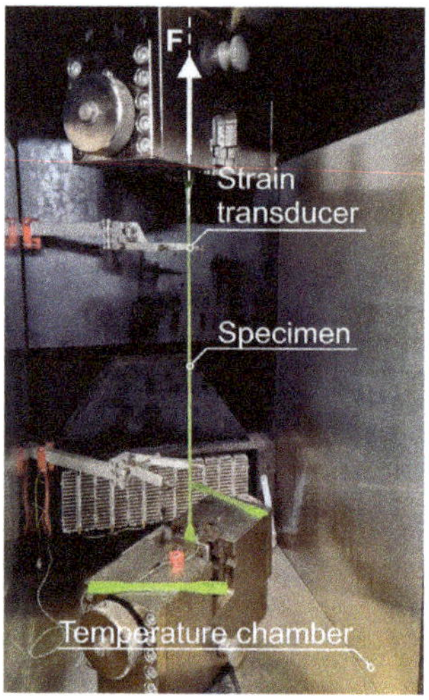

Figure 5. Tensile test according to DIN EN ISO 527 [8] using a temperature chamber.

The tests are carried out on a uniaxial *Zwick Roell Z100*, ZwickRoell GmbH and Co. KG, Ulm, Germany, testing machine. Deformation quantities are recorded locally in the parallel measuring area of the shoulder specimen using two longitudinal extensometers. These are also used for displacement-controlled regulation at 50 mm/min in accordance with [7]. A preload of 10 N is applied prior to testing.

The characteristic values of strength, fracture strain and modulus of elasticity are determined on the basis of the experimentally collected data. The latter is determined in the stress interval [1 MPa; 4 MPa]. This range is particularly suitable due to the relatively linear behavior of the material. The corresponding strains are between 0.5% and 0.25%, an interval that is usually used to determine the modulus.

2.2. Accelerated Aging Procedures

2.2.1. Sanding Method

Due to the location of the Fehmarnsund Bridge, the HDPE coating of the bridge cables is constantly abraded by sand during its service life, which is carried to the bridge from the surrounding beaches and shore areas, especially in strong winds (sand abrasion).

Selected HDPE samples are spattered with sand for testing. The sanding method for glass and plastics according to DIN 52348 [40] is used for this purpose. To increase the abrasion of the HDPE, the test is carried out at twice the drop height of 3.0 m (Figure 6). A total of 3 kg of quartz sand (grain size 0.5/0.71) falls evenly through a special outlet nozzle and two sieve levels onto the sample, which is mounted on a rotary plate to ensure homogeneous abrasion.

2.2.2. Climatic Boundary Conditions and Compatibility

Three basic types of climatic exposure are analyzed as part of the accelerated aging process: temperature cycles, salt spray and UV irradiation. The exposure is combined in order to consider possible, mutually reinforcing influences on the mechanical properties of the HDPE. The combinations of the aging processes carried out can be seen in Figure 7.

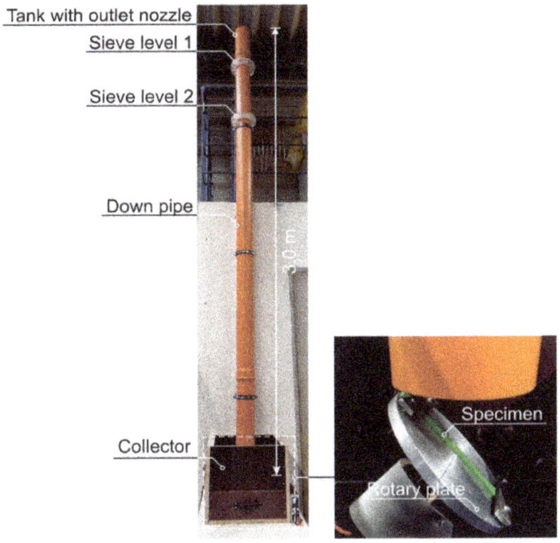

Figure 6. Sanding device based on DIN 52348 [40].

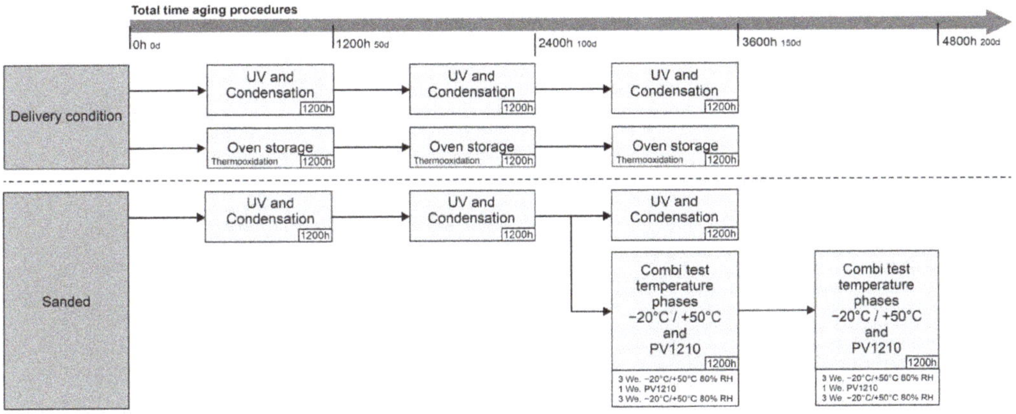

Figure 7. Overview of the accelerated aging procedures carried out.

The selected temperature range of −20 °C to +50 °C corresponds to the extreme temperatures to be expected at the Fehmarnsund Bridge. Due to the green color of the bridge cables and the maritime climate of the Baltic Sea, a surface temperature of more than +50 °C is not to be expected even on sunny summer days. The temperature range of −20 °C to +50 °C is used as the upper and lower test temperature for both the accelerated aging methods and the uniaxial tensile test. The upper service temperature of HDPE without mechanical stress should not exceed 90 °C to 100 °C for short periods and 80 °C to 90 °C for long periods [11].

To investigate the influence of UV radiation on the material, aging tests are carried out with artificial irradiation in accordance with DIN EN ISO 4892 Part 1 [41] and Part

3 [42] using a *UVTest*® by Atlas Material Testing Technology GmbH, Linsengericht, Germany. Cycle 1 of procedure A with a UVA-340 lamp is selected. This contains drying and condensation phases.

As already discussed, a correlation between real and accelerated aging for plastics and HDPE in particular has not been described or cannot be determined experimentally for the most part due to a large number of simultaneously acting influences. With regard to UV exposure, however, a comparative calculation can be used to make a rough estimate of the artificial UV irradiation and the UV exposure occurring on the building. Using data from the German Weather Service (DWD), the average annual sum of global radiation on Fehmarn can be determined as 1040 kWh/m^2. According to DIN EN ISO 4892-1 [41] the proportion of UV radiation is approx. 6.8% of the total radiation. With the power of the UV source (0.76 W/m^2), a duration of approx. 39 days corresponds to the real UV exposure of a year on Fehmarn. If it is also taken into account that the UV lamp is switched off according to cycle 1, method A, during the wet phases, the duration is extended to approx. 50 days. The total duration of the artificial irradiation of 3 × 50 days thus corresponds approximately to a UV exposure of three years. It must also be taken into account that the constant change between wet and dry phases puts more strain on the material than in a constantly dry climate.

Samples are collected after every 50 days (1200 h) and tested in a uniaxial tensile test. In addition, further samples are collected at one-week intervals in order to be able to analyze possible degradation with a higher resolution.

After 2400 h of UV irradiation, some samples are taken and subjected to a corrosion and aging test. This consists of the modified aging test PV1200 (−20 °C/+50 °C with 80% RH) in a *C600-70* climate chamber by Weiss Technik GmbH, Reiskirchen, Germany and the condensation and salt spray test PV1210 (both tests according to VW AG specifications) in a *SC/KWT 1000* salt spray chamber by Weiss Technik GmbH, Reiskirchen,, Germany. A 1200 h block is constructed according to the following aging regime. A detailed description of the two aging tests can be found in Table 1 and Figure 8. The modified aging test is also extended with additional samples in order to resolve the degradation of the mechanical properties more precisely. Details can be found in the general overview in Table 2.

Table 1. Aging regime sequence from climate change test PV1200 (−20 °C/+50 °C) and salt spray test PV1210.

Test Cycles	Duration	Stages	Conditions
3 weeks modified aging test PV1200	1 h	Heating phase	+50 °C, 80% RH
	4 h	Cooling phase	+50 °C, 80% RH
	2 h	Cooling phase	−20 °C, from T < 0 °C non-regulated RH
	4 h	Holding phase	−20 °C, 80% non-regulated RH
	1 h	Heating phase	+23 °C, ab T = 0 °C r. L 30%
1 week salt spray test PV1210	4 h	Salt spray	
	4 h	Ambient climate	
	16 h	Condensing water climate, 40 °C	
3 weeks modified aging test PV1200	1 h	Heating phase	+50 °C, 80% RH
	4 h	Cooling phase	+50 °C, 80% RH
	2 h	Cooling phase	−20 °C, from T < 0 °C non-regulated RH
	4 h	Holding phase	−20 °C, 80% non-regulated RH
	1 h	Heating phase	+23 °C, ab T = 0 °C r. L 30%

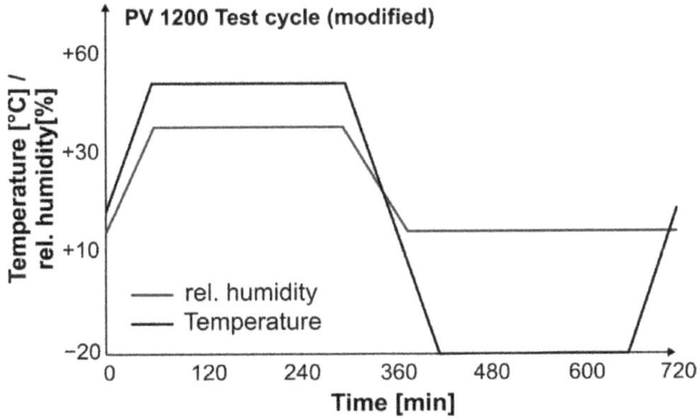

Figure 8. PV1200 test cycle.

Table 2. Overview of sample series and accelerated aging procedure.

Test Series			Accelerated Aging Procedures
	sanded	Non-sanded	
Manufacturing: Extrusion			
Reference	–	X	Non-aged reference samples
Pipe cut RWTH	–	X	Samples cut from HDPE pipe at the RWTH
UV 1200 h	X	X	UV irradiation + condensation, duration: 1200 h/50 d
UV 2400 h	X	X	UV irradiation + condensation, duration: 2400 h/100 d
UV 3600 h	X	X	UV irradiation + condensation, duration: 3600 h/150 d
TO 1200 h	–	X	Oven storage at +60 °C, duration: 1200 h/50 d
TO 2400 h	–	X	Oven storage at +60 °C, duration: 2400 h/100 d
TO 3600 h	–	X	Oven storage at +60 °C, duration: 3600 h/150 d
UV 2400 h + KWT 1200 h	X	–	UV irradiation + condensation, duration: 2400 h/100 d +
UV 2400 h + KWT 2400 h	X	–	Aging test PV1200 & PV1210, duration: 1200 h/50 d
Cyclically conditioned	X	-	UV irradiation + condensation, duration: 2400 h/100 d +
Manufacturing: Injection molding			
Reference	X	X	Non-aged reference samples
UV 1	–	X	UV irradiation + condensation, duration: 7 d
UV 2	–	X	UV irradiation + condensation, duration: 14 d
UV 3	–	X	UV irradiation + condensation, duration: 21 d
UV 4	–	X	UV irradiation + condensation, duration: 28 d
UV 5	–	X	UV irradiation + condensation, duration: 35 d
UV 6	–	X	UV irradiation + condensation, duration: 42 d
KWT 1	–	X	21 d KWT (−20/+50 °C 80% RH)
KWT 2	–	X	21 d KWT (−20/+50 °C 80% RH) + 2 d NK + 5 d PV1210 (salt spray test)
KWT 3	–	X	21 d KWT (−20/+50 °C 80% RH) + 2 d NK + 5 d PV1210 (salt spray test) + 2 d NK + 21 d KWT (−20/+50 °C 80% RH)
KWT 4	–	X	2 cycles: 21 d KWT (−20/+50 °C 80% RH) + 2 d NK + 5 d PV1210 (salt spray test) + 2 d NK + 21 d KWT (−20/+50 °C 80% RH)
Rope filler	–	–	Storage in rope filler, duration: 1200 h/50 d, RT
Mineral oil	X	–	Storage in mineral oil, duration: 1200 h/50 d, RT

KWT: aging test modified PV1200, NK: normal climate, RH: relative humidity.

In addition to the climatic influences, the compatibility of the HDPE with rope filler and mineral oil is also analyzed. For this purpose, samples are stored fully enclosed in both media at room temperature for 1200 h (50 days). The filling agent is a solvent-based rope lubricant containing zinc dust. Table 2 summarizes the descriptions and properties of all the sample series studied.

2.2.3. Cyclical Conditioning

Some plastics exhibit material softening under repeated loading. This can manifest itself in the form of a reduced stiffness after initial loading (Mullins effect) or in the form of reduced strength.

During the manufacturing and installation of the bridge cables, the HDPE sheath is also subjected to repeated cyclical stress. The rope is coiled up after the sheathing has been extruded, uncoiled before the rope core is socketed, coiled up again after socketing and finally uncoiled during installation. With a winding diameter of 2.20 m, this results in a maximum compression of approx. 3.1% on the inside and maximum elongation of approx. 3.0% on the outside of the HDPE sheath.

In order to investigate the influence of this load cycle, uniaxial tensile tests are carried out after a defined cyclic conditioning with subsequent destructive testing. For a conservative estimate, a total of 20 loading and unloading cycles up to a 3% elongation and then a further 20 loading and unloading cycles up to 5% are performed. This is immediately followed by the quasi-static test until failure.

2.3. *Sample Manufacturing*

Two different manufacturing processes are examined: injection molded samples (hereinafter referred to as "injection molded") and samples cut from an HDPE pipe manufactured using an extrusion process (hereinafter referred to as "pipe cut"). Pipe cut samples were provided by the manufacturer ("pipe cut manufacturer") on the one hand and produced from offcuts ("pipe cut RWTH") on the other. All injection molded samples were provided by the manufacturer.

All mechanical tests are carried out using uniaxial tensile tests on a dog bone specimen in accordance with ISO 527-2 [8].

2.4. *Differential Scanning Calorimetry and Oxidation Induction*

DSC is an analytical measurement method used to investigate the thermal properties of materials, such as phase transitions [43]. It is usually carried out in an inert gas atmosphere to prevent oxidative decomposition reactions in polymers. However, if the oxidation stability of a polymer or the quantity (quality) of the added stabilizers is to be determined, the oxidation induction time (OIT) can be used, which is also determined using DSC measurements. In this process, heating is first carried out in the nitrogen flow. When the specified temperature is reached, the atmosphere is replaced by the oxygen atmosphere at the same flow rate. The test specimen is then held at a constant temperature until the oxidation reaction is indicated by an exothermic deviation of the DSC heat flow curve.

It is known from the literature from measurements of the oxidation induction temperature that the start of the oxidation of HDPE can be above 250 °C [44]. However, temperatures from 200 °C are also found for polyethylene itself. In this article, 200 °C was therefore chosen as the test temperature in order to achieve the best possible resolution of the various OITs. Tests were performed with a *DSC 214 Polyma* by NETZSCH-Gerätebau GmbH, Germany.

3. Results

3.1. Mechanical Behavior and Surface Conditions

The comparison of the mechanical values shown in Table 3 clearly shows the influence of sample production on the mechanical behavior. Compared to extruded samples, injection molded samples have virtually no yield potential. After reaching the maximum stress, failure occurs relatively quickly. The sample tapers considerably before a crack forms, which then leads to cross-sectional failure. In the case of extruded samples cut from the HDPE pipe, on the other hand, a pronounced yielding of the material can be seen, which is accompanied by a cross-sectional tapering that occurs over the entire gauge length (Figure 9).

Table 3. Comparison of the mechanical characteristic values between injection molding and pipe cut sample types.

		Tensile Strength [MPa]	Fracture Strain [%]	Modulus of Elasticity [MPa]
Injection molded	M	25.4	20.5	1031
	SD	0.97	5.2	73
Pipe cut manufacturer	M	22.2	151	1031
	SD	0.61	82	33
Pipe cut	M	23.7	479	1306
	SD	0.28	242	159

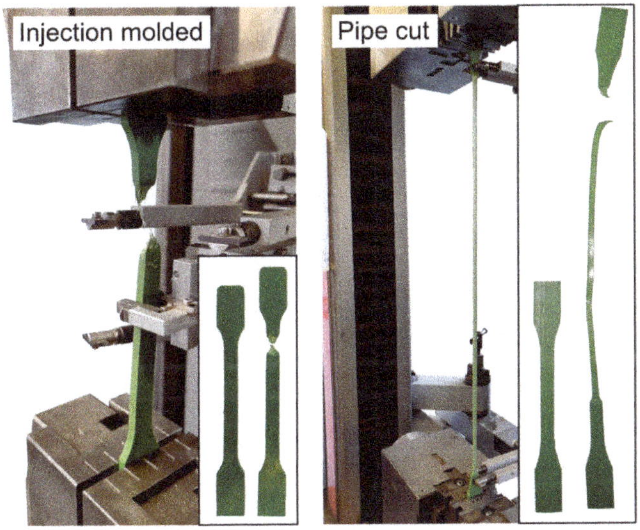

Figure 9. Comparison of the deformation and failure behavior of injection molded and extruded specimens.

The influence of the manufacturing process with different cooling processes and the differences resulting from the formation of different superstructures can also be clearly seen in images taken using a scanning electron microscope (SEM) and reflected light microscope, Figure 10. Compared to the extruded samples, the surface texture of the injection molded samples is much smoother. The extruded samples have a granular-structured surface, which indicates a bigger spherulite formation.

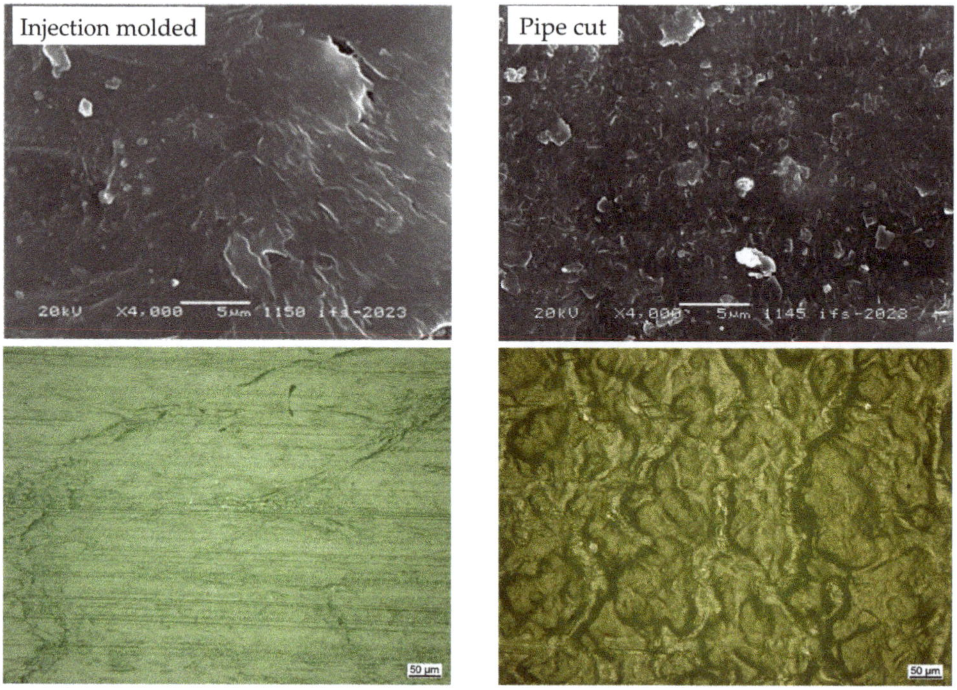

(a) (b)

Figure 10. Comparison of SEM (top) and reflected light microscope images (bottom) of (**a**) injection molded and (**b**) extruded samples.

3.2. Sanding Method

Irradiating the samples with quartz sand damages the surface of the material. The microscope images shown in Figure 11 reveal a clear difference between the unsanded and sanded sample surfaces. There are more light-colored, punctual defects and scratches, which can negatively influence the strength.

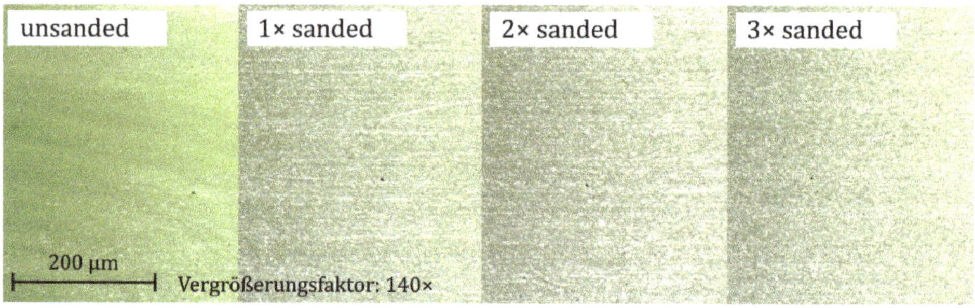

Figure 11. Microscope images of differently sanded samples (injection molding).

Additional sanding cycles lead to an optical homogenization of the surface. Therefore, the samples are only sanded once to avoid homogenization.

A series of tests carried out on injection molded, sanded samples show no significant influence of sanding on the tensile strength, fracture strain or modulus of elasticity,

Figure 12. However, since a damaged surface can influence degradation by increasing the area of attack during the accelerated aging process, sanding is carried out on a large number of samples.

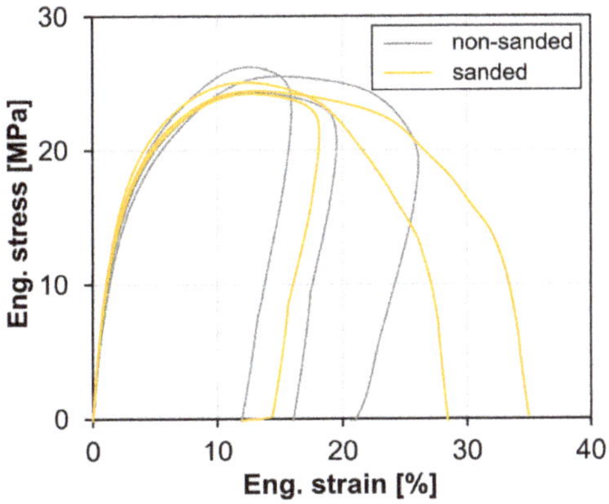

Figure 12. Tensile stress–strain curves of sanded and unsanded samples (injection molded).

3.3. Cyclical Conditioning

The cyclic conditioning of the samples causes the formation of hysteresis curves (Figure 13). Due to the viscoelastic behavior of HDPE, a residual strain remains after returning to the force zero point, which increases degressively with increasing cycles. This behavior is evident for both strain levels investigated. There is no difference between injection molded and extruded samples. Overall, the behavior of both sample types in the linear-elastic range is almost identical, which also demonstrates the good agreement of the hysteresis curves.

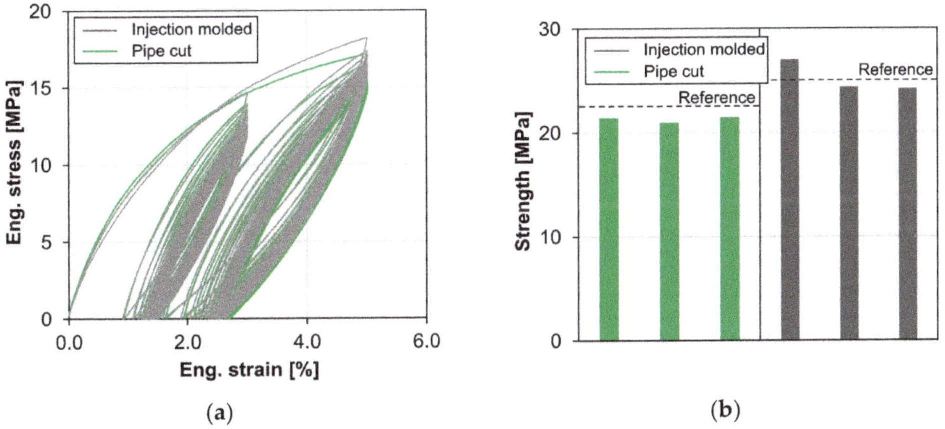

Figure 13. (**a**) Tensile stress–strain curves of cyclic conditioning (pipe cut and injection molded); (**b**) subsequently determined tensile strengths (pipe cut and injection molded, right).

After completion of the cycle runs at 3% and 5% elongations, the samples were visually inspected for localized defects and compared with control images of the unloaded condition. No defects or indications of damage were found.

The tensile strengths determined in subsequent quasi-static tests are reduced by approx. 1 MPa for the pipe cut samples compared to the non-conditioned reference samples (Figure 13b). The average strength of the injection molded samples, on the other hand, is at the level of the corresponding reference samples. The moduli of elasticity and the fracture strain show no irregularities in comparison to non-conditioned samples, considering the manufacturing process and the scattering.

3.4. Temperature Tests

The mechanical behavior of thermoplastics is strongly dependent on the temperature. In the differential thermal analysis (DSC) of the pipe cut material, Figure 14, there are still slight differences in the melting peak in the first heating due to the material history of the individual sample; in the second heating, a peak at 131 °C is obtained in both measurements. The comparative dynamic mechanical thermal analysis (DMTA), Figure 15, shows a melting point at 125 °C (evaluation of the tan Δ) with a constantly slightly decreasing shear modulus. The crystallization range ends in the DSC when cooling down to approx. 60 °C; the melting range of the first micro crystallites starts when heating up from 70 °C.

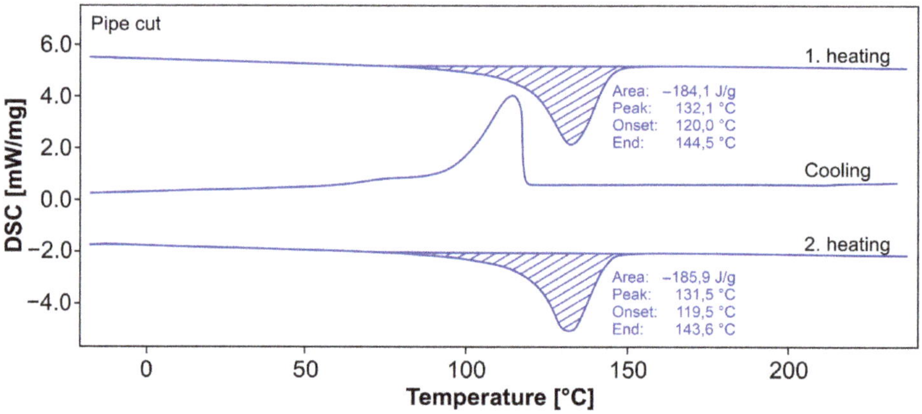

Figure 14. DSC analysis (pipe cut).

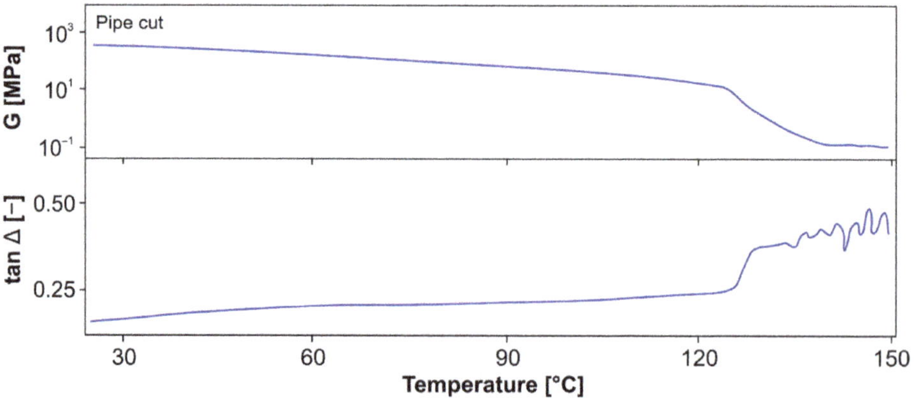

Figure 15. DMTA analysis (pipe cut).

Using the average enthalpy of fusion of two samples with ΔH_m of 189.3 J/g and $\Delta H_{m\infty}$ of 287.3 J/g at 100% crystallinity of PE results, according to Mirabella and Bafna [45], in a degree of crystallinity of 65.9% (and 65.3% using ΔHm_∞ of 290 J/g like in [14]) for the pipe cut material used, with $\alpha = \Delta H_m / \Delta H_{m\infty} \times 100\%$.

A comparison of the tensile stress–strain curves shown in Figure 16 illustrates the influence of test temperatures well below the melting range on the material. Figure 17 compares the material characteristics.

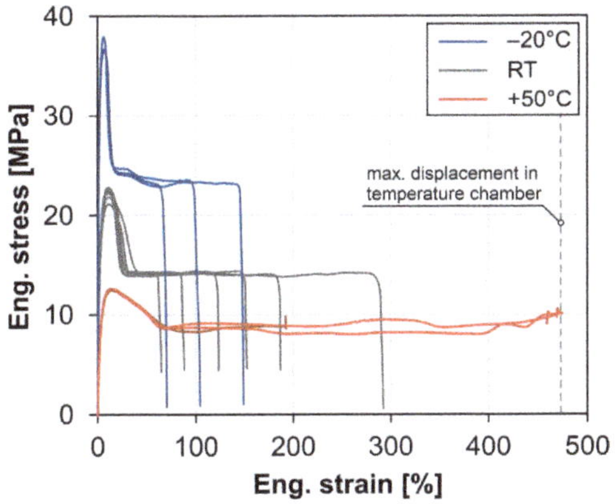

Figure 16. Stress–strain curves of the temperature tests (pipe cut).

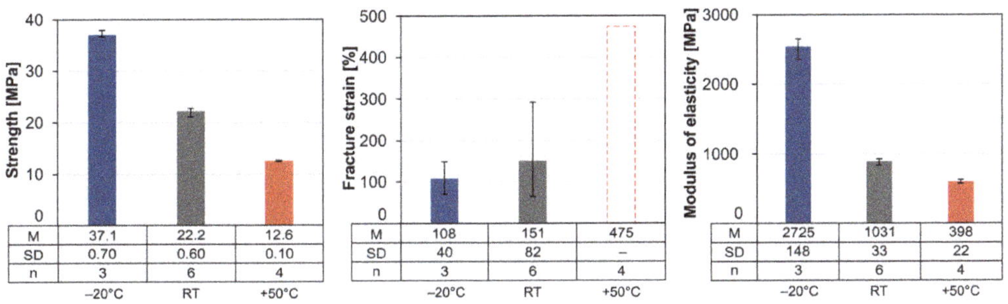

Figure 17. Average strength, fracture strain and modulus of elasticity of the temperature tests (pipe cut).

Poisson's ratio is determined at room temperature and lies in a range from 0.40 to 0.45, depending on the evaluation method.

According to fib Bulletin [7], requirements for strength at RT and fracture strain at −20 °C and RT are specified for the pipe material. The requirements for strength and fracture strain at −20 °C are met, but the required minimum fracture strain is not met at RT.

3.5. Accelerated Aging

The results of the various accelerated aging methods are compared below. For reasons of clarity, the individual tensile stress–strain curves are not shown. Initially, only the results of the extruded pipe cut samples are shown.

Considering the tensile strength first, it can be seen that the average strengths determined are between 21.3 MPa and 23.7 MPa (Figure 18). Only the samples from the

1200 h UV exposure are significantly below the other average strengths, with a strength of 16.8 MPa and 16.6 MPa, respectively. After a further 1200 h, the strength of the unaged samples can be observed again. An additional 1200 h then has no further influence on the strength. The experimental investigations were unable to provide reliable evidence to explain this phenomenon. However, various polymer–chemical effects are conceivable, which initially lead to debonding and subsequent re-polymerization. These include, for example, the destruction of chain bonds by UV radiation and the formation of free oxygen radicals (photodegradation). With increasing durations of UV irradiation, opposing effects initiated by the UV light, such as the formation of oxidative groups or the post-crosslinking of polymer chains, can lead to renewed polymerization. It is also conceivable that internal stresses are reduced during initial storage or that moisture is absorbed. Due to a lack of sample material, however, this could not be investigated further analytically. From the literature, this effect could not be observed in the pure temperature storage at 125 °C (i.e., without UV and without the influence of moisture) of thermoplastic blends made of LLDPE/HDPE. Here, there was only a reduction in tensile strength over a 210-day observation period [46]. In comparison, the long-term storage of HDPE geogrids at 25 °C, 45 °C and 75 °C in leachate for up to one year shows no changes in tensile strength [47]. However, this study shows an increase in fracture strain over time when stored in leachate at 75 °C.

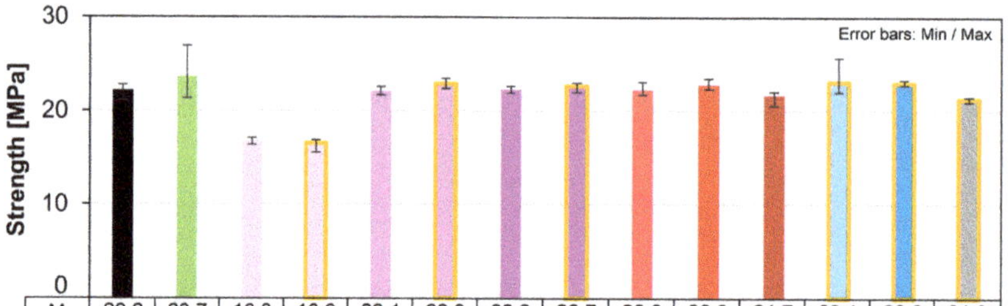

	Reference	Pipe cut RWTH	UV 1200h	UV 1200h + sanded	UV 2400h	UV 2400h + sanded	UV 3600h	UV 3600h + sanded	TO 1200h	TO 2400h	TO 3600h	UV 2400h + KWT 1200h + sanded	UV 2400h + KWT 2400h + sanded	Cyclically conditioned
M	22.2	23.7	16.8	16.6	22.1	22.9	22.3	22.7	22.2	22.8	21.7	23.1	23.0	21.3
SD	0.61	2.08	0.53	0.57	0.63	0.38	0.50	0.40	0.73	0.59	0.73	1.49	0.23	0.28
n	6	5	2	5	2	5	2	5	3	3	4	5	5	3

Figure 18. Comparison of the strength of artificially aged samples (pipe cut).

In the present study, oven storage at +60 °C only had a marginal effect on the strength. Only after 3600 h is the average strength reduced by approx. 1 MPa and is therefore outside the scatter band of the pre-series (TO 2400 h). The effect of thermal oxidation is discussed in more detail in the context of differential scanning calorimetry (DSC) (see Section 3.9).

The aging test (KWT), on the other hand, does not lead to a reduction in strength. After cyclic conditioning, the strength is slightly reduced but is still within the scatter band of the reference samples. Sanded samples show a slightly increased strength compared to non-coated samples of the same series.

The evaluation of the fracture strain of extruded pipe cut samples shows an extreme volatility in the determined data. The lowest fracture strain is 55%, the highest value 700%. The enormous scatter that can be observed within series and across series makes it impossible to compare the various accelerated aging processes. As the geometric deviations between the samples are very small and the test and storage conditions within the series are identical, the cause of the high scatter can only be assumed to be in the material itself. For example, samples may have different local defects and degrees of crystallization and superstructures, which influence the yield capacity and thus also the fracture strain achieved. An indication of this would be the deviating melting behavior in the first heating run [44]. Since only a very limited number of samples were available, it was not possible to carry out comparative measurements on untested samples.

The similarity between the moduli of elasticity of different aging processes largely corresponds to the distribution of their strengths, but overall, all aging processes result in an increase in modulus compared to the reference (Figure 19). Nevertheless, the reduction in the average modulus after 1200 h of UV irradiation can also be observed here. After a further irradiation period of 1200 h, it increases again and is at the level of the other series. The cyclic conditioning results in a reduced modulus of elasticity compared to the other aging scenarios.

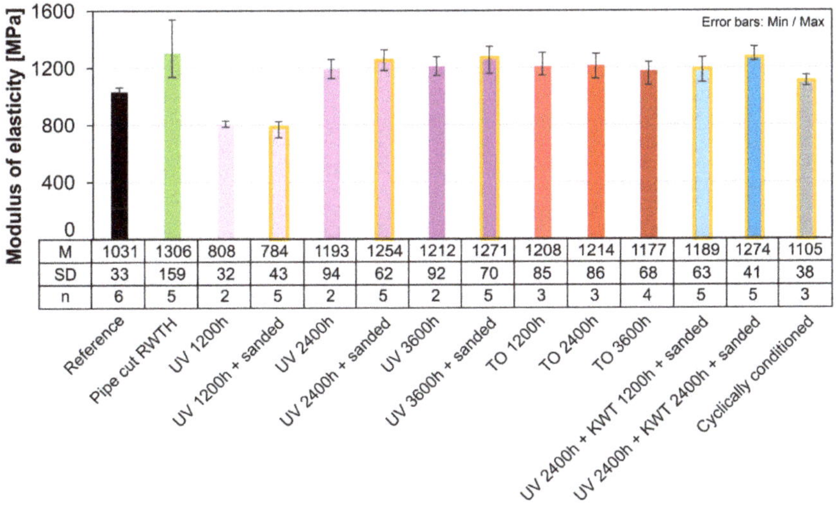

Figure 19. Comparison of the modulus of elasticity of artificially aged samples (pipe cut).

The strength requirements according to fib Bulletin [7] are met for all the accelerated aging processes investigated. The requirements for minimum fracture strain are only partially met due to the very high scatter.

Additional aging tests are carried out with further injection molded samples. For UV irradiation with condensation water phases, these are carried out over a period of six weeks. Three samples are taken and tested every week. The aim here is to achieve a more precise resolution of the initial aging. The additional aging tests have a total duration of 14.5 weeks. For comparison purposes, the respective results of the extruded pipe cut are also shown (Figure 20).

As with the extruded samples, accelerated aging has no effect on the strength of the injection molded samples. This is approx. 24 MPa for all series tested. The scatter is also extremely small and lower than for the extruded samples.

As already discussed, the total fracture strain determined is much lower than for the extruded samples due to the lack of yield potential. However, the results here vary greatly, albeit less extremely than for the extruded samples. It is therefore not possible to make a statement regarding the effects of aging on the fracture strain. Only the aging test KWT shows a tendency towards an increase in the average fracture strain. The influence on the modulus of elasticity is also very low (Figure 21).

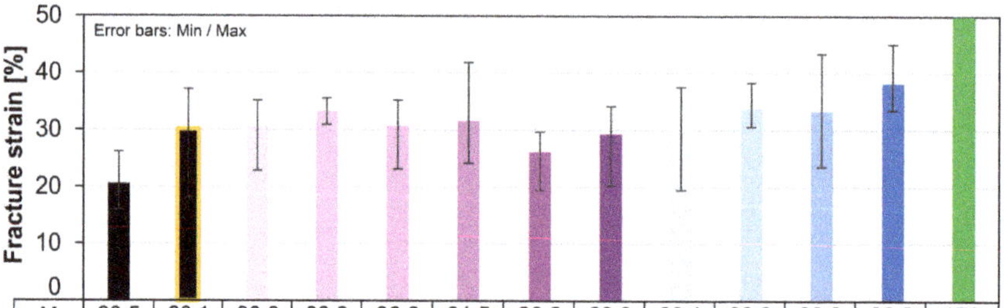

	Reference	sanded	UV1	UV2	UV3	UV4	UV5	UV6	KWT1	KWT2	KWT3	KWT4	Pipe cut (RWTH)
M	20.5	30.1	30.6	33.3	30.8	31.5	26.2	29.3	30.1	33.9	33.3	38.2	479
SD	5.2	10.4	6.7	2.3	6.7	9.1	5.9	6.4	6.6	2.9	7.1	4.4	242
n	3	3	3	3	3	3	3	4	5	5	5	5	6

Figure 20. Comparison of the strength of artificially aged samples (injection molded).

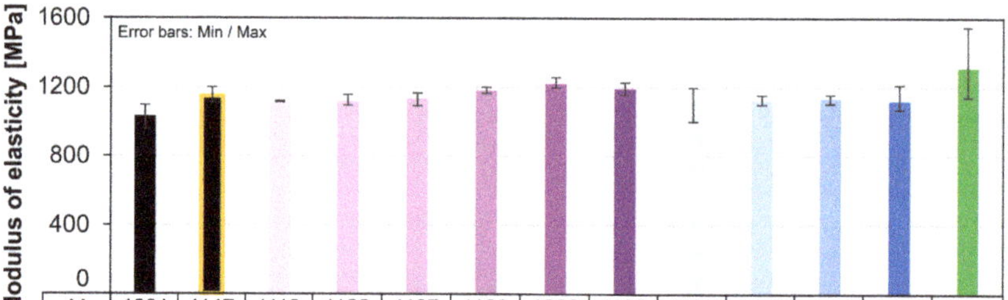

	Reference	sanded	UV1	UV2	UV3	UV4	UV5	UV6	KWT1	KWT2	KWT3	KWT4	Pipe cut (RWTH)
M	1031	1147	1116	1122	1137	1183	1222	1192	1118	1126	1138	1122	1306
SD	73	46	4	32	39	18	31	31	72	20	25	54	159
n	3	3	3	3	3	3	3	4	5	5	5	5	6

Figure 21. Comparison of the modulus of elasticity of artificially aged samples (injection molded).

3.6. Compatibility

The results of the compatibility tests with rope filler and mineral oil (*TruLub A11*) are shown below (Figure 22). Storing samples in rope filler for fifty days reduces the average strength by 1 MPa. Storage in mineral oil also reduces the average strength by around 1 MPa compared to the sanded reference sample. At the same time, exposure to both substances cause an increase in fracture strain. However, the generally high scattering of this parameter should be noted here. Storage in the filling agent causes an increase of approx. 100 MPa, in mineral oil a reduction of approx. 100 MPa, of the modulus of elasticity. In principle, HDPE is resistant to mineral oils [48]. However, due to the immersion storage and a similar (non-)polarity of HDPE and mineral oil, it is conceivable that small quantities migrate into the HDPE and thus lead to the observed softening of the polymer.

No permanent discoloration occurred. No incompatibility with the cured WIRELOCK® socketing compound was found.

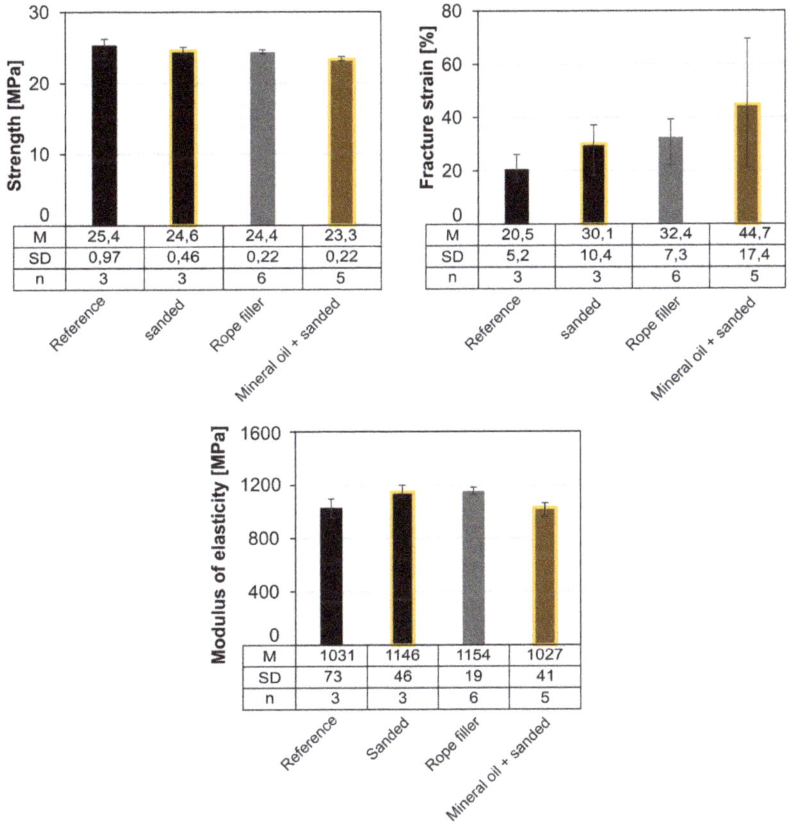

Figure 22. Average strength, fracture strain and modulus of elasticity of the compatibility tests (injection molded).

3.7. Knife Cut and Groove

During the intended service life and in particular during the installation of the bridge cables, defects in the HDPE sheathing can occur. In order to investigate the influence of notches and scratches on the load-bearing capacity, samples were provided with a measured cut orthogonal to the load direction. Three cutting depths of 0.5 mm, 1.0 mm and 1.5 mm were examined (Figure 23a). At this point, it should be noted that the thermoplastic properties of the polymer make it easy to repair surface defects.

In addition, samples were provided with a groove (width 2 mm, depth 2 mm) to simulate a larger cross-sectional recess (Figure 23b). Such a recess can be used, for example, for a force-fit connection of HDPE and potting compound within the rope sleeve.

Both types of tests are intended to clarify whether cross-section defects lead to an excessive reduction in the load-bearing capacity of the HDPE compared to fully intact cross-sections. For this purpose, the maximum loads determined are plotted against the net cross-section (Figure 24). The reference curve shown is based on the reference samples without cross-sectional weakening. As all results are on or above the reference curve, it can be shown that the weakening of the cross-section does not lead to an excessive reduction in the load-bearing capacity or strength of the material.

(a) (b)

Figure 23. (**a**) Fracture patterns of the knife-edge samples; (**b**) sample and fracture patterns of the groove test.

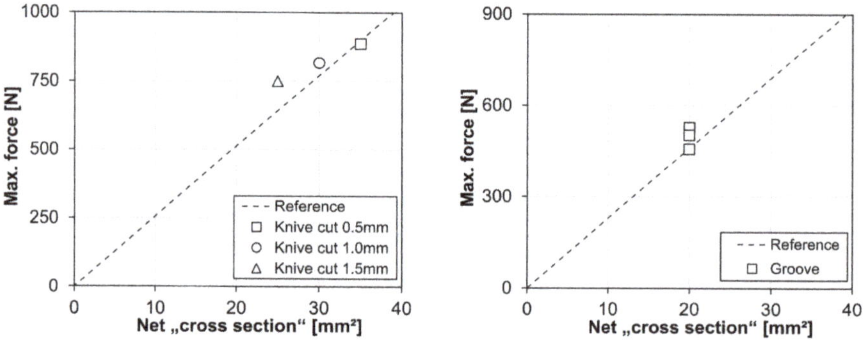

(a) (b)

Figure 24. Influence of cross-section reductions on the load-bearing capacity: (**a**) knife cut (injection molded); (**b**) groove (pipe cut).

3.8. Relaxation

The HDPE sheath has an extremely low stiffness compared to the bridge cables. As the HDPE is force-fitted to the outer Z-wire layer by extrusion, the deformation of the sheath corresponds to that of the rope. Therefore, creep tests were carried out to estimate the relaxation behavior of the polymer under imposed deformation. The two strain levels of 3% and 5%, which were already considered in the cyclic conditioning, were investigated. Injection molded samples were used. Since the selected strain levels are in the elastic range and this is virtually identical for injection molded and extruded samples, it can be assumed that the results are transferable to extruded HDPE.

The stress curves shown in Figure 25 demonstrate the viscoelastic character of the polymer. Under constant deflection, the stress decreases immediately. This is due to the movement and reorientation of the molecular chains [49] and a degressively decreasing process. A large part of the stress reduction takes place in the first few minutes after the deformation is imposed. After 12 h, the stress level for an applied strain of 5% has already fallen to 41.6% of the initial level; after 120 h, the material relaxes by only a further 0.3%. At this point, thermal effects already have a greater influence, as the results presented show. As the relaxation tests were not carried out at controlled temperatures, the jagged curves can be attributed to the different temperatures during the day and night. The reduction of the stresses and secant moduli can be seen in Table 4.

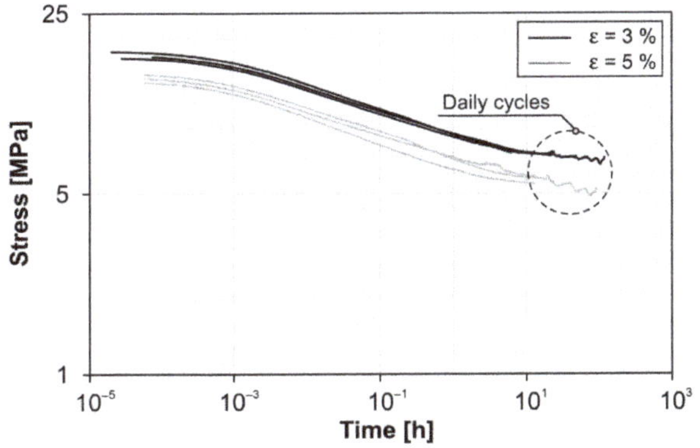

Figure 25. Relaxation tests with two different strain levels (injection molded).

Table 4. Tensile stress and secant modulus as a function of the test time.

	t = 0 h		t = 12 h			t = 24 h			t = 120 h		
ε [%]	σ_0 [MPa]	E [MPa]	σ [MPa]	σ/σ_0 [%]	E [MPa]	σ [MPa]	σ/σ_0 [%]	E [MPa]	σ [MPa]	σ/σ_0 [%]	E [MPa]
3.0	13.97	466	5.68	40.6	189	5.41	37.4	180	5.19	35.9	173
5.0	17.17	343	7.14	41.6	143	7.10	41.4	142	7.05	41.1	141

3.9. Differential Scanning Calorimetry and Oxidation Induction

Isothermal measurements at 200 °C were used to investigate the influence of oven storage and UV irradiation on the OIT of extrusion samples. The curves shown in the following figures represent the heat flow of the material over time. The OIT is reached when this changes significantly. A special tangent method is used to determine the point in time [50]. For a better overview, the curves, which actually overlap each other, are drawn apart.

For oven storage at +60 °C, a significant reduction in OIT can only be seen for samples stored for 3600 h or 150 days (Figure 26). Compared to the reference samples, the duration decreases here by approx. 48%. The shorter storage times in the oven for 1200 or 2400 h show no influence here.

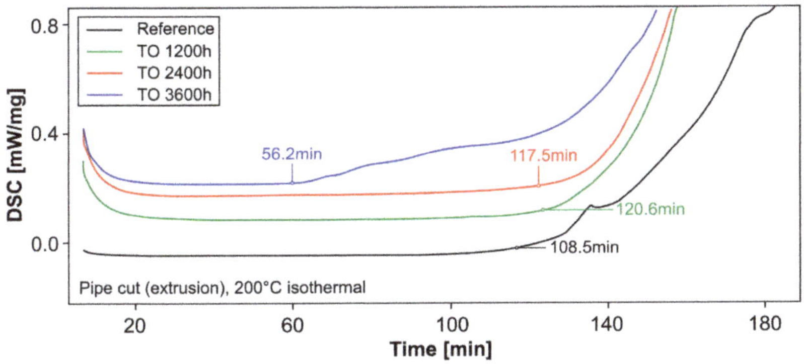

Figure 26. DSC: Influence of oven storage (pipe cut).

According to Arrhenius [51], chemical processes take place approximately twice to four times as fast when the reaction temperature is increased by +10 °C. If a reference temperature of 20 °C is selected and it is also taken into account that the test atmosphere of the DSC consists of 100% oxygen, this results in an oxidation stability of the HDPE pipes (reference, OIT = 108.5 min) in a normal atmosphere (21% oxygen) of approx. 258 years as a very rough estimation, without taking into account the effects of weathering, UV exposure and loss of stabilizers due to migration [52].

The results shown in Figure 27 show the influence of UV radiation in combination with moisture on the material. The test specimens are injection molded samples. The reference of the OIT corresponds to approx. one and a half times the reference of the OIT of the extrusion samples. The oxidation stability is continuously reduced by the artificial UV exposure and moisture. After five weeks, the effectiveness of the added antioxidants is still approx. 68% of the reference sample. However, it is still at a comparatively high level, higher than the reference samples of the pipe cut.

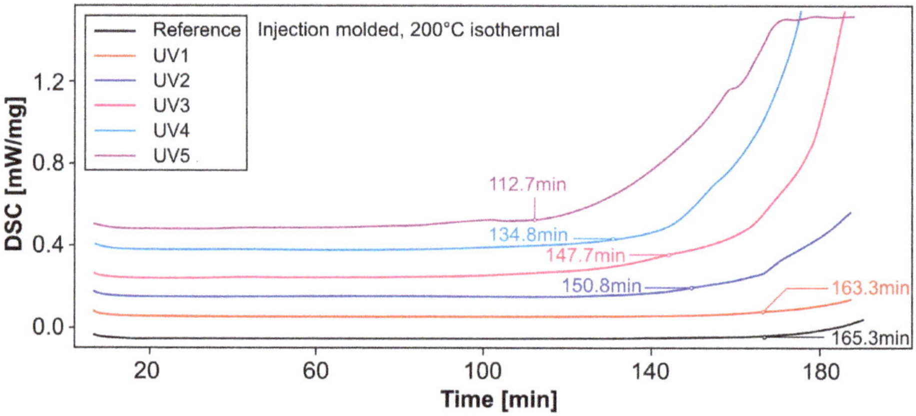

Figure 27. DSC: Influence of UV radiation (injection molded).

It should be mentioned at this point that reaching the OIT is not directly associated with a defect or loss of the mechanical, optical or chemical properties of the material. It merely indicates the point in time at which oxidation processes begin significantly. This point in time depends on the type and amount of stabilization of the material by the antioxidants. The influence on the mechanical properties cannot be derived directly from this.

4. Discussion

4.1. Manufacturing Process

The sample production was complex, as the HDPE coatings to be analyzed are extruded directly onto the ropes. The material tests had to be carried out before the ropes were manufactured, but industrial extrusion only takes place for the finished ropes with the corresponding color pigments and material thicknesses. For this reason, small batches of HDPE were injection molded, and a few samples were extruded in a small series prior to rope production, which then had to be compared with each other.

Our results show, in accordance with the literature [14,15], that the difference between the two manufacturing processes is particularly evident in the fracture strain achieved and in the yield behavior. This is most likely due to the cooling process of the molding compound during the respective manufacturing processes. In injection molding, the molecules have less time to orient and arrange themselves due to the rapid cooling within the surrounding mold, and the structures from crystallites and lamellae to spherulites are on nanoscale level. In contrast, the extrusion process allows for slower cooling, especially on the inner side of the material, and continuous movement of the polymer, which can lead to a different super structure due to bigger spherulites in microscale, as the SEM pictures show, and higher yield properties.

Due to the limited availability of samples cut from the pipe, not all aging processes can be investigated in the same way as with injection molded samples. However, as the strength and elongation at yield point of both sample types differs only slightly, it can be assumed that the influences on the material are comparable.

An important result with regard to the characterization of the material and the quality control of the product properties is the manufacturing process of the accompanying HDPE test specimens. Samples from injection molding did not exhibit any yield potential in the tests carried out here. Compared to pipe cut samples made from extruded material, the fracture strain is therefore many times lower. The average strength is at a similar level, but the modulus at a slightly reduced level compared to the extruded samples.

4.2. Accelerated Aging Tests

The influence of a variety of different accelerated aging processes on the material was also investigated. This proved to be very robust against all methods.

Parameters such as UV irradiation and temperature, water and salt had no significant influence on the strength and modulus of elasticity within a time frame of 3600 h for pipe cut samples. Compared to the results from the literature, a deterioration in the parameters would have been expected here. Even the influence of sand, which is blown onto the surface by wind and mechanically loads it, is not recognized in the mechanical characteristic values of our tests, even if the surface is clearly attacked. The reasons for this certainly lie in the experimental design. In many cases in the literature, especially in studies on the resistance of polymers, influencing variables are designed for the greatest possible damage. The influence of water/sea water in particular is often modeled using immersion, like in [22], which does not correspond to the application at hand here, as bridge cables always have the possibility of re-drying. There is also no permanent exposure to light (due to the night phases) like in [24], nor is the temperature permanently in high ranges like in [25]. The fact that a reduction in the characteristic values is visible in these literature contributions can be explained by the permanent exposure to the influencing factor. Due to the more realistic parameters we used, we do not come into a damage range. This emphasizes once again how important the selection of influences and the evaluation of results from accelerated aging tests are.

The service life should not be estimated at this point using the DSC-OIT. As also described in [26], misinterpretations can occur if a parameter (such as the oxygen partial pressure in the measuring system) does not correspond to the real influencing conditions when the test is carried out. However, the method is well suited to comparing a course, e.g., the need for stabilizers over time under different influences.

Compatibility with mineral oil and rope filler was demonstrated. When stored for 50 days in a mineral oil bath, the strength and modulus of elasticity are slightly reduced and fracture strain is slightly increased. This is most probably due to the migration of components of the oil into the HDPE. However, this is not considered to be critical, especially when considering the analysis in [39] in motor oil, because the sharp drop in yield strength of >30% in HDPE for water pipes after a storage period of 12 weeks or even much higher modulus losses in [15] for immersion in crude oil after one week exposure is not observed in our tests. No incompatibility with the WIRELOCK® socketing compound was found either. We are unable to draw a comparison with results from the literature at this point, as this special filling material is not generally taken into account in compatibility studies in scientific articles.

Unfortunately, due to the limited material, we were not able to investigate the amount or migration of stabilizers and the separation of inner and outer layers (regarding, e.g., crystallinity, modulus) of the pipe sections in more detail. A detailed analysis of the oil absorption as in [15] in comparison to the structure and the mechanical parameters could have contributed further to our understanding and must be considered in further work. The same applies to the influence of moisture and UV exposure, as well as the determination of diffusion parameters. The low resolution of possible effects in our experiments, which are due to the inner and outer layer of the pipe sections and which would have to be separated individually, could be explained because in our case each sample represents the cross-section across the material, and differences can therefore be averaged out.

The mean fracture strain of our tests fluctuates very strongly in some cases. However, high scattering can be observed for all sample series including the reference series and regardless of the manufacturing process, so that the influence of accelerated aging is not a possible cause. As discussed in the section before, the causes for this are to be found in the material itself.

The reduction in strength and modulus of extruded samples after 1200 h of UV irradiation with condensation water phases is the only significant deviation from the other sample series. However, this effect could not be reproduced in a second study on injection molded samples. Due to the lack of sample material, no further analytical tests could be carried out.

In the opinion of the authors, other chemical environmental influences like solvents or cleaning agents are not decisive for the Fehmarnsund Bridge or comparable bridge structures.

Guo's investigations have shown a Mullins effect due to shearing of the crystallites under cyclic loading [14]. In Figure 13, hysteresis behavior was observed under cyclic conditioning; this cannot be clearly attributed to the Mullins effect but can be explained by the viscoelastic material behavior. Further investigations would be necessary for this. However, here too there was no significant difference between pipe cut and injection molded samples in the linear-elastic range.

Overall, the results show the robust character and high weather resistance of the polymer material. The HDPE coating is a sensible alternative to the corrosion protection coatings previously used. As the HDPE is applied to the bridge cables at the factory, there is no need for a time-consuming on-site work step after the cables have been installed, which is a great advantage, especially in highly exposed locations. Local defects can be repaired with little effort due to the thermoplastic nature of the polymer. By adding appropriate colorants, the bridge cables can be manufactured in any color.

4.3. *Influence of Climate Zones and Temperatures*

The mechanical behavior of HDPE at different temperatures has been investigated, for example, in the literature sources [26–29] in Chapter 1.2.2, but mainly on HDPE pipes or geomembranes. The investigation of the HDPE sheathing of bridge cables has so far only been examined very sporadically and with a focus on specific influences [31,33,36,37]. In this article, we have attempted to take into account all the main influences that affect the sheathing of the cables in a bridge structure in order to draw a full picture. In the authors' opinion, there is generally no reason why HDPE should not be used in different climatic zones; HDPE can even be used in very hot or very cold regions, taking into account the change in mechanical properties, see Section 3.4. The effectiveness of HDPE as corrosion protection is not affected by the temperatures.

The presented investigations were carried out during the planning of the cable replacement of the Fehmarnsund Bridge. The cable replacement is necessary to extend the service life of the bridge and to adapt the bridge to new conditions. Therefore, the selected aging regimes, temperatures and UV loads are adapted to the specific location. However, due to the exposed maritime location, these are very harsh corrosive conditions that do not apply to many bridges in Europe. Therefore, this is a component-related analysis, but the results can be transferred with caution to similar bridge structures. However, the level of UV exposure would have to be re-evaluated for other structures.

This point addresses the fundamental discussion of the transferability of accelerated aging tests in the construction industry.

4.4. Exceptional Effects

The behavior of the material was not investigated under extreme conditions such as an explosion or an impact/accident (increased temperatures + dynamic shock load), as this can be ruled out due to the roadway layout and bridge construction.

While the present investigations are tailored to the Fehmarnsund Bridge, the conceptual approach of the investigations can be transferred to other bridges in other climate zones at any time. It can be seen that the sample production has a strong influence on the results, and therefore the transferability to other production processes is limited. In addition, it is imperative that samples for reliable results are taken from original components.

5. Conclusions and Outlook

Full-locked coil ropes are provided with a multi-layered, epoxy resin and polyurethane-based corrosion protection coating in accordance with the state of the art according to ZTV-ING. This is applied after installation of the bridge cables and must be renewed at regular intervals. An alternative, innovative option is the use of an HDPE coating, which is extruded onto the rope during production. The material is also already being used for strand bundle ropes, where HDPE tubes are applied during assembly. HDPE-sheathed FLCRs have so far only been used on two bridges in Germany and require a certificate of applicability from the building authorities.

The mechanical behavior of the thermoplastic HDPE used for the cladding was comprehensively characterized on the basis of a large number of tests. Various accelerated aging procedures were derived based on the general conditions of the Fehmarnsund Bridge. The material proved to be robust against all the aging regimes carried out and showed no incompatibilities with substances with which it could come into contact during its service life. However, the method of sample production has a decisive influence on the yield behavior of the plastic.

Due to its resistance to climatic influences, the HDPE coating is a competitive alternative to conventional corrosion protection coatings for FLCRs.

Author Contributions: Conceptualization, B.S.; Data curation, B.S., B.A. and E.S.; Funding acquisition, M.F.; Investigation, B.S., B.A. and E.S.; Methodology, M.F. and K.D.; Project administration, M.F.; Supervision, M.F.; Validation, B.A. and E.S.; Writing—original draft, B.S. and E.S.; Writing—review and editing, B.A., E.S., M.F. and K.D. All authors have read and agreed to the published version of the manuscript.

Funding: The research activities were carried out as part of an overarching bridge project, in which the authors were commissioned by an engineering firm as a subcontractor as part of the renovation of the Fehmarnsund Bridge.

Institutional Review Board Statement: Not applicable.

Informed Consent Statement: Not applicable.

Data Availability Statement: The original contributions presented in this study are included in the article. Further inquiries can be directed to the corresponding author(s).

Acknowledgments: The results presented in this article were compiled on behalf of Deutsche Bahn AG. We would like to take this opportunity to thank them for their cooperation.

Conflicts of Interest: The authors declare no conflicts of interest.

References

1. Friedrich, H.; Hamme, M. Brückenseile. In *Stahlbau Kalender 2021*; Kuhlmann, U., Ed.; Ernst & Sohn; John Wiley & Sons Inc., Berlin: Hoboken, NJ, USA, 2021. [CrossRef]
2. Bundesministerium für Digitales und Verkehr: ZTV-ING Teil 4: Stahlbau, Stahlverbundbau, Abschnitt 5: Korrosionsschutz von Brückenseilen. Available online: https://technical-regulation-information-system.ec.europa.eu/en/notification/26195/text/D/DE (accessed on 19 December 2024).
3. Feyrer, K.; Wehking, K.-H. *FEYRER—Drahtseile, Bemessung, Betrieb, Sicherheit*; Springer Vieweg: Berlin/Heidelberg, Germany, 2017; ISBN 978-3-642-54295-4.
4. Friedrich, H. Bridge Ropes for Road Bridges—Comparison Between Fully Locked Coil Ropes and Bundles of Parallel Strands. Berichte der Bundesanstalt für Straßenwesen, Heft B 98 Brücken- und Ingenieurbau. 2014. Available online: http://bast.opus.hbz-nrw.de/volltexte/2014/767/pdf/B98b.pdf (accessed on 19 December 2024).
5. *German Version DIN EN 1993-1-11:2006+AC:2009*; Design of Steel Structures—Part 1-11: Design of Structures with Tension Components. Beuth Verlag GmbH: Berlin, Germany, 2010.
6. Bundesministerium für Digitales und Verkehr: Technische Lieferbedingungen und Technische Prüfvorschriften für Ingenieurbauten. TL/TP-ING (2022) TL/TP-VVS, Part 4 Section 4 'Technical Delivery Conditions and Technical Test Specifications for Fully Locked Bridge Cables'. Available online: https://www.bast.de/DE/Publikationen/Regelwerke/Ingenieurbau/Baudurchfuehrung/TL_TP-Gesamt.pdf (accessed on 19 December 2024).
7. *fib Bulletin No. 30. Acceptance of Stay Cable Systems Using Prestressing Steals*; International Federation for Structural Concrete (fib): Lausanne, Switzerland, 2005; ISBN 978-2-88394-070-3.
8. *German Version EN ISO 527-2:2012*; Plastics—Determination of Tensile Properties—Part 2: Test Conditions for Moulding and Extrusion Plastics. Beuth Verlag GmbH: Berlin, Germany, 2012.
9. Millfield Enterprises (Manufacturing) Limited (2023) Wirelock—Technical Data Manual; V10-1123. Available online: https://www.wirelock.com/ (accessed on 19 December 2024).
10. *German Version EN 13411-4:2021*; Terminations for Steel Wire Ropes—Safety—Part 4: Metal and Resin Socketing. German Version EN ISO 13411-4:2022; Beuth Verlag GmbH: Berlin, Germany, 2022.
11. Ehrenstein, G. *Polymer Werkstoffe, Struktur—Eigenschaften—Anwendungen, 3. Auflage*; Carl Hanser Verlag: München, Germany, 2011; ISBN 978-3-446-42283-4.
12. Ehrenstein, G.; Pongratz, S. *Beständigkeit von Kunststoffen. Band 1*; Carl Hanser Verlag: München, Germany, 2007; ISBN 978-3-446-21851-2.
13. Elleuch, R.; Taktak, W. Visocelastic behavior of HDPE Polymer using Tensile and Compressive Loading. *J. Mater. Eng. Perform.* **2006**, *1*, 111–116. [CrossRef]
14. Guo, H.; Rinaldi, R.G.; Tayakout, S.; Broudin, M.; Lame, O. Characterization of the spherulitic deformation in equatorial region and cavitation in HDPE materials submitted to mixed-mode oligo-cyclic tensile loading. *Polym. Test.* **2021**, *99*, 107208. [CrossRef]
15. Chaoui, K.; Ghabeche, W.; Azzouz, S. Study of Crude Oil Interaction with High Density Polyethylene Pipe Surface Material. 24ème Congrès Français de Mécanique. Available online: https://www.researchgate.net/publication/335994735_STUDY_OF_CRUDE_OIL_INTERACTION_WITH_HIGH_DENSITY_POLYETHYLENE_PIPE_SURFACE_MATERIAL (accessed on 19 December 2024).
16. *DIN EN ISO 179-1*; Plastics—Determination of Charpy Impact Properties—Part 1: Non-Instrumented Impact Test. Beuth Verlag GmbH: Berlin, Germany, 2023.

17. *DIN EN ISO 6259-1*; Thermoplastics Pipes—Determination of Tensile Properties—Part 1: General Test Method. Beuth Verlag GmbH: Berlin, Germany, 2015.
18. *DIN EN ISO 6259-3*; Thermoplastics Pipes—Determination of Tensile Properties—Part 3: Polyolefin Pipes. Beuth Verlag GmbH: Berlin, Germany, 2015.
19. *D1505-18*; Standard Test Method for Density of Plastics by the Density-Gradient Technique. ASTM International: West Conshohocken, PA, USA, 2018.
20. *D1238-23a*; Standard Test Method for Melt Flow Rates of Thermoplastics by Extrusion Plastometer. ASTM International: West Conshohocken, PA, USA, 2013.
21. Albozahid, M.A.; Diwan, A.A.; Diwan, M.A.; Alansari, L.S. Effect of Weathering on Tensile Properties of Low and High Density Polyethylene. *IOP Conf. Ser. Mater. Sci. Eng.* **2021**, *1094*, 012140. Available online: https://iopscience.iop.org/article/10.1088/1757-899X/1094/1/012140 (accessed on 19 December 2024).
22. Elkori, R.; Lamarti, A.; Salmi, H.; El Had, K.; Hachim, A.; Yamari, I. Stady of the Effekt of Accelerated Aging by Sea Water on the Mechanical and Chemical Properties of High Density Polyethylene Bottles. Available online: https://www.researchsquare.com/article/rs-1893605/v1 (accessed on 19 December 2024).
23. Hsueh, H.-C.; Kim, J.H.; Orski, S.; Fairbrother, A.; Jacobs, D.; Perry, L.; Hunston, D.; White, C.; Sung, L. Micro and macroscopic mechanical behaviors of high-density polyethylene under UV irradiation and temperature. *Polym. Degrad. Stab.* **2020**, *174*, 109098. [CrossRef]
24. Carrasco, F.; Pages, P.; Pascual, S.; Colom, X. Artificial aging of high-density polyethylene by ultraviolet irradiation. *Eur. Polym. J.* **2001**, *37*, 1457–1464. [CrossRef]
25. Zhao, B.; Zhang, S.; Sun, C.; Guo, J.; Yu, Y.X.; Xu, T. Aging behavior and properties evaluation of high-density polyethylene (HDPE) in heating-oxygen environment. *IOP Conf. Ser. Mater. Sci. Eng.* **2018**, *369*, 012021. [CrossRef]
26. Mueller, W.; Jakob, I. Oxidative resistance of high-density polyethylene geomembranes. *Polym. Degrad. Stab.* **2003**, *1*, 161–172. [CrossRef]
27. Ewais, A.; Rowe, R.K. Effect of aging on the stress crack resistance of an HDPE geomembrane. *Polym. Degrad. Stab.* **2014**, *109*, 194–208. [CrossRef]
28. Rowe, R.K.; Ewais, A. Ageing of exposed geomembranes at locations with different climatological conditions. *Can. Geotech. J.* **2015**, *3*, 326–343. [CrossRef]
29. Guermazi, N.; Elleuch, K.; Ayedi, H.F. The effect of time and aging temperature on structural and mechanical properties of pipeline coating. *Mater. Des.* **2009**, *6*, 2006–2010. [CrossRef]
30. Saul, R.; Nützel, O. Neuartige Sanierung der Tragkabel einer Hängebrücke in Norwegen. *Stahlbau* **2014**, *2*, 96–100. [CrossRef]
31. Yu, Y.; Man, M.; Zhao, F.; Lin, S.; Guo, F. Corrosive degradation evaluation of semi-parallel wire cables with high-density polyethylene sheath breaks. *Eng. Fail. Anal.* **2020**, *116*, 104714. [CrossRef]
32. Huang, R.; Yu, J.; Yang, Q.; Chen, J.; Song, Q. Discussion on the Wear Durability of HDPE Outer Sheath for Bridge Cable. *Prestress. Technol.* **2009**, *5*, 23–26. [CrossRef]
33. Liu, S.; Su, H.; Xu, J.; Zhou, Z.; Hao, H.; Wei, L.; Du, J. Experimental study on tensile properties of HDPE sheath for parallel wire suspender after photo-oxidative aging. *Constr. Build. Mater.* **2024**, *416*, 135304. [CrossRef]
34. Mkacher, I.; Brument, Y.; Murin, V.; Sellier, I.; Colin, X. Methodology for evaluating the durability of HDPE outer sheaths of underground electric cables. In Proceedings of the 21st International Conference on Electricity Distribution (CIRED), Frankfurt, Germany, 6–9 June 2011. Paper No 0779.
35. Lin, B.; Zhang, C. Photo-aging performance of high-density polyethylene sheath in dry environment. In *Advances in Civil Engineering: Structural Seismic Resistance, Monitoring and Detection*; CRC Press: Boca Raton, FL, USA, 2022.
36. Dan, D.; Cheng, W.; Sun, L.; Guo, Y. Fatigue durability study of high density polyethylene stay cable sheathing. *Constr. Build. Mater.* **2016**, *111*, 474–481. [CrossRef]
37. Dan, D.; Sun, L.; Guo, Y.; Cheng, W. Study on the Mechanical Properties of Stay Cable HDPE Sheathing Fatigue in Dynamic Bridge Environments. *Polymers* **2015**, *8*, 1564–1576. [CrossRef]
38. Lai, J.; Bakker, A. Analysis of the non-linear creep of high-density polyethylen. *Polymer* **1994**, *1*, 93–99. [CrossRef]
39. Rezakalla, A.I.; Gennadyevech, D.A. Effect of Certain Hydrocarbon Compounds on High-density Polyethylene Water Pipes. *Mater. Plast.* **2022**, *59*, 91–99. [CrossRef]
40. *DIN 52348*; Testing of Glass and Plastics—Abrasion Test—Sand Trickling Method. Beuth Verlag GmbH: Berlin, Germany, 1985.
41. *German Version EN ISO 4892-1:2016*; Plastics—Methods of Exposure to Laboratory Light Sources—Part 1: General Guidance. Beuth Verlag GmbH: Berlin, Germany, 2016.
42. *German Version EN ISO 4892-3:2016*; Plastics—Methods of Exposure to Laboratory Light Sources—Part 3: Fluorescent UV Lamps. Beuth Verlag GmbH: Berlin, Germany, 2016.

43. Grellmann, W.; Seidler, S. *Kunststoffprüfung;* Hanser: München, Wien, 2015; ISBN 978-3-446-44350-1.
44. Qiao, Z.; Wu, W.; Wang, Z.; Zhang, L.; Zhou, Y. Space Charge Behavior of Thermally Aged Polyethylene Insulation of Track Cables. *Polymers* **2022**, *11*, 2162. [CrossRef]
45. Mirabella, F.M.; Bafna, A. Determination of the crystallinity of polyethylene/α-olefin copolymers by thermal analysis: Relationship of the heat of fusion of 100% polyethylene crystal and the density. *J. Polym. Sci. Part B Polym. Phys.* **2002**, *15*, 1637–1643. [CrossRef]
46. Lunzhi, L.; Jinghui, G.; Lisheng, Z.; Kai, Z.; Xiaohan, Z. Aging phenomena in non-crosslinked polyolefin blend cable insulation material: In: Electrical treeing and thermal aging. *Front. Chem.* **2022**, *10*, 903986. [CrossRef]
47. Kiersnowska, A.; Fabianowski, W.; Koda, E. The Influence of the Accelerated Aging Conditions on the Properties of Polyolefin Geogrids Used for Landfill Slope Reinforcement. *Polymers* **2020**, *9*, 1874. [CrossRef]
48. Professional Plastics (2024) HDPE and LDPE Resistance Chart by Chemical. Available online: https://www.professionalplastics.com/professionalplastics/HDPE-LDPEChemicalResistanceChart.pdf (accessed on 15 October 2024).
49. Rösler, J.; Harders, H.; Bäker, M. *Mechanisches Verhalten der Werkstoffe;* Springer Vieweg: Wiesbaden, Germany, 2019; ISBN 978-3-8348-1818-8.
50. *Kunststoffe—Dynamische Differenzkalorimetrie (DSC)–Teil1: Allgemeine Grundlagen (ISO11357-1:2023); Deutsche Fassung EN ISO11357-1:2023;* Beuth Verlag GmbH: Berlin, Germany, 2023.
51. Binnewies, M.; Finze, M.; Jäckel, M.; Schmidt, P.; Willner, H.; Rayner-Canham, G. *Allgemeine und Anorganische Chemie;* Springer Spektrum: Berlin, Germany, 2016; ISBN 978-3-662-45067-3.
52. Troughton. *Handbook of Plastics Joining: A Practical Guide;* William Andrew: Norwich, NY, USA; TWI The Welding Institute: Cambridge, UK, 2008; ISBN 978-0-8155-1581-4.

materials

MDPI

Article

Precision of Fungal Resistance Test Method for Cereal Husk-Reinforced Composite Construction Profiles Considering Mycelium Removal Techniques

Ewa Sudoł [1,*], Ewa Szewczak [2], Mariia Goron [1] and Ewelina Kozikowska [1]

[1] Construction Materials Engineering Department, Instytut Techniki Budowlanej, 00-611 Warsaw, Poland; m.goron@itb.pl (M.G.); e.kozikowska@itb.pl (E.K.)
[2] Group of Testing Laboratories, Instytut Techniki Budowlanej, 00-611 Warsaw, Poland; e.szewczak@itb.pl
* Correspondence: e.sudol@itb.pl

Abstract: Many test methods used in the assessment of construction products are characterised by poor precision, which is reflected in the discrepancies of results obtained by different laboratories. The test procedure for fungal resistance of natural particle-reinforced composite construction profiles has not been fully specified, which may lead to such discrepancies and erroneous utility assessment. In this study, the precision of the method and the influence of the mycelium removal on the fungal resistance were assessed based on the flexural strength and modulus of elasticity test results obtained for millet- and oat husk-reinforced PVC composites exposed to *Coniophora puteana*. The study revealed low precision of the test method, the coefficient of variation, in which, based on the standard deviation of interlaboratory reproducibility for one of the tested composites, was even higher than 20%. Additionally, it was found that the method of mycelium removal can significantly (at the level of 16% difference between flexural strength results) affect the resistance test results. This indicates the need to modify the test method and clearly specify the recommended method of cleaning samples after exposure to fungi.

Keywords: repeatability; reproducibility; polymer composites; fungi; flexural properties

1. Introduction

The development of new materials and products in the construction sector requires the development of new test methods. However, not all test methods, both those used so far and those developed recently, provide reliable, credible results. The standardisation of test methods is intended to provide a uniform system for assessing products, including construction products, and enable the comparison of their performance properties. The basic metrological condition that must be met to achieve this goal is the comparability of the results. According to VIM 3 [1], the measurement results are metrologically comparable when they are both metrologically traceable to the same measurement unit, such as the meter. It is relatively simple to compare the measurement results, but it should be noted that laboratory test results cannot always be treated in the same way as the measurement results. The already withdrawn document EA-4/16 [2] aptly presented the substance of the difference between the terms measurement and test. First, it distinguished the results of these activities into a 'measurand' in the case of measurement and a 'characteristic' in the case of the test. In practice, a measurand is a physical quantity that has a 'true value', independent of the test method. Typical examples of measurement are linear dimensions, mass, and temperature. In the case of testing, the result is determined by the test method itself, and there is no well-defined 'true value' that can be obtained using another test method. Tests (reaction to fire, fire resistance, impact resistance, resistance to climatic conditions, water absorption, adhesion, etc.) are predominant in most activities aimed at assessing building materials and elements. In the case of tests, even if the results are expressed numerically and are comparable owing to the same measurement unit, they cannot always be compared with each other as they may depend on the test method used [3].

As research [4–7] indicates, even the standardisation of test methods may not ensure uniform product evaluations. Uncertainty plays a major role when assessing the compliance of the test results with the criteria [8]. However, the uncertainty of measurement, which is estimated by the laboratory based on the results of calibrations of its equipment and possible in-laboratory experiments, most often does not include a component related to the reproducibility of the test method. In many cases, the level of reproducibility of the methods developed by standard committees is unsatisfactory and does not allow a reasonable assessment of compliance with the criteria, especially in the case of multi-stage methods [5,9]. Often, this is owing to the underspecification of the test procedures or their stages, which, in the sense of the paradigm presented by Walker [10], represents uncertainty in the structure of the test model.

One of the most important tests for building materials and elements is their resistance to environmental factors. With regard to natural fibre-reinforced composites, as well as other composites reinforced with natural particles, such as cereal husk-based composites, the level of susceptibility to environmental factors is of crucial interest [11,12]. Studies on the resistance to environmental factors have paid a lot of attention to the change in properties as a result of artificial ageing involving UV light [13–15], water, high temperature and hydro-thermal cycles involving water, low and high temperature [14,16–18], and microorganisms [19–31]. It was found that artificial ageing has the most significant effect on bending properties, followed by fungi, and hydro-thermal cycles the least [12]. The hydrophilic nature of the plant filler means that, in contact with water, it swells easily, causing cracks to form in the hydrophobic polymer matrix. The interaction between the lignocellulosic particles and the polymer also deteriorates, reducing the stress transfer capacity between the fibres and the matrix and resulting in a deterioration of the composite performance [11,17]. Composites are susceptible to biotic agents, particularly fungi of the class *Basidiomycetes* [18,19], classified in modern taxonomy as the subfamily *Agaricomycotina*. The polymer matrix is characterised by rather low susceptibility to biodegradation processes. The fungal resistance of a composite depends mainly on the type, amount, and size of the filler and its degree of dispersion in the matrix [17,18,24]. The development of microorganisms is made possible by the nutrients present in plant particles, lignin, and cellulose. Composites reinforced with low-absorbing fibres are less susceptible to fungi than those using high-absorbing fibres [23]. It has been established that the fibres in composites reach up to 70% moisture content, which, with the right temperature and ph, provides optimal conditions for the growth of microorganisms [22]. White decay fungi, such as *Coriolus versicolor*, primarily attack lignin, while brown decay fungi, such as *Coniophora puteana*, primarily degrade cellulose [26]. Proper fibre dispersion in the matrix is an ally of bioresistance. It promotes good interfacial adhesion by reducing voids, which act as channels for fungal movement through the material, facilitating enzyme transport [20]. The resistance to microorganisms decreases as the amount of filler increases [31]. Previous studies have recorded varying susceptibility to particular strains of fungi. The most intensive growth was observed for *Trametes versicolor* and *Coniophora puteana*, with weight loss not exceeding 5% and no significant change in flexural modulus [22–24]. Nevertheless, there are known cases of decreases in flexural strength of up to 30%, modulus of elasticity of up to 40% [24,29,31], and impact strength of up to 16% [30].

Tests of resistance to biotic factors comprise several stages, such as specimen preparation, exposure to microorganisms, and post-exposure property tests, including changes in morphological, physical, or mechanical properties and comparison with the initial state.

Several sources of uncertainty that affect the final result can be identified in each step.

Differences in microbial activity are important in tests of resistance to biotic factors. This impact can be reduced by controlling the strain quality and culture conditions prior to test sample exposure [19,20]. The results can also be influenced by the method of mycelium removal as well as the methods used to assess the performance properties. During long-term exposure to fungi, mycelia tend to overgrow in the samples. The mycelium can be harmful to human health and interfere with the performance of the test equipment; therefore, it should be removed after completing the incubation and before further testing. This stage of the procedure is often overlooked in descriptions of test methods, including test standards [32]. Some authors have proposed a cleaning method. The samples can be cleaned with a soft brush [30], pressurised water [31], or sandblasting [22]. The test procedure for natural fibre composite fungal resistance [33] has also not been fully defined in this area, which may lead to discrepancies in the results obtained by different laboratories.

As indicated by the results of previous evaluations of metrological properties of some test methods [5], their precision seems to be insufficient for the purposes of uniform evaluation of construction products. In this study, the precision of the resistance to biotic factors test method was analysed. There is no similar analysis in the literature.

The study consisted of exposing samples of natural fibre-reinforced composites to the action of microorganisms and then subjecting them to flexural properties tests. The changes in flexural strength and modulus of elasticity occurring under the influence of fungi exposure were assessed. Two methods were used to remove mycelium from samples after exposures, which is also a novelty of this work. Based on the results of these tests, in the first stage, the precision of the fungus resistance test method was assessed, independently for both methods of mycelium removal. The precision analysis was based on the repeatability variance and intra-laboratory reproducibility variance (tests performed by two independent researchers). Then, the influence of the mycelium removal technique on the results of fungi resistance tests of cereal husk-reinforced polymer construction profiles was analysed, which has not been previously studied. The results of the statistical analysis of the significance of differences in flexural properties after using different mycelium removal methods were also supported by the analysis of the microstructure observed using a scanning electron microscope (SEM). The problem was studied in the case of millet- and oat husk-reinforced PVC composites exposed to *Coniophora puteana*. The results obtained allow for the assessment of the level of precision of the method and the influence of the sample cleaning step, which were not included in the standard test procedure, due to possible discrepancies in the results.

2. Materials and Methods

2.1. Samples

The profiles for outdoor terrace flooring were also tested. Polyvinyl chloride (PVC) containing chalk ($CaCO_3$) was used as the matrix, and pulverised oats and millet husks were used as fillers. The oat husk filler was dominated by slender particles, up to 50 μm wide and 100 μm or more long (Figure 1a), which can be considered fibres. Pulverised millet husks took on a plate-like shape, with a width of up to 100 μm (Figure 1b). No millet-based fibres were noted.

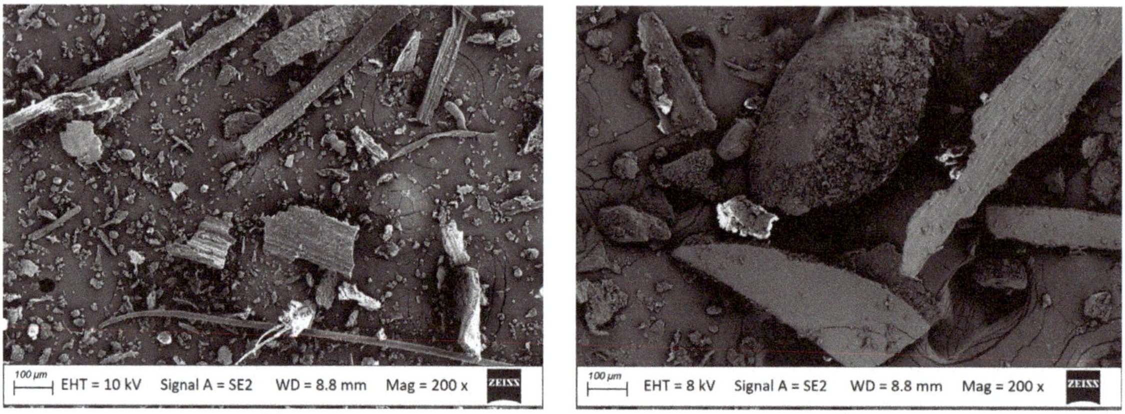

(**a**) (**b**)

Figure 1. Pulverised (**a**) oat husks and (**b**) millet husks, SEM images, magnification 200×.

The compositions of composites are listed in Table 1.

Table 1. The composition of cereal husk-reinforced PVC composites.

Series Name	Matrix	Mineral Filler	Cereal Husk Filler	Other
M	PVC (100 phr)	$CaCO_3$ (50 phr)	pulverised millet husks (30 phr)	impact modifiers (4.5 phr) flow modifiers (1 phr) stabiliser (4.5 phr) wax (2 phr)
O			pulverised oat husks (30 phr)	

The profiles were manufactured under industrial conditions by using an extrusion method. They had a chamber shape, width of 125 mm, height of 22 mm, and a wall thickness of 5 mm. Their face surfaces were flat without milling, but the factory was treated with a metal brush to give them a slightly rough wood-like texture (Figure 2).

Figure 2. Appearance of profiles from which samples were taken for testing.

Test specimens were cut from the face surface to preserve the original wall thickness. They were obtained from the central part of the chamber, parallel to the ribs. They had dimensions of 80 mm × 10 mm × 5 mm. From both the millet husk-reinforced PVC composite (M) and the oat husk-reinforced PVC composite (O), 46 samples were prepared. Of these, six samples were seasoned under laboratory conditions, and the remaining samples were aged in water for 2 weeks [32]. Distilled water was used, which was changed nine times.

2.2. Exposition to Fungi

Lyophilised strain of the fungus *Coniophora puteana* Schumach. BAM Ebw. Fifteen were used to prepare the parental cultures. According to the recommendations of the supplier, the fungus was rehydrated with 1 mL of sterile distilled water. After rehydration, which lasted for 2 h under normal conditions, the fungus was transplanted onto a sterile malt extract agar medium to grow the parental cultures. A medium consisting of 40 g malt extract, 35 g agar, and 1000 mL of water was used. The medium was sterilised in an autoclave at 121 ± 2 °C and a pressure of 1 bar. The fungus was cultured at 22 ± 1 °C and 70 ± 5% humidity for 2 weeks and then transplanted onto freshly prepared malt extract agar medium. After cultivation for one week under the same conditions, subcultures were obtained.

Kolle flasks were prepared to the specification of the test samples and fungi. Fifty mL of malt extract agar medium was poured into each flask, sterilised, and left to solidify horizontally. Subcultures of *Coniophora puteana* with well-developed mycelia were used. Each flask was inoculated with two pieces of mycelia from subcultures under sterile conditions. The flasks were left in an incubation chamber at 22 ± 1 °C and 70 ± 5% until the surface of the medium was completely grown using mycelium. The activity of the fungi was verified in a concurrent test on the sapwood of *Pinus sylvestris*, resulting in a mass loss of 45% after 16 weeks.

Samples previously aged in water were placed on the mycelium with a usable face surface in direct contact. A total of 40 samples of millet husk-reinforced PVC composite and 40 samples of oat husk-reinforced PVC composite were exposed to *Coniophora puteana*. Flasks with samples were held in an incubation chamber at 22 ± 1 °C and 70 ± 5% for 16 weeks. The degree of mycelial growth on the samples was assessed visually according to the following scale: grade 0—no mycelium, grade 1—scarcely developed mycelium, grade 2—medium-developed mycelium, grade 3—well-developed mycelium, and grade 4—very thick mycelium. After, the mycelium was removed.

Mycelium removal was performed independently by two researchers with several years of experience in microbiological testing of building products. Each used two cleaning methods. Half of the samples from the given composite were manually cleaned using a brush. A soft nylon bristle brush with a diameter of 0.18–0.20 mm was used. The brush head measured 30 × 15 mm with four rows of fibre bundles arranged in 10 columns. The other half was cleaned using a water jet. A pressure washer (Kärcher, Poland, Kraków) was used. Water at 23 ± 2 °C was distributed using a lance at 12 bar pressure, at an intensity of 9.2 L/min. No detergent was used for either method. Cleaning was performed carefully until the mycelium was completely removed and visually controlled. After cleaning, all samples were dried for one week at 40 ± 2 °C in a forced-air laboratory dryer. Figure 3 shows the flowchart of the experimental steps.

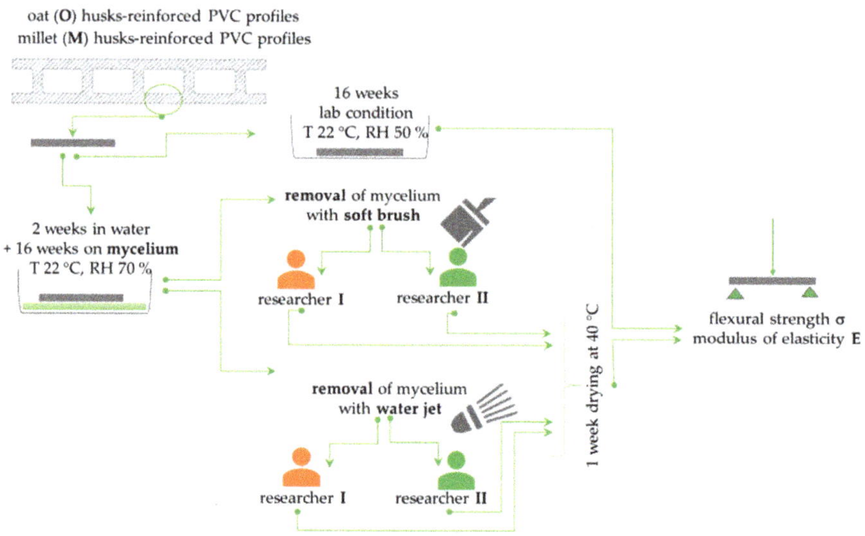

Figure 3. Experimental flowchart.

2.3. Microstructure Analysis

The microstructure was examined with Sigma 500 VP cold-field emission scanning electron microscope (Carl Zeiss Microscopy GmbH, Köln, Germany). The tests were carried out at the accelerating voltage of 5 KeV, 8 KeV, or 10 KeV inductive electron beam, using an SE detector. The observation was carried out on samples in initial state and after exposure to *Coniophora puteana*, both unpurified and after removal of the mycelium. All samples were dried for 7 days at T 40 ± 2 °C in a chamber with forced-air circulation before observation, and then a layer of gold was sputtered. SEM observations of the mycelium were performed at 500× and 5000× magnification, while the surfaces of the composites in the initial state and after the fungus culture were examined at 100× magnification. Microscopic observations were carried out on the usable surface of the profiles. Brittle fracture after fungus cultivation was also analysed.

2.4. Flexural Properties Test

The flexural strength σ and modulus of elasticity E, were determined, according to ISO 178 [34], using a class 1 strength testing machine (Instron, Darmstadt, Germany). Three-point bending, similar to that in [14,35], was performed. Supports with a 5 mm radius were used, spaced every 64 mm, and a 5 mm radius pressing element was placed in the middle of the span. The samples were then provided with free support. The load was applied to the ground surface at a constant rate of 2 mm/min until it was destroyed. The flexural strength, σ, was calculated according to Equation (1) [34] and expressed in MPa.

$$\sigma = \frac{3FL}{2bh^2} \quad (1)$$

where the following are defined: F—maximum force, in N; L—support spacing, in mm; b—sample's width, in mm; and h—sample thickness, in mm.

A load–deflection curve was recorded during bending in a linearly elastic range, including the force and deflection values corresponding to strains ε_{f1} = 0.0005 and ε_{f2} = 0.0025. The deflection values of f_1 and f_2 were calculated using Equation (2) [34].

$$f_1 = \frac{\varepsilon_{f1} L^2}{6h};\ f_2 = \frac{\varepsilon_{f2} L^2}{6h} \tag{2}$$

where the following are defined: L—spacing of supports, in mm; and h—sample thickness, in mm.

The force values recorded when ε_{f1} and ε_{f1} strains occurred were used to determine the values of the σ_{f1} and σ_{f2} normal stresses. The E modulus was calculated according to Equation (3) [34] and is expressed in MPa.

$$E = \frac{\sigma_{f2} - \sigma_{f1}}{\varepsilon_{f2} - \varepsilon_{f1}} \tag{3}$$

where the following are defined: σ_{f1}, σ_{f2}—maximum normal stress corresponding to f_1 and f_2 stress, determined according to (2).

The series of both the millet husk-reinforced PVC composite and the oat husk-reinforced PVC composite are described in Table 2. The initial results were sets of six results obtained by one researcher, and the sets of results after exposure and mycelium removal consisted of ten results for each of the two researchers.

Table 2. Test results series.

Properties	State of samples				
	Initial	After exposure to fungi			
		mycelium removal method			
		soft brush		water jet	
		researcher			
		I	II	I	II
Flexural strength	σ_i	σ_{bI}	σ_{bII}	σ_{wI}	σ_{wII}
Modulus of elasticity	E_i	E_{bI}	E_{bII}	E_{wI}	E_{wII}

2.5. Statistical Analysis Methods

The precision analysis of the method was based on the estimates of the values of the repeatability variance, between-researcher variance, and reproducibility variance obtained for all properties Y: σ_b, σ_w, E_b, E_w (Table 2) according to the methodology presented in ISO 5725-2 [36]. The assessment of precision should be preceded by the detection of outliers, and the method recommended by ISO 5725-2 is the Grubbs test, which is based on the assumption of normality of the result distributions. Therefore, first of all, all series of results described in Table 2 were subjected to Shapiro–Wilk tests for normality, which is characterised by a relatively high power. Then, to detect a single deviating value in a dataset, a Grubbs test was performed for all result series.

- The repeatability variance, s_r^2, is calculated for all properties Y according to Equation (4).

$$s_{rY}^2 = \frac{s_{rYI}^2 + s_{rYII}^2}{2} \tag{4}$$

where

s_{rYI}—the sample standard deviation of the first operator's results for quantity Y;

s_{rYII}—the sample standard deviation of the second operator's results for quantity Y.

For the initial result series σ_i, and E_i, the standard deviation of the repeatability was assumed to be the standard deviation of the sample of six results

- The between-researcher variance, s_L^2, is calculated for all properties Y according to Equation (5).

$$s_{LY}^2 = \frac{s_{dY}^2 - s_{rY}^2}{\overline{\overline{n}}_Y} \tag{5}$$

where

s_{LY}—between-researcher standard deviation for quantity Y

and

$$s_{dY}^2 = n_{YI}\left(\overline{Y_I} - \overline{\overline{Y}}\right)^2 + n_{YII}\left(\overline{Y}_{II} - \overline{\overline{Y}}\right)^2 \tag{6}$$

where

n_{YI}, n_{YII}—number of samples for quantity Y and operator I and II, respectively;
$\overline{Y}_I$, $\overline{Y}_{II}$—average result for quantity Y and operator I and II, respectively;
$\overline{\overline{Y}}$—general mean calculated according to Equation (7).

$$\overline{\overline{Y}} = \frac{n_{YI}\,\overline{Y_I} + n_{YII}\,\overline{Y_{II}}}{n_{YI} + n_{YII}} \tag{7}$$

and

$$\overline{\overline{n}}_Y = (n_{YI} + n_{YII}) - \frac{(n_{YI}{}^2 + n_{YII}{}^2)}{(n_{YI} + n_{YII})} \tag{8}$$

- The within-laboratory reproducibility variance, s_R^2, is calculated for all properties Y according to Equation (9).

$$s_{RY}^2 = s_{rY}^2 + s_{LY}^2 \tag{9}$$

where

s_{RY}: within-laboratory reproducibility standard deviation for quantity Y.

All Equations (4)–(9) are derived from the equations given in standard ISO 5725-2 [36].

To ensure the comparability of the variability of the results for different test result values, all standard deviations were converted to coefficients of variation v_Y according to Equations (10) or (11):

$$v_Y = \frac{100 s_Y}{\overline{\overline{Y}}}, \% \tag{10}$$

or for initial tests

$$v_{Yi} = \frac{100 s_{Yi}}{\overline{Y}}, \% \tag{11}$$

The relatively large dispersion of test results (shown in Section 3.3.1 of this article) requires statistical evaluation of whether the difference in means for given result groups (in the case of the described experiment of result sets for the same operator and the same method of sample cleaning) is significant. A typical test used to analyse the statistical significance of differences between result groups is the analysis of variance (ANOVA), which is based on the comparison of variance within result groups and intergroup variance related to the differences between means. One-way ANOVA was used to analyse whether the differences in test results resulting from different surface cleaning methods are statistically significant.

The F-test was used to draw some conclusions regarding differences in intra-group variances of the obtained results. (The test concerns two groups of results from normal distributions. For a larger number of groups, other tests are used, e.g., Bartlett's).

3. Results and Discussion

3.1. Surface Assessment

After 16 weeks of exposure to *Coniophora puteana*, the degree of mycelial growth on the samples was assessed visually. For the millet husk-reinforced composite, well-developed mycelium, corresponding to grade 3, was observed (Figure 4a). For oat husk-reinforced composite, medium mycelium coverage was noted, which corresponds to grade 2 (Figure 4b). The millet husk-reinforced composite showed more fungal overgrowth, which indicates that the oat husk-reinforced composite is more resistant to fungi [12,13].

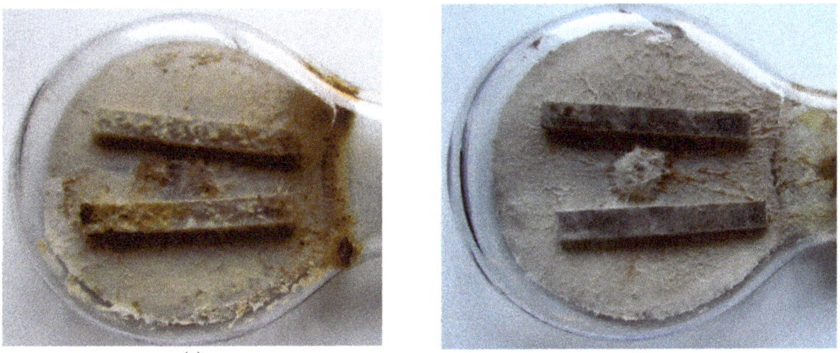

(a) (b)

Figure 4. Appearance of specimens after exposure to *Coniophora puteana*: (**a**) millet husk-reinforced composite and (**b**) oat husk-reinforced composite.

The surface of samples before mycelium removal was subjected to SEM analysis. It showed a developed mycelium with distinct spores (Figure 5a) characteristic for the *Coniophora puteana* strain. Mycelium images on the millet husk-reinforced composite were similar to those observed on the oat husk-reinforced composite. Due to the degree of mycelium development completely obscuring the composite, it was not possible to assess the surface morphology of the samples before cleaning. A brittle fracture test was also conducted to see if the mycelium enters inside the composite structure. The microstructure analysis showed that *Coniophora puteana* penetrates inside the walls of the profiles. In the images (Figure 5b), it is visible in the form of a dense network of thin threads, indicating that the fungus was present throughout the composite.

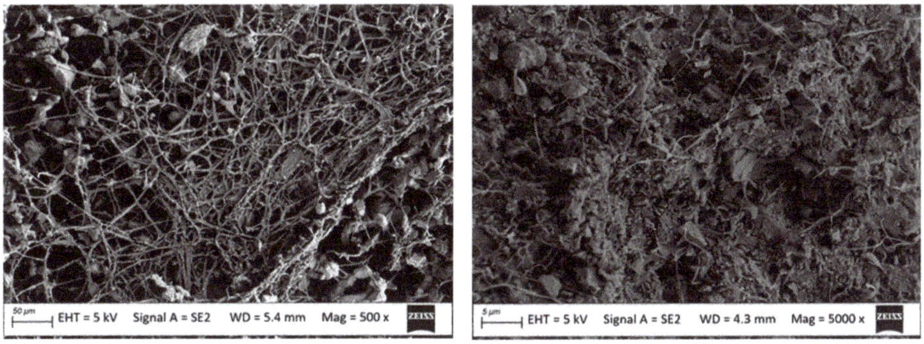

(a) (b)

Figure 5. *Coniophora puteana* mycelium (**a**) on the surface before cleaning and (**b**) in cross-section of the oat husk-reinforced composite, magnification, respectively, 500× and 5000×.

Cleaning of the samples with both a brush and wet jet resulted in complete removal of the mycelium. There were no differences in the surface appearance of the samples owing to the cleaning method. However, there was a significant difference in the surface appearance of the samples with millet husks compared to those with oat husks. In visual assessment, the surface of the millet husk-reinforced composite was noticeably rougher. In many places, filler particles were observed. Single points with holes in the PVC matrix, indicating filler leaching [31], were also noted. In the case of the oat husk-reinforced composite, the filler remained covered in the matrix after exposure. No uncovering particles were observed.

SEM analysis of the cleaned surfaces confirmed the macroscopic observations. The surface of both composites in the initial state had longitudinal grooves, in accordance with the mechanical treatment applied, but without exposing the filler (Figure 6a,b). After exposure to *Coniophora puteana* and removal of the mycelium with the brush in the case of the oat husk-reinforced composite, single, small fragments of filler particles not surrounded by the matrix became visible (Figure 6c). In the case of the millet husk-reinforced composite, large voids in the matrix with smooth surfaces (Figure 6d) were observed after the same cleaning process, indicating that interfacial bonds were lost, resulting in the removal of filler from the polymer matrix [11]. The surface morphology after water jet cleaning was similar (Figure 6e,f), with the millet husk composite showing filler loss of a smaller area than after brush cleaning. It was also revealed that there were non-matrix-covered millet husk particles. The above may indicate that not only the action of *Coniophora puteana*, but also the cleaning technique, is responsible for the filler loss.

3.2. Flexural Properties

In the initial state, the oat husk-reinforced composite had a flexural strength σ_i close to 44 MPa, whereas the millet husk-reinforced composite had it slightly greater than 30 MPa (Figure 7a). The modulus of elasticity E_i reached approximately 3800 MPa and 2900 MPa, respectively (Figure 7b).

The analysis of the dispersion of the results and their impact on the significance of the differences in mean results are presented in detail in Section 3.3, therefore, error bars are not shown.

The above differences between oat and millet husk-based product may be due to differences in the shape of the filler [11,25]. The mechanical parameters of the composite depend, among other things, on the quantity, size, and shape of the filler and the degree of its dispersion in the matrix [11,14]. The oat husk particles were predominantly in the shape of long, slender fibres (Figure 1a), favourable in terms of bending and tensile properties [17]. Millet husk particles were plate-like shaped (Figure 1b), which makes proper dispersion in the matrix difficult [11]. Flexural properties are similar to those obtained for composites with rice husk filler [14], although lower than those reported for composites reinforced with sisal [37] or hemp fibres [11,12].

The flexural properties of composites decreased after exposure to *Coniophora puteana*. The flexural strength (the average value of σ_{bI}, σ_{bII}, σ_{wI}, and σ_{wII}) of the oat husk-reinforced composite reduced by 8% (Figure 7a); whereas, that of the millet husk-reinforced composite reduced by 10% (Figure 7b). The modulus of elasticity decreased by approximately 50% and 35%, respectively. These changes are illustrated by the flexural curves shown for selected specimens in Figure 7c,d. After exposure, lower values of the maximum force were noted. A flattening of the flexural curve and thus a change in the tangent angle were also observed, especially for the millet husk-reinforced composite.

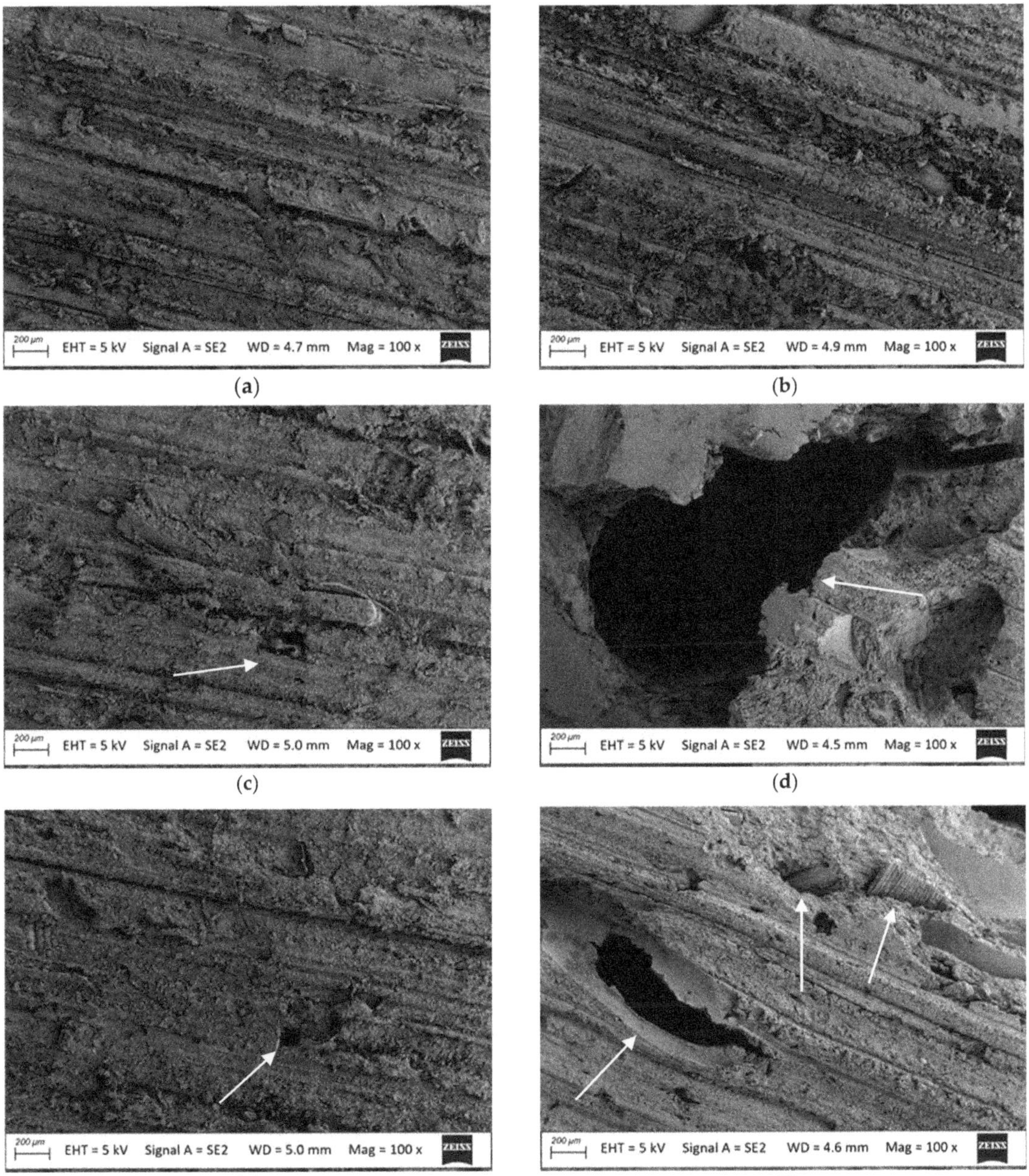

Figure 6. Surface morphology (**a**,**b**) in inital state, oat and millet husk-reinforeced composite, respectively, (**c**,**d**) after mycelium removal with brush, oat and millet husk-reinforeced composite, respectively, (**e**,**f**) after mycelium removal with water jet, oat and millet husk-reinforeced composite, respectively; magnification 100×.

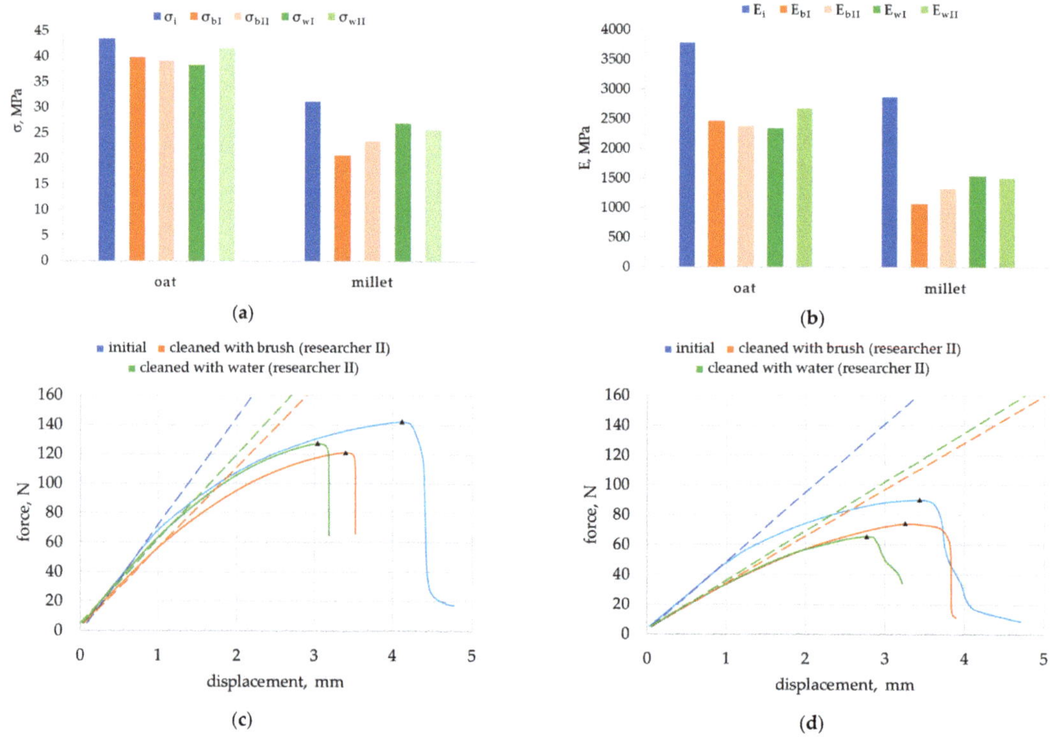

Figure 7. Test results of (**a**) flexural strength σ; (**b**) modulus of elasticity E of oat husk-reinforced composite and millet husk-reinforced composite in initial state (σ_i, E_i), after exposure to *Coniophora puteana* and cleaning with brush performed by researcher I (σ_{bI}, E_{bI}) and by researcher II (σ_{bII}, E_{bII}) or with water jet, also by researcher I (σ_{wI}, E_{wI}) and by researcher II (σ_{wII}, E_{bII}); (**c**) flexural test curves (examples) for oat husk-reinforced composite; and (**d**) flexural test curves (examples) for millet husk-reinforced composite; ▲ maximum force, the dotted line shows the tangent.

After exposure and mycelium removal, fragments of filler not surrounded by the matrix were noted (Figure 6c,f) as well as voids in the matrix (Figure 6d–f). This indicates that the interaction between the lignocellulosic particles and the PVC matrix was weakened after exposure [35]. This may be related to the profile surface treatment used. Although SEM images (Figure 6a,b) did not show exposed filler in initial state, as previous studies show, mechanical surface treatment reduces the protection of hydrophilic lignocellulosic particles by the matrix [38]. Even a partially exposed filler is more susceptible to swelling, which leads to a decrease in interfacial adhesion [11,12]. The reduction in the interphase interaction on the composite surface can determine the modulus of elasticity much more than the flexural strength. Weakening of the top layer significantly increases the susceptibility to strain [17]. However, it should be noted that recorded changes are comparable to observed for lignocellulose fibre-reinforced polymer composites exposed to fungi and moisture [19–21,26]. The statistical significance of flexural property changes is shown in Section 3.3.2.

3.3. Statistical Analyses

3.3.1. Metrological Properties of the Test Method

For all series of results, the null hypothesis for the Shapiro–Wilk test of normality of distribution was confirmed, which means that it can be assumed that the populations from which the result sets described in Table 2 and point 2.4 have a normal distribution.

Grubbs tests showed no single outliers; therefore, all results obtained were used in further analysis.

First, the standard deviation and coefficients of variation of repeatability and reproducibility were estimated for all series of results. Figure 5 shows the results obtained.

The repeatability and reproducibility results are presented in Figure 8 in comparison with the repeatability coefficients of variation for the results of the initial flexural properties σ_i and E_i. The repeatability and reproducibility coefficients of variation in the flexural strength and modulus results of both composites after exposure to *Coniophora puteana* and after cleaning were in most cases several times higher than the coefficients in the initial tests. The statistical significance of differences in variances before and after exposure was confirmed using the F-test for the strength variance for the oat composite after cleaning with brush, for the millet composite after cleaning with water and with brush, and for the modulus variance for the millet composite.

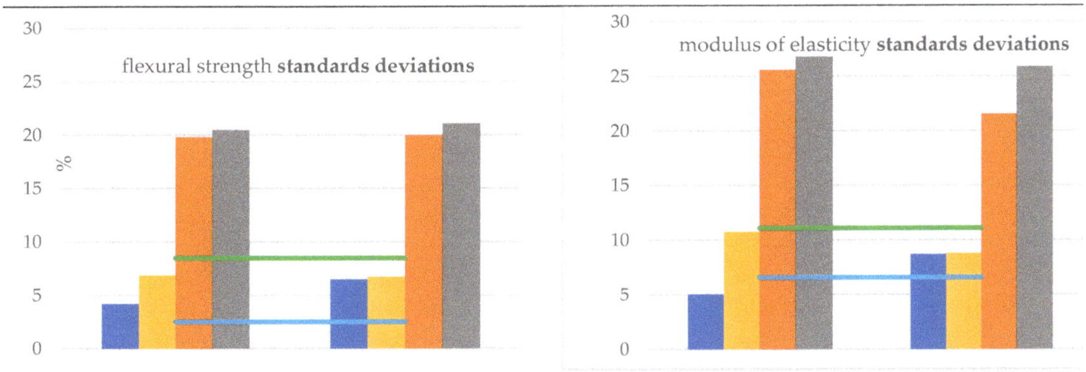

		water $S_{\sigma w}$	brush $S_{\sigma b}$				water S_{Ew}	brush S_{Eb}
$v_r\sigma$, %		4.2	6.5	oat husks	v_{rE}, %		5.0	8.7
$v_R\sigma$, %		6.9	6.7	oat husks	v_{RE}, %		10.7	8.8
$v_r\sigma_i$, %	2.5				v_{rEi}, %	6.6		
$v_r\sigma$, %		19.8	20.0	millet husks	v_{rE}, %		25.6	21.5
$v_R\sigma$, %		20.5	21.1	millet husks	v_{RE}, %		26.8	25.9
$v_r\sigma_i$, %	8.5				v_{rEi}, %	11.1		

(a) (b)

Figure 8. Values of repeatability and within-laboratory reproducibility coefficients of variation for millet husk-reinforced PVC composite and oat husk-reinforced PVC composite samples. v_r- repeatability coefficient of variation, v_R—reproducibility coefficients of variation, $v_r\sigma_i$, v_{rEi} repeatability coefficients of variation values for initial flexural strength and initial modulus of elasticity, respectively, and v_{Yw}, v_{Yb}-, values of coefficients of variation (v_r and v_R) for the Y quantity after exposure to fungi and cleaning with water jet and soft brush, respectively. (**a**) Flexural strength results and (**b**) modulus of elasticity results.

Considering that exposure and cleaning can introduce additional sources of variability in the results beyond those obtained from flexural testing alone, larger coefficients of variation seem obvious.

The relative expanded uncertainty of the test method assuming a 95% coverage interval and a normal distribution can be approximately estimated as $2v_R$ and can reach 40–50% of the result value, which poses a high risk when evaluating the product. Considering the possibility that the coefficients of variation in reproducibility will be higher in interlaboratory tests than in intra-laboratory tests, it must be assumed that this uncertainty can be even larger.

It is also evident that the variability in the results expressed by the values of the coefficients of variation differed between the composites. Already in the initial tests, a more than two-fold difference can be observed in the case of $v_{r\sigma i}$ and almost two-fold in the case of the v_{rEi} The differences deepened after exposure and cleaning and reached a four-fold difference between the composites. Therefore, it can be concluded that the uncertainty of the test method depends on the tested material.

Table 3 presents the results of the coefficient of variation between researchers, which was calculated based on the interlaboratory standard deviation S_L. Considering that the experiment was conducted under conditions of within-laboratory reproducibility, the S_L component of the uncertainty can reach significantly higher values in the case of tests in different laboratories.

Table 3. Results of between-researchers coefficient of variation.

Composite	Water Jet Cleaning	Soft Brush Cleaning	Water Jet Cleaning	Soft Brush Cleaning
	$v_{L\sigma w}$, %	$v_{L\sigma b}$, %	v_{LEw}, %	v_{LEb}, %
Oat husk-reinforced	5.4	1.8	9.5	1.2
Millet husk-reinforced	5.1	6.6	8.0	14.4

3.3.2. Differences Between the Water- and Brush-Cleaned Samples

To assess the statistical significance of the differences between the flexural parameters of the samples subjected to different procedures, a one-way ANOVA analysis was performed. Table 4 shows the differences in the results, which were calculated using Equation (12).

$$100\frac{\overline{\overline{Y}}_1 - \overline{Y_i}}{\overline{Y_i}} \quad (12)$$

where the following are defined:

$\overline{Y_i}$—result in initial state (mean strength or mean modulus);

$\overline{\overline{Y}}_1$—general mean of results (strength or modulus) after exposure and cleaning, calculated according to Equation (7).

To calculate the difference in results for the water jet-cleaned and soft brush-cleaned samples, Equation (13) was used:

$$100\frac{\overline{\overline{Y}}_b - \overline{\overline{Y}}_w}{\overline{\overline{Y}}_w} \quad (13)$$

where the following are defined:

$\overline{\overline{Y}}_b$—general mean of results (strength or modulus) after exposure and cleaning with brush;

$\overline{\overline{Y}}_w$—general mean of results (strength or modulus) after exposure and cleaning with water.

Table 4 also shows the F-statistic values obtained from the analysis of variance. In all cases, statistical significance was found for the differences in flexural parameters before and after exposure and cleaning (Table 4). It was also observed that in the case of the oat husk composite, the difference in flexural parameters between the water jet- and soft brush-cleaned samples was not statistically significant. These values were 1.3% for flexural strength and 3.6% for modulus. In contrast, in the case of the millet husk composite, the difference was statistically significant. The loss of flexural strength was 16% greater and the loss of modulus was 21% greater with brushing than with water jet cleaning. Therefore, the method of mycelium removal may have caused significant differences in the test results.

Table 4. Differences between values of flexural strength and modulus of elasticity for oat husk composites and millet husk composites before exposure (initial tests), after exposure to *Coniophora puteana*, cleaning with water, and after exposure to *Coniophora puteana* and cleaning with a brush. Statistically significant differences are highlighted in grey. F-statistics values resulting from analysis of variance: F_{str}—strength results., F_{mod} is the modulus value, and F_{crit} is the critical value for the ANOVA test.

		Oat Husk-Reinforced Composite		**Millet Husk-Reinforced Composite**	
		Initial Test Results	**Cleaning with Water Jet**	**Initial Test Results**	**Cleaning with Water Jet**
Cleaning with water jet	Difference between values of flexural strength, %	−8		−14	
	Difference between values of modulus, %	−34		−45	
	F_{crit}	4.26		4.26	
	F_{str}	12		4.73	
	F_{mod}	153		64	
Cleaning with soft brush	Mean difference between values of flexural strength, %	−9	−1.3	−28	−16
	Mean difference between values of modulus, %	−36	−3.6	−57	−21
	F_{crit}	4.26	4.10	4.26	4.10
	F_{str}	14	0.45	21	7.7
	F_{mod}	181	1.85	151	9.4

Given that different laboratories may use different cleaning methods, the standard deviation of the reproducibility of the test method may increase. The estimation of the standard deviation of reproducibility for all results after exposure, regardless of the cleaning method, gives the composite with millet husks the coefficient of variation values of 22% for flexural strength and 28% for modulus of elasticity. Assuming a 95% coverage interval and a normal distribution, the expanded uncertainty value of the results can reach 44% for the flexural strength and 56% for the modulus of elasticity. These values may be even higher because the results of this study were obtained based on intra-laboratory reproducibility. When the study is performed by different laboratories, additional factors may also increase the coefficients of variation. This raises the question of whether a test method with such low precision can be used for the assessment of construction products.

4. Conclusions

The test procedure for the fungal resistance of construction profiles made of natural fibre composites was analysed in terms of its precision and the influence of a factor not considered in the method model, namely the procedure for cleaning samples after exposure to fungi.

The results showed that the precision of the method is relatively poor. The coefficient of variation related to intra-laboratory reproducibility can take values above 20%, which can lead to uncertainty values of the results of the order of 50% and higher if interlaboratory reproducibility is considered.

It has been shown that different mycelium removal methods can produce different results in resistance tests. Composites with different properties and microstructures may show different susceptibility to changes in the method of mycelium removal.

Considering the above findings, it would be necessary to clarify the test method with respect to the mycelium removal step and other aspects as well.

The issue of the suitability of test methods for assessing the performance of construction products is extremely important and research into the precision and uncertainty associated with testing should be carried out in multiple directions.

Considering the dynamic development of natural fibre composites and their growing significance in civil engineering, further studies are planned. Future studies should cover other biotic factors, including algae and composites, with other fillers.

Author Contributions: Conceptualisation, E.S. (Ewa Sudoł); methodology, E.S. (Ewa Szewczak), E.S. (Ewa Sudoł), M.G. and E.K.; investigation, E.S. (Ewa Sudoł), E.S. (Ewa Szewczak) and E.K.; resources, E.S. (Ewa Sudoł), E.S. (Ewa Szewczak) and M.G.; data curation, E.S. (Ewa Szewczak); writing—original draft preparation, E.S. (Ewa Szewczak) and M.G.; writing and editing: E.S. (Ewa Sudoł), E.S. (Ewa Szewczak) and E.K.; visualisation, E.S. (Ewa Sudoł), E.S. (Ewa Szewczak), M.G. and E.K.; supervision, E.S. (Ewa Sudoł); project administration, E.S. (Ewa Szewczak). All authors have read and agreed to the published version of the manuscript.

Funding: This research was funded by Ministry of Science and Higher Education, Poland, grant number ITB L-2/2024, NZM-068 and NZK-080.

Institutional Review Board Statement: Not applicable.

Informed Consent Statement: Not applicable.

Data Availability Statement: The original contributions presented in this study are included in the article. Further inquiries can be directed to the corresponding author.

Acknowledgments: We would like to thank Jarosław Sówka for technical support.

Conflicts of Interest: The authors declare no conflicts of interest. The funders had no role in the study's design; in the collection, analyses, or interpretation of data; writing of the manuscript; or in the decision to publish the results.

References

1. BIPM, IEC, IFCC, ILAC, ISO, IUPAC, IUPAP, and OIML. International Vocabulary of Metrology—Basic and General Concepts and Associated Terms (VIM). Joint Committee for Guides in Metrology, JCGM 200:2012. (3rd edition). Available online: https://www.bipm.org/en/doi/10.59161/jcgm200-2012 (accessed on 15 December 2024). [CrossRef]
2. The European Co-Operation for Accreditation (EA). *EA-4/16 G: 2003 EA Guidelines on the Expression of Uncertainty in Quantitative Testing;* The European Co-Operation for Accreditation (EA): Paris, France, 2003.
3. Simonet, B.M.; Lendl, B.; Valcárcel, M. Method-Defined Parameters: Measurands Sometimes Forgotten. *TrAC Trends Anal. Chem.* **2006**, *25*, 520–527. [CrossRef]

4. Szewczak, E.; Winkler-Skalna, A.; Czarnecki, L. Sustainable Test Methods for Construction Materials and Elements. *Materials* **2020**, *13*, 606. [CrossRef] [PubMed]
5. Szewczak, E. Does Standardisation Ensure a Reliable Assessment of the Performance of Construction Products? *Standards* **2022**, *2*, 260–275. [CrossRef]
6. Stancu, C.; Michalak, J. Interlaboratory Comparison as a Source of Information for the Product Evaluation Process. Case Study of Ceramic Tiles Adhesives. *Materials* **2022**, *15*, 253. [CrossRef]
7. Michalak, J. Standards and Assessment of Construction Products: Case Study of Ceramic Tile Adhesives. *Standards* **2022**, *2*, 184–193. [CrossRef]
8. BIPM, IEC, IFCC, ILAC, ISO, IUPAC, IUPAP, and OIML. Evaluation of Measurement Data—The Role of Measurement Uncertainty in Conformity Assessment. Joint Committee for Guides in Metrology, JCGM 106:2012. Available online: https://www.bipm.org/en/doi/10.59161/JCGM106-2012 (accessed on 15 December 2024). [CrossRef]
9. Schabowicz, K. Testing of Materials and Elements in Civil Engineering. *Materials* **2021**, *14*, 3412. [CrossRef] [PubMed]
10. Walker, W.E.; Harremoës, P.; Rotmans, J.; van der Sluijs, J.P.; van Asselt, M.B.A.; Janssen, P.; Krayer von Krauss, M.P. Defining Uncertainty: A Conceptual Basis for Uncertainty Management in Model-Based Decision Support. *Integr. Assess.* **2003**, *4*, 5–17. [CrossRef]
11. Vercher, J.; Fombuena, V.; Diaz, A.; Soriano, M. Influence of fibre and matrix characteristics on properties and durability of wood–plastic composites in outdoor applications. *J. Thermoplast. Compos. Mater.* **2020**, *33*, 477–500. [CrossRef]
12. Friedrich, D.; Luible, A. Investigations on ageing of wood-plastic composites for outdoor applications: A meta-analysis using empiric data derived from diverse weathering trials. *Constr. Build. Mater.* **2016**, *124*, 1142–1152. [CrossRef]
13. Stark, N.M.; Matuana, L.M. Ultraviolet weathering of photostabilised wood-flour-filled high-density polyethylene composites. *J. Appl. Polym. Sci.* **2003**, *90*, 2609–2617. [CrossRef]
14. Sudoł, E.; Kozikowska, E.; Choińska, E. The Utility of Recycled Rice Husk-Reinforced PVC Composite Profiles for Façade Cladding. *Materials* **2022**, *15*, 3418. [CrossRef] [PubMed]
15. Qin, J.; Jiang, J.; Tao, Y.; Zhao, S.; Zeng, W.; Shi, Y.; Xiao, M. Sunlight tracking and concentrating accelerated weathering test applied in weatherability evaluation and service life prediction of polymeric materials: A review. *Polym. Test.* **2021**, *93*, 106940. [CrossRef]
16. Pilarski, J.M.; Matuana, L.M. Durability of wood flour-plastic composites exposed to accelerated freeze-thaw cycling. Part I. Rigid PVC Matrix. *J. Vinyl Addit. Technol.* **2005**, *11*, 1–8. [CrossRef]
17. Sethi, S.; Ray, B.C. Environmental effects on fibre reinforced polymeric composites: Evolving reasons and remarks on interfacial strength and stability. *Adv. Colloid Interface Sci.* **2015**, *217*, 43–67. [CrossRef]
18. Ladaci, N.; Saadia, A.; Belaadi, A.; Boumaaza, M.; Chai, B.X.; Abdullah MM, S.; Al-Khawlanif, A.; Ghernaout, D. ANN and RSM Prediction of Water Uptake of Recycled HDPE Biocomposite Reinforced with Treated Palm Waste *W. filifera*. *J. Nat. Fibers* **2024**, *21*, 2356697. [CrossRef]
19. Barton-Pudlik, J.; Czaja, K.; Grzymek, M.; Lipok, J. Evaluation of wood-polyethylene composites biodegradability caused by filamentous fungi. *Int. Biodeterior. Biodegrad.* **2017**, *118*, 10–18. [CrossRef]
20. Altuntas, E.; Yilmaz, E.; Salan, T.; Alma, M.H. Biodegradation properties of wood-plastic composites containing high content of lignocellulosic filler and zinc borate exposed to two different brown-rot fungi. *BioResources* **2017**, *12*, 7161–7177. [CrossRef]
21. Xu, K.; Feng, J.; Zhong, T.; Zheng, Z.; Chen, T. Effects of volatile chemical components of wood species on mould growth susceptibility and termite attack resistance of wood plastic composites. *Int. Biodeterior. Biodegrad.* **2015**, *100*, 106–115. [CrossRef]
22. Hosseinaei, O.; Wang, S.; Taylor, A.M.; Kim, J.W. Effect of hemicellulose extraction on water absorption and mold susceptibility of wood-plastic composites. *Int. Biodeterior. Biodegrad.* **2012**, *71*, 29–35. [CrossRef]
23. Yap, S.Y.; Sreekantan, S.; Hassan, M.; Sudesh, K.; Ong, M.T. Characterization and Biodegradability of Rice Husk-Filled Polymer Composites. *Polymers* **2021**, *13*, 104. [CrossRef]
24. Rajak, D.K.; Pagar, D.D.; Menezes, P.L.; Linul, E. Fiber-Reinforced Polymer Composites: Manufacturing, Properties, and Applications. *Polymers* **2019**, *11*, 1667. [CrossRef] [PubMed]
25. Fabiyi, J.S.; McDonald, A.G.; Morrell, J.J.; Freitag, C. Effects of wood species on durability and chemical changes of fungal decayed wood plastic composites. *Compos. Part A Appl. Sci. Manuf.* **2011**, *42*, 501–510. [CrossRef]
26. Catto, A.L.; Rosseto, E.S.; Reck, M.A.; Rossini, K.; da Silveira RM, B.; Santana RM, C. Growth of white rot fungi in composites produced from urban plastic waste and wood. *Macromol. Symp.* **2014**, *344*, 33–38. [CrossRef]
27. Camarero, S.; Martínez, M.J.; Martínez, A.T. Understanding lignin biodegradation for the improved utilization of plant biomass in modern biorefineries. *Biofuels Bioprod. Biorefining* **2014**, *8*, 615–625. [CrossRef]
28. Catto, A.L.; Montagna, L.S.; Almeida, S.H.; Silveira, R.M.; Santana, R.M. Wood plastic composites weathering: Effects of compatibilization on biodegradation in soil and fungal decay. *Int. Biodeterior. Biodegrad.* **2016**, *109*, 11–22. [CrossRef]

29. Mankowski, M.; Morrell, J.J. Patterns of fungal attack in wood-plastic composites following exposure in a soil block test. *Wood Fiber Sci.* **2000**, *32*, 340–345. Available online: https://ir.library.oregonstate.edu/concern/articles/fn106z43g (accessed on 8 November 2024).
30. Ashori, A.; Behzad, H.M.; Tatmain, A. Effects of chemical preservative treatments on durability of wood flour/HDPE composites. *Compos. Part B* **2013**, *47*, 308–313. [CrossRef]
31. Naumann, A.; Seefeldt, H.; Stephan, I.; Braun, U.; Noll, M. Material resistance of flame retarded wood-plastic composites against fire and fungal decay. *Polym. Degrad. Stab.* **2012**, *97*, 1189–1196. [CrossRef]
32. *ENV 12038:2022*; Durability of Wood and Wood-Based Products. Wood based panels. Method of Test for Determining the Resistance Against Wood-Destroying Basidiomycetes. European Committee for Standardization (CEN): Brussels, Belgium, 2002.
33. *EN 15534-1:2017*; Composites Made from Cellulose-Based Materials and Thermoplastics (Usually Called Wood-Polymer Composites (WPC) or Natural Fibre Composites (NFC)). Part 1: Test Methods for Characterisation of Compounds and Products. European Committee for Standardization (CEN): Brussels, Belgium, 2017.
34. *ISO 178:2019*; Plastics. Determination of Flexural Properties. International Organization for Standardization ISO: Geneva, Switzerland, 2019.
35. Sudoł, E.; Kozikowska, E.; Szewczak, E. Artificial Weathering Resistance Test Methods for Building Performance Assessment of Profiles Made of Natural Fibre-Reinforced Polymer Composites. *Materials* **2022**, *15*, 296. [CrossRef]
36. *ISO 5725-2:2019*; Accuracy (Trueness and Precision) of Measurement Methods and Results. Part 2: Basic Method for the Determination of Repeatability and Reproducibility of a Standard Measurement Method. International Organization for Standardization ISO: Geneva, Switzerland, 2019.
37. Prasad, A.; Rao, K. Mechanical properties of natural fibre reinforced polyester composites: Jowar, sisal and bamboo. *Mater. Design* **2011**, *32*, 4658–4663. [CrossRef]
38. Stark, N.M.; Matuana, L.M.; Clemons, C.M. Effect of processing method on surface and weathering characteristics of wood-flour/HDPE composites. *J. Appl. Polym. Sci.* **2004**, *93*, 1021–1030. [CrossRef]

materials

MDPI

Article

Assessment of Cracking Development in Concrete Precast Crane Beams Using Optical and Deep Learning Methods

Marek Słoński

Faculty of Civil Engineering, Cracow University of Technology, ul. Warszawska 24, 31-155 Kraków, Poland; marek.slonski@pk.edu.pl; Tel.: +48-12-628-2562

Abstract: The longevity and safety of concrete precast crane beams significantly impact the operational integrity of industrial infrastructure. Assessment of surface cracks development in concrete structural elements during laboratory tests is performed mainly by applying standard tools such as linear-variable-differential transformers and strain gauges. This paper presents a novel assessment methodology combining deep convolutional neural network for image segmentation with digital image correlation method to evaluate the structural health of precast crane beams after more than fifty years of service. The study first outlines the adaptation of the deep learning U-Net architecture for detecting and segmentation of surface cracks in crane beams. Concurrently, DIC technique is employed to measure surface strains and displacements under load. The integration of these technologies enables a non-destructive, accurate, and detailed analysis, facilitating early detection of deterioration that may compromise structural safety. Initial results from field tests validate the effectiveness of our approach, demonstrating its potential as a tool for predictive maintenance of aging industrial infrastructure.

Keywords: concrete; precast; crane beam; crack; digital image correlation; convolutional neural network; U-Net; segmentation

1. Introduction

The assessment of concrete precast crane beams using advanced imaging and machine learning techniques represents a significant change from traditional methods in structural health monitoring. This shift is driven by the need to address the inherent limitations of manual inspection methods, which have long been the standard practice for assessing the structural integrity of such critical components.

Manual inspections, while invaluable for the detailed assessment they provide, are fraught with several limitations that compromise their effectiveness and efficiency. Manual inspections are heavily based on the experience and judgment of the inspectors. This subjectivity can lead to inconsistencies in the assessment outcomes, with potential cracks being overlooked or non-critical anomalies being overemphasized. These inspections are labor-intensive and time-consuming, often requiring scaffolding, facility closure, and extensive human resources, which can lead to significant operational disruptions. The need for physical access to inspect structural elements, especially in large or complex installations such as crane beams, poses safety risks to inspection personnel, particularly in environments with high or difficult-to-reach structures. Manual methods do not scale well with the increasing number and complexity of infrastructures. They become impractical as the frequency of inspections increases or as more detailed and frequent data collection is required for predictive maintenance. The time taken to schedule, conduct, and analyze manual inspections can lead to delays in identifying and addressing structural problems, potentially exacerbating problems that could have been mitigated if detected earlier.

Given these significant drawbacks, there is a pressing need for more reliable, efficient and objective methods to monitor structural health. This research addresses these challenges by leveraging digital image correlation (DIC) and deep learning models, specifically U-Net variants, to provide a non-destructive, automated, and precise method for detecting and analyzing cracks in concrete beams. Incorporating these technologies improves both the accuracy and effectiveness of damage identification while significantly minimizing operational interruptions and safety hazards linked to conventional manual inspections.

Assessment of concrete precast crane beams using optical techniques such as digital image correlation (DIC) and deep learning methods such as a convolutional neural network (CNN) involves leveraging advanced image processing and machine learning methods to assess structural integrity. This approach is part of a broader trend in structural health monitoring, where non-contact methods are increasingly used to assess and monitor infrastructure. The integration of DIC and CNN can enhance the precision and efficiency of detecting and quantifying damage to crane beams, which are critical components in construction and industrial settings.

Concrete structural elements, such as beams, columns and slabs located in industrial building facilities, have been experiencing rapid deterioration due to load and environmental impact [1]. For example, a large proportion of precast post-tensioned structures [2], especially those made of segments in several European countries, have been in service for more than 50 years [3–5]. As these structures age or are exposed to various environmental factors, cracks often develop in them, which if not addressed promptly, can compromise the structural integrity and safety of the entire structure.

Traditional methods of evaluating and characterizing these cracks often involve manual inspections, which can be time-consuming, labor intensive, and prone to human error. Furthermore, the subjective nature of manual inspections can lead to inconsistencies in crack assessment, potentially leading to overlooked threats or unnecessary repairs [6–8].

The detection and characterization of surface defects, such as cracks in concrete, plays a crucial role in vision-based structural health monitoring systems and in the evaluation of the condition of civil infrastructure. These defects can serve as indicators of ongoing severe degradation processes within a structure. To address this, several strategies have been proposed that use computer vision and image processing methods for the automatic detection of surface defects [9–11].

In recent years, rapid advances in computer vision and deep learning architectures for semantic segmentation [12] have opened new avenues for automating and improving the concrete crack detection and evaluation process. Among the various deep learning architectures, the U-Net model has emerged as a promising tool for image segmentation tasks, which makes it particularly suitable for delineating and quantifying cracks in concrete surfaces. By automating the process, we can achieve faster, more consistent, and potentially more accurate assessments of concrete cracks. This approach not only prolongs the lifespan and ensures the safety of structures, but it also results in reduced costs for maintenance and repairs.

Furthermore, as urban areas continue to expand and the number of concrete structures increases, the need for efficient and reliable crack detection systems becomes even more pressing. Using the capabilities of the U-Net model, we can address this growing demand and pave the way for smarter and safer urban development.

To bridge the gap between conventional methods and modern needs, this paper introduces a hybrid approach that combines advanced imaging techniques and machine learning models. This approach aims to improve the precision, efficiency and safety of structural assessments.

This paper is organized as follows. In Section 2, the related works are reviewed and presented. Section 3 presents the novel methodology developed in this work. In Section 4, a comparison of U-Net-based models for the semantic segmentation of cracks is given. Section 5 presents the application of the methodology for the assessment of the concrete precast crane beams. Finally, in Section 6, the most relevant conclusions are drawn.

2. Related Works

The evolution of techniques for the detection and characterization of concrete cracks has been marked by significant advancements in computer vision and image processing technologies. This section reviews various methods, highlighting the transition from traditional manual approaches to sophisticated automated systems that employ deep learning for enhanced accuracy and efficiency. To facilitate a clearer understanding of the diverse methodologies and their respective contributions to the field, a tabular summary is provided below.

Standard image-based approaches to the detection and characterization of concrete cracks on the surface of structural elements are based on computer vision methods and image processing techniques integrated into custom-designed methods tailored to specific types of images. These approaches have been extensively presented and evaluated in numerous recent survey articles, for example in [10,11,13,14]. Despite saving time, these approaches often require manual recalibration for different datasets due to their algorithmic rules, which limit their generalizability. This recalibration process could be both laborious and complex, as shown in [15].

On the other hand, a variety of alternative techniques rooted in deep learning have been proposed. These methods can autonomously learn rules from datasets, addressing a key limitation of traditional analysis techniques. Furthermore, adapting an algorithm to incorporate new data generally entails merely augmenting the training set with new samples, while most other training and model parameters remain unchanged. Numerous deep learning approaches have been introduced to segment concrete cracks, employing various methods such as patch-based techniques and semantic segmentation techniques, among others. Islam et Kim in [16] propose an autonomous crack identification method based on a convolutional neural network (CNN) model. It comprises a fully convolutional neural network for crack segmentation, and VGGNet is utilized as the backbone. The results show that the method developed is very efficient for the classification of concrete cracks, attaining a performance of approximately 92% for the recall. Mousavi et al. in [17] used an optimized U-shaped convolutional neural network with a novel training strategy for the segmentation of concrete cracks. The proposed model is an encoder-decoder model that uses EfficientNet-B7 as the encoder and the modified expansion path of U-Net as the decoder.

Recently, ED-CNNs have been developed for semantic image segmentation [18]. Motivated by these accomplishments, numerous recent investigations have devised ED-CNN-based models aimed at automatic semantic segmentation of concrete cracks [19,20]. Dung and Anh [21] proposed a segmentation method using a VGG16-based encoder. Yang et al. [22] applied the fully convolutional network (FCN) to enhance the effectiveness of crack detection at the image level and the cell level of the grid. The findings indicate that the use of FCN is both practical and adequate for detecting and assessing cracks. In the paper by Pu et al. [23] research on recognizing crack patterns in concrete through images is presented using a convolutional neural network and an encoder–decoder based component for semantic segmentation. Shen et al. in their paper [24] propose an enhanced method based on the open source Deeplabv3+ model and name it Deeplabv3+ BDF in line with the optimization strategy employed. Deeplabv3+ BDF initially replaces the original Xception backbone with MobileNetv2 and then substitutes replace all traditional convolutional layers with depthwise separable convolutions to create a lightweight model. Razveeva et al. in [25] present an examination of the geometric properties of cracks and delamination in aerated concrete products utilizing convolutional neural networks. Guo et al. in [26] describe the monitoring and automatic characterization of cracks in a hardening-resistant cementitious compound (SHCC) by intelligently analyzing photographs. Xu et al. in [27] propose a data set benchmark for the segmentation of concrete cracks at the pixel level, using a deep learning model. The suggested benchmark for the concrete crack dataset offers engineering importance and can enhance model efficiency by carefully choosing suitable techniques, such as expanding the dataset size or refining the model design, while also reducing the time needed for dataset annotation. Forest et al. in [28] propose applying the explainable artificial intelligence (XAI) methodology to segment and monitor surface cracks. Ahmadi et al. in [29] evaluated the effectiveness of two deep learning models, SAM and U-Net, to identify cracks in concrete structures. The findings suggest that each model has distinct strengths and weaknesses in detecting various types of cracks.

König et al. in [30] reviewed recent studies in this research area. In particular, semantic segmentation techniques have exhibited state-of-the-art (SOTA) performance in crack segmentation tasks on concrete surface images [31–33]. Zhou et al. in [34] review recent developments in deep learning-based crack segmentation methods and investigate their performance under the impact of different image types. Deng et al. in [35] conducted a literature review, focusing mainly on qualitative damage evaluation or damage segmentation.

Most approaches for semantic segmentation of concrete cracks utilize the encoder-decoder deep neural network structure inspired by the superior accuracy of the U-Net architecture for medical applications [36]. This architecture improves the original fully convolutional network (FCN) methodology [37]. Other popular encoder-decoder architectures like Segnet [38] and DeepLab [39] were also introduced.

In U-Net, skip connections facilitate gradient flow and enable information exchange between the down-sampling and up-sampling paths by linking the corresponding encoder and decoder layers. U-Net processes an input image and assigns each pixel to one of several classes, thus segmenting the image into distinct regions. For surface concrete crack segmentation, these classes differentiate between crack and non-crack areas. The result is a pixel-wise map from the U-Net, which provides precise segmentation of the target regions.

Various adaptations have been created to improve the effectiveness of the initial U-Net design. Dense U-Net links the outputs of all convolutional blocks to the inputs of the following blocks within each level, promoting feature reuse and mitigating vanishing gradients. The Inception U-Net uses multiple convolutional kernel sizes in parallel paths at each level, providing robustness to images of varying scales.

Attention U-Net integrates an attention module within each skip connection, allowing the network to focus more on important spatial regions in an image and less on less relevant ones. Rao et al. in [40] propose a novel deep learning framework to detect cracks and then estimate crack widths in images of concrete surfaces. The proposed framework uses Attention Recurrent Residual U-Net (Attention R2U-Net) with Random Forest regressor to predict the crack width. Hang et al. in [41] present an attention-based feature fusion network (AFFNet), with a backbone residual network (ResNet101) enhanced with two modules of the attention mechanism for automatic detection of concrete cracks at the pixel level. Evaluations against other leading models demonstrate the proposed method high efficiency and precision.

Lastly, residual U-Net incorporates residual learning by adding shortcut connections in each layer, which improves gradient flow and operates under the principle that learning a residual mapping is more straightforward than learning the original one.

Shamsabadi et al. in [42] introduced a vision transformer (ViT)-oriented framework to detect cracks on asphalt and concrete surfaces. In contrast to CNN-based models such as DeepLabv3+ and U-Net, our TransUNet model, which incorporates a CNN-ViT architecture, demonstrated enhancements in mean Intersection over Union (IoU) by approximately 61% and 3.8% on the original images of the corresponding datasets. This improvement was observed particularly among images exhibiting small and multi-scale crack features.

In a study [43], Wang and Su introduce an innovative SegCrack model designed for pixel-wise crack segmentation, utilizing a Transformer encoder with a hierarchical structure to produce multiscale features. This model employs a top-down pathway enriched with lateral connections to incrementally up-sample and integrate features from the encoder's deepest layer.

Past research on concrete crack detection has introduced several adaptations of the U-net architecture using Transformer architecture to enhance performance. Xiang et al. in [44] proposed a dual encoder network that combines transformers and convolutional neural networks (DTrC-Net). DTrC-Net architecture was designed to effectively capture both the local features and the global context of the crack images. To improve the merging of features between adjacent layers and the coding-decoding stages, the network incorporates a feature fusion module alongside a residual path module. In general, DTrC-Net showed superior performance in crack segmentation accuracy and exhibited exceptional generalization capabilities in intricate environments compared to other state-of-the-art networks.

However, the specific impact of these modifications on the segmentation accuracy of concrete crack images remains uncertain due to the lack of an impartial comparison. Shi et al. in [45] compared semantic segmentation networks for crack detection in construction materials. Li et al. in [46] made a survey of deep learning-based pixel-level crack image segmentation methods. They grouped the methods into 10 topics based on the backbone network architecture. The comparison of the application of convolutional neural networks for concrete crack detection is presented in [47].

This study addresses this gap by evaluating three prominent U-net variants in the SDNET2018 dataset, aiming to discern their influence on segmentation performance and balancing these findings against training, evaluation time, and computational complexity.

To facilitate a clearer understanding of the diverse methodologies and their respective contributions to the field, a tabular summary is provided in Table 1.

Table 1. Summary of methods used for concrete surface crack detection.

Method	Key Features	Benefits	Drawbacks
Manual Inspection	Visual examination by experts	High detail accuracy	Subjective, time-consuming, safety risks
Linear Variable Differential Transformers (LVDT)	Mechanical measurement of displacement	Precise, reliable	Limited to visible, accessible areas
Strain Gauges	Measures deformation under load	Direct measurement of physical properties	Requires physical contact, limited coverage

Table 1. *Cont.*

Method	Key Features	Benefits	Drawbacks
Digital Image Correlation (DIC)	Optical method for full-field displacement and strain	Non-contact, high resolution	Requires high-quality imaging, complex setup
Fully Convolutional Networks (FCN)	Deep learning for semantic segmentation	Automates crack detection, scalable	Requires large labeled datasets, intensive computation
U-Net	Encoder-decoder architecture with skip connections	Precise segmentation, good with small datasets	Computationally intensive
Attention U-Net	Integrates attention mechanisms	Improves focus on relevant features	More complex than standard U-Net
TransUNet	Combines Transformers with U-Net	Captures global and local context	Very high computational resources needed

3. Materials and Methods

3.1. Overview

The methodology depicted in Figure 1 involves steps that facilitate a comprehensive evaluation of cracking in the beams through advanced image processing techniques. Here is a detailed description of each step illustrated in the diagram.

1. Setup of a U-Net based model for crack segmentation.
   This step involves off-line training of selected U-Net based deep learning models for semantic segmentation of surface cracks using available datasets from the Internet. The trained best U-Net based model is set up to identify and segment cracks in images of crane beams on-line during the laboratory experiments.
2. Setup of the vision system in laboratory.
   The vision system, which includes high-resolution DSLR cameras and appropriate lighting setups, is installed in a laboratory environment. This setup is crucial to capture detailed images of the crane beams under examination. The system is calibrated to ensure that the images obtained are suitable for both U-Net segmentation and DIC analysis.
3. Raw images acquisition and storage.
   High-quality raw images of the crane beams are captured using the laboratory vision system for subsequent processing. The image acquisition is done under controlled lighting to minimize noise and ensure that the finest cracks can be detected.
4. Image pre-processing.
   The raw images (optionally) undergo pre-processing, which includes noise reduction, contrast enhancement, and scaling to prepare the images for analysis by both the U-Net model and the DIC software. Concrete images may contain noise due to the conditions under which the photos are taken. The application of a Gaussian blur filter smooths out the images, reducing pixel-level noise. It may help minimize false positives in crack detection. The contrast enhancement technique was used to improve the visibility of the cracks by sharpening the intensity differences between the cracks and the surrounding concrete. The images were resized to ensure consistent input size for the models.
5. Segmentation and estimation of cracks density.

The preprocessed images are fed into the U-Net model, which performs segmentation of the cracks online. Following segmentation, the density of the cracks is calculated, providing quantitative data on the extent of cracking, which is critical for assessing the structural capacity of the beams.

6. Computation of full field of deformation.
   Using 2D DIC software, the deformation of the crane beams under load is analyzed off-line. This software computes the displacement fields across the beam's surface from the images, allowing for a detailed understanding of how the beams deform under loading.
7. Computation of cracks width development
   The final step involves using the DIC results to calculate the development of crack widths over time or under varying load conditions. This analysis helps to understand the behavior of cracks during the loading process and contributes to predicting the future behavior of these structural elements.

This method combines U-Net with DIC to improve the identification and segmentation of structural flaws and to deliver an in-depth evaluation of the mechanical performance of the beams, thereby presenting a strong strategy for monitoring structural health in civil engineering.

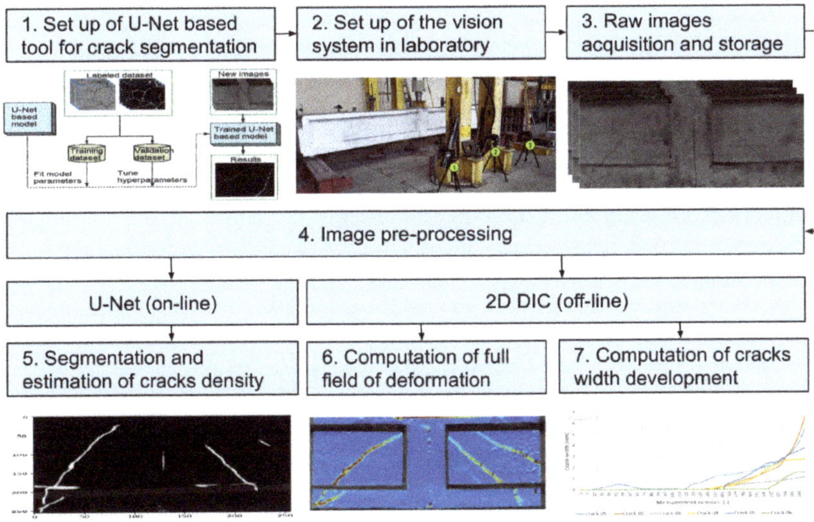

Figure 1. A diagram illustrating the experimental–computational framework employed in this study.

3.2. Digital Image Correlation

Digital Image Correlation (DIC) is a very useful method for evaluating structural element deformations [48–50]. DIC method was first introduced at University of South Carolina at the beginning of the 1980s as described in [51,52]. DIC method works by examining successive images recorded in the process of the deformation of an element under loading and establishing a correspondence between the pixel coordinates of the undeformed image and the pixel coordinates of the deformed image through correlation analysis, which gives the maximum correlation between the reference image and the deformed image. This mapping is utilized for determining full-field strains [53].

The nodes of the grid based on the defined image regions are paired and recognized as that corresponding to the maximum value of the correlation coefficient. This correlation measure is computed between the reference subset "f" and the target subset "g", with identical dimensions and are M × N pixels employing the zero-mean normalized cross-correlation formula specified in Equation (1). The basic principle of digital image correlation is illustrated in Figure 2.

$$CC^{ZMN} = \frac{\sum_{i=1}^{M}\sum_{j=1}^{N}\left[\left(f(i,j) - u_f\right) \times \left(g(i,j) - u_g\right)\right]}{\sqrt{\sum_{i=1}^{M}\sum_{j=1}^{N}\left(f(i,j) - u_f\right)^2 \times \sum_{i=1}^{M}\sum_{j=1}^{N}\left(g(i,j) - u_g\right)^2}}, \quad (1)$$

where u_f is the intensity of the reference subset and u_g is the intensity of the target subset form.

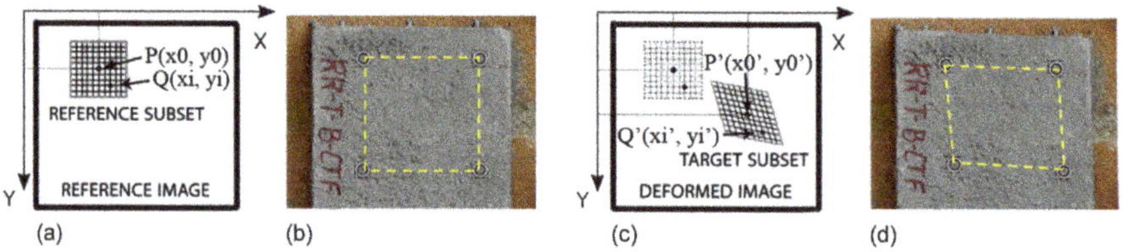

Figure 2. Illustration of the basic principle of the digital image correlation method: Sketch of the subset (**a**,**c**) and measurement points on the surface of a specimen (**b**,**d**) before (**a**,**b**) and after (**c**,**d**) deformation [54].

3.3. Deep Convolutional Encoder-Decoder Architectures

A deep convolutional encoder-decoder architecture is a type of neural network that is commonly used for computer vision tasks [55]. It consists of two main components: An encoder and a decoder. The encoder component of a deep convolutional encoder-decoder network is responsible for extracting meaningful features from the input data. It typically consists of multiple convolutional layers followed by pooling layers to reduce the spatial dimensions of the feature maps. The encoder gradually learns to encode the input data in a lower-dimensional representation, capturing important information in the process.

The decoder component of the network takes the encoded representation and reconstructs the output data. It usually consists of upsampling layers to increase the spatial dimensions of the feature maps, followed by convolutional layers to refine the reconstructed output. The decoder gradually learns to decode the encoded representation back into the original input data, generating the desired output. By combining the encoder and decoder components, a deep convolutional encoder-decoder network can effectively perform tasks such as image segmentation, image-to-image translation, and image generation, by learning to encode and decode the relevant information in the input data.

In this paper, 3 architectures are applied: U-Net, Attention U-Net and TransUNet. They are briefly described in the following subsections. The number of parameters and the number of convolutional layers are presented in Table 2.

Table 2. Number of parameters and number of convolutional layers.

Model Name	Number of Parameters [$\times 10^6$]	Number of Conv Layers
U-Net	31.5	13
Attention U-Net	37.3	17
TransUNet	72.8	24

3.3.1. U-Net Architecture

U-Net is a convolutional neural network architecture for fast and precise image segmentation. It was proposed by Olaf Ronneberger, Philipp Fischer, and Thomas Brox in 2015 [36]. The network is based on the fully convolutional network (FCN) which was introduced by Long et al. in 2014 in [37]. Its architecture was modified and extended to work with fewer training images and yield more precise segmentations. U-Net is widely used in the field of medical image segmentation, where it has achieved state-of-the-art performance.

U-Net architecture is comprised of a contracting path to capture local details and an expanding path for global context. The contracting path includes convolutional and pooling layers, while the expanding path features upsampling and convolutional layers. Furthermore, U-Net uses skip connections to maintain spatial detail and enhance segmentation precision. The U-Net architecture is presented in Figure 3.

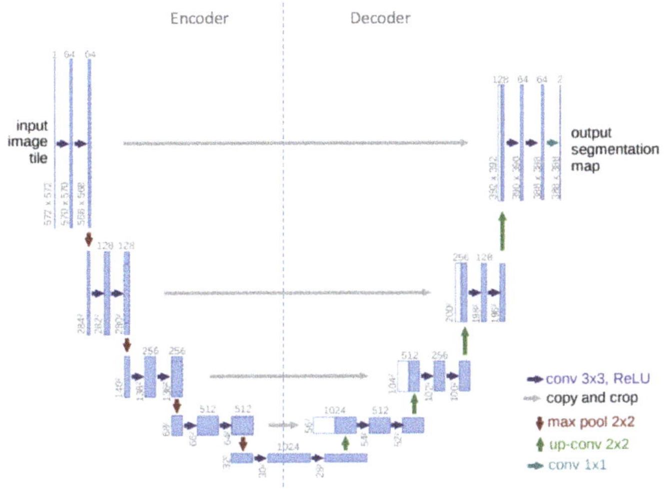

Figure 3. U-Net architecture, taken from the original paper [36].

3.3.2. Attention U-Net

The Attention U-Net architecture is an extension of the traditional U-Net architecture that incorporates the concept of attention mechanisms introduced by Bahdanau et al. in 2014 in [56]. The Attention U-Net architecture was proposed by Ozan Oktay et al. in 2018 in [57]. It is a variant of the original U-Net architecture that incorporates attention mechanisms. The architecture is presented in Figure 4 and the attention gate diagram is shown in Figure 5.

It was designed to improve the performance of U-Net in tasks where capturing long-range dependencies and focusing on relevant image regions is crucial, such as image segmentation and object detection. Attention U-Net was introduced to the research community to enhance the model's ability to handle fine-grained details and complex patterns.

The inclusion of an attention mechanism in Attention U-Net enhances the ability of the model to concentrate on essential areas of the input image, thereby increasing its segmentation precision. This approach is modeled on the human visual system, which prioritizes certain parts of an image over others. Integrating attention mechanisms within the U-Net framework has led to notable improvements in model performance in a range of medical image segmentation applications, including tumor and organ segmentation.

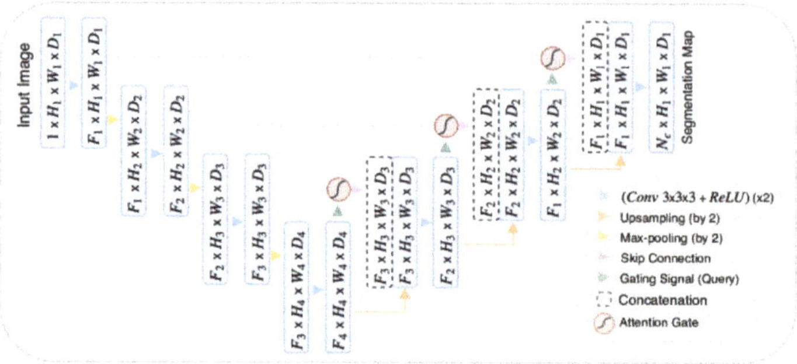

Figure 4. Attention U-Net architecture, taken from the original paper [57].

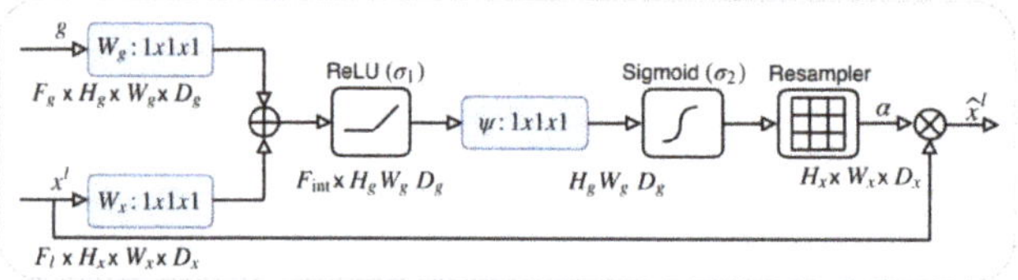

Figure 5. The block diagram of attention gate in Attention U-Net architecture, taken from the original paper [57].

3.3.3. TransUNet

The TransUNet architecture combines the strengths of transformer-based models introduced by Vaswani et al. in 2017 in [58] (for context and feature extraction) with the architecture of a U-Net (for spatial segmentation tasks). This architecture was proposed by Chen et al. in 2021 in [59]. It can be used for tasks such as semantic segmentation, object detection, or medical image analysis, where capturing fine-grained spatial details and global context is essential. The architecture is presented in Figure 6.

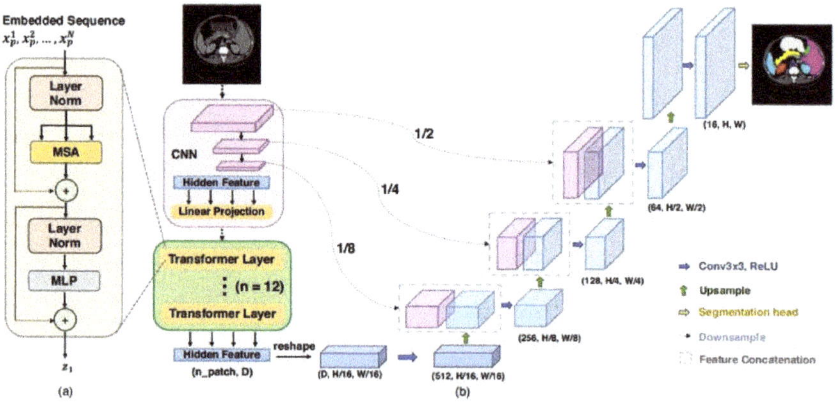

Figure 6. TransUNet architecture: Transformer layer framework (**a**), TransUNet layer structure (**b**), taken from the original paper [59].

The choice of U-Net variants - specifically the standard U-Net, Attention U-Net, and TransUNet - for this study was guided by their distinct architectural strengths, which are particularly beneficial for the complex task of segmenting concrete cracks. Each of these models offers unique advantages in processing images characterized by subtle and intricate crack patterns, which are common in aged concrete structures.

The standard U-Net model is renowned for its efficiency and effectiveness in image segmentation tasks, particularly when dealing with limited data. Its architecture leverages a symmetric encoder-decoder structure with skip connections that help preserve spatial hierarchies between image features at different scales. This capability is crucial for detecting and delineating fine cracks, ensuring that even minor discrepancies on the surface of the concrete are captured and analyzed.

Attention U-Net extends the standard U-Net by incorporating attention mechanisms, which are particularly adept at refining the focus on relevant features while suppressing less useful information. This model improves the precision of segmentation in images where cracks are surrounded by complex textures or noise, a common scenario in deteriorating infrastructure. By directing the model's focus towards salient features, Attention U-Net is better equipped to isolate and characterize cracks with higher accuracy, making it ideal for detailed structural assessments where every millimeter counts.

TransUNet integrates the robust feature extraction capabilities of Transformers with the spatial context preservation of U-Nets. This hybrid model excels at capturing both the global and local context, a necessary feature for comprehensive crack analysis across varying scales and intensities. Transformer components address the challenge of long-range dependencies, allowing for more consistent segmentation in larger crack zones and improving the generalizability of the model to different types of concrete surfaces and damage patterns.

The models were chosen for their technical strengths, supported by initial tests that showed that they outperformed other deep learning frameworks in crack segmentation tasks. For example, TransUNet demonstrated exceptional capability in maintaining high recall rates, crucial for ensuring that no potential crack is missed, which is paramount in predictive maintenance and safety assessments. Attention U-Net showed higher precision, which is critical in minimizing false positives, thus avoiding unnecessary inspections and repairs.

Through these enhanced capabilities, the selected variants of U-Net facilitate a more accurate, reliable, and cost-effective way to monitor and maintain the aging infrastructure, in line with the goals of advanced structural health monitoring systems. Thus, this strategic selection directly contributes to the overarching aim of the investigation: to improve the precision and reliability of concrete crack detection in support of structural integrity assessments.

3.4. Dataset

The dataset used for this study consists of high-resolution images of concrete surfaces. These images were manually annotated to provide ground-truth labels for crack regions.

Before feeding the images into the model, several preprocessing steps were applied. These include resizing the images, normalization, and data augmentation techniques to improve the model's robustness.

To evaluate the performance of the models, we us ed 800 images of cracked concrete elements randomly selected from the open source data set called SDNET2018 [60]. SDNET2018 contains more than 56,000 images of cracked and non-cracked concrete bridge decks, walls, and pavements. The data set includes cracks as narrow as 0.06 mm and as wide as 25 mm. The data set also includes images with a variety of obstructions, including shadows, surface roughness, scaling, edges, holes, and background debris. Four images of cracked and uncracked concrete pavements from the benchmark dataset are shown in Figure 7.

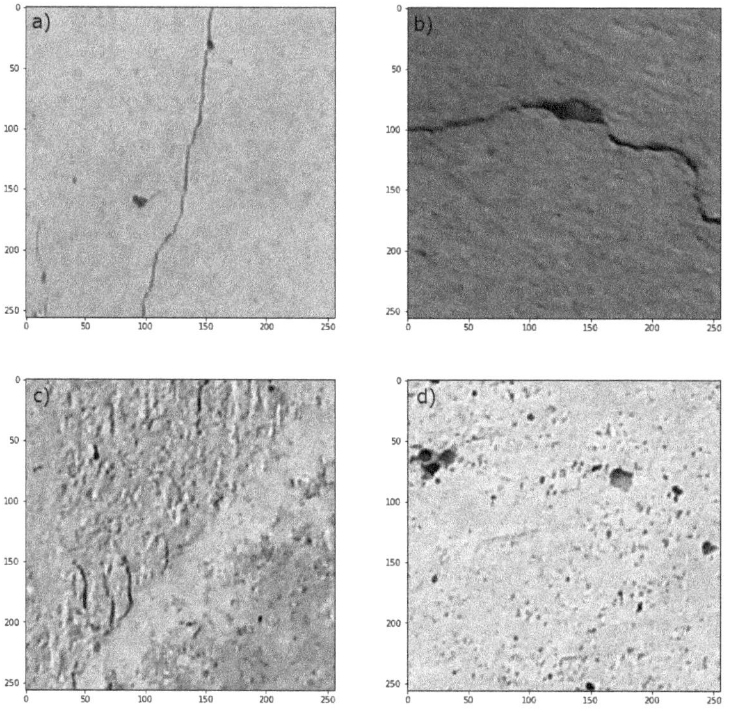

Figure 7. Images of cracked (**a**,**b**) and uncracked concrete pavements (**c**,**d**), sampled from SDNET2018 dataset [60].

While SDNET2018 is a widely used benchmark, it has several shortcomings. The dataset predominantly features images captured under controlled conditions, which may not represent the variability encountered in real world scenarios. This limitation could potentially introduce a bias toward specific types of cracks or surface conditions, affecting the generalizability of the model. The images in SDNET2018 do not adequately capture the diverse environmental impacts that real-world structures undergo, such as weathering or chemical degradation. This exclusion can lead to models that perform well in laboratory settings, but less so in harsh or outdoor environments.

The are some challenges in annotating the dataset (inaccuracies and subjectivity) that can influence the accuracy and effectiveness of the resulting models. These challenges were addressed using an automated tool to pre-annotate images based on common crack patterns, which were then manually refined. Cracks were annotated using the Segments-ai application [61].

In Figure 8, an example of a new image and a mask obtained from the original image and mask by horizontal flip are shown. Two examples of new images and masks obtained by vertical flip and transpose are presented in Figure 9.

The data set was divided into a training data set (2560 images) and a validation data set (640 images). The created dataset (accessed on 20 January 2024) is available at: https://www.kaggle.com/datasets/jakubniemiec/concrete-crack-images.

Data augmentation (horizontal, vertical, and flipping) was used to increase the number of labeled images to 3200 images.

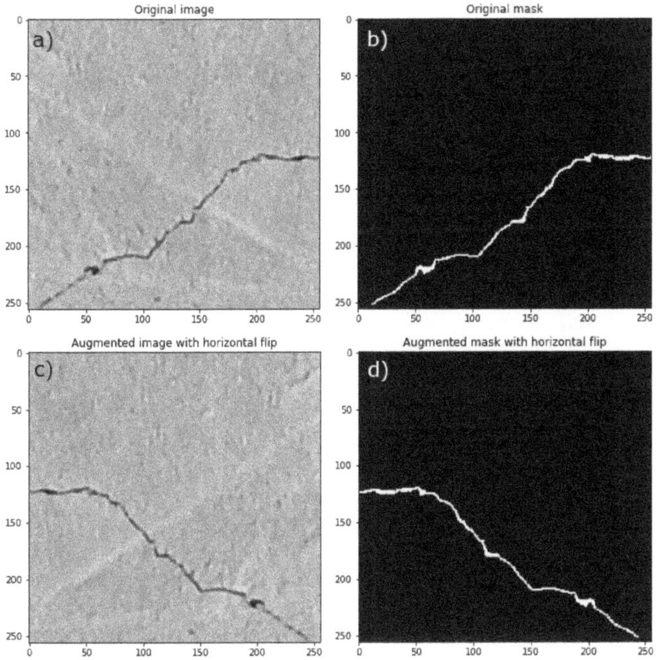

Figure 8. Example of new image and mask (**c**,**d**) obtained from original image and mask (**a**,**b**) by horizontal flip, taken from [62].

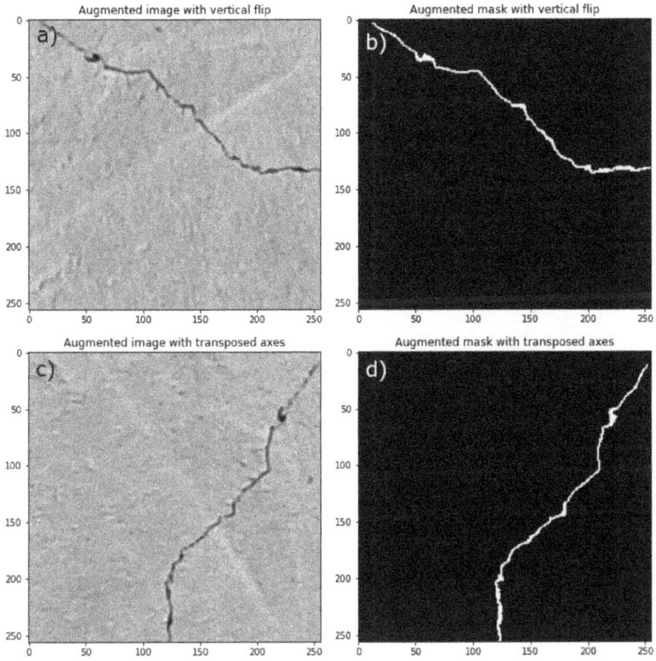

Figure 9. Examples of new images and masks obtained by vertical flip (**a**,**b**) and by transpose (**c**,**d**), taken from [62].

3.5. Evaluation Metrics for Crack Segmentation

The true measure of the efficacy of any model lies in its evaluation against robust and relevant metrics. In this section, we introduce and elucidate the key metrics used to gauge the performance of the U-Net model in the context of concrete crack assessment. The selection of these metrics is based on both their statistical robustness and their applicability to real-world scenarios, guaranteeing that our evaluation is both theoretically robust and practically significant. From assessing the model's accuracy in detecting minute cracks to its precision in characterizing their parameters, each metric provides a unique lens through which to view the model's capabilities. By delving into these evaluation metrics, we aim to provide a transparent and comprehensive understanding of the standards against which our research findings have been benchmarked.

The effectiveness of crack detection algorithms in segmentation was evaluated using a confusion matrix, as detailed in Table 3 and represented graphically in Figure 10. This matrix is composed of two rows and two columns that display the pixel-wise classification results on the test dataset, where the rows indicate the actual classes and the columns denote the predicted classes. Specifically, TP signifies crack pixels that are correctly classified as cracks, TN indicates non-crack pixels accurately recognized as non-cracks, FP refers to non-crack pixels incorrectly predicted as cracks, and FN represents crack pixels that the model fails to identify, misclassifying them as non-cracks.

Table 3. Confusion matrix for pixel-wise crack segmentation.

	Predicted Non-Crack	Predicted Crack
Actual non-crack	TN	FP
Actual crack	FN	TP

To evaluate the performance of the model, four metrics were used:

- Precision—measures the proportion of true positives out of all predicted positives,
- Recall (or Sensitivity)—measures the proportion of true positives out of all actual positives,
- Dice Score Coefficient (DSC) or F1 score is the harmonic mean of precision and recall, balancing both metrics,
- Mean Intersection over Union (mIoU) is a common metric in semantic segmentation tasks, measuring the overlap between predicted and ground truth labels.

These metrics provide a comprehensive evaluation of the model's ability to accurately segment concrete cracks.

$$Precision = \frac{TP}{TP + FP}, \tag{2}$$

$$Recall = \frac{TP}{TP + FN}, \tag{3}$$

$$DSC = F1 = \frac{2|A \cap B|}{|A| + |B|} = \frac{2 \times Precision \times Recall}{Precision + Recall} = \frac{2TP}{2TP + FP + FN}, \tag{4}$$

$$IoU = \frac{|A \cap B|}{|A \cup B|}. \tag{5}$$

where:

A—original mask,

B—predicted mask.

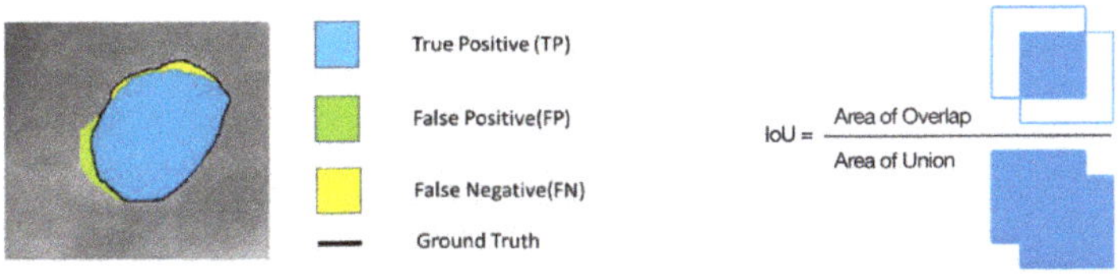

Figure 10. Graphical representation of: confusion matrix (**left**) and Equation (5) (**right**).

3.6. Experimental Settings

The models were trained using the supervised learning approach and the cost function is defined as the Dice loss introduced in [63], using the Dice coefficient; see the Equation (4):

$$Dice\ loss = 1 - DSC = 1 - \frac{2|A \cap B|}{|A| + |B|}. \tag{6}$$

The implementation of the models was based on the Keras-Unet-Collection [64]. The experiments were carried out using Google Colab Pro services [65]. The specification description of the Colab environment applied consists of GPU (NVIDIA Tesla A100 Tensor Core), CPU (Intel® Xeon® E5-2630) and RAM (27.3 GB). This advanced hardware set-up facilitated effective experimentation with different deep learning techniques for crack segmentation. To improve the crack segmentation performance of the model, we applied a series of targeted training methodologies. Initially, we used ImageNet for pretraining to initialize the base model parameters, offering general-purpose features as an initial foundation. During simulation, we adjusted the weights in both the convolutional and up-sampling layers, scaling them to enhance the learning of distinctive crack features.

In Table 4, the total number of parameters, the number of trainable parameters and the training time of the three networks is shown. The TransUNet model has substantially more parameters and a longer time than the other two models.

Table 4. Number of parameters and training time for each model.

Model Name	# of Parameters [$\times 10^6$]	# of Trainable Parameters [$\times 10^6$]	Training Time [min]
U-Net	28.0	13.3	10
Attention U-Net	28.4	13.7	10
TransUNet	146.0	131.4	40

4. Comparison of Three U-Net Based Models for Semantic Segmentation of Concrete Surface Cracks

After training, the models were used to classify each pixel in the crack image of the concrete surface as cracked or non-cracked. For example, Figure 11 presents the segmentation results for the trained U-Net model. The leftmost image shows the original grayscale concrete surface with a visible crack. The middle image presents the prediction results of the U-Net model. It uses a heatmap style where red indicates higher probabilities that pixels belong to a crack, transitioning to blue for lower probabilities. This visualization clearly shows how the model has identified the crack path with varying confidence levels along its length. The rightmost image is the ground truth (labeled data) used to train the model. It marks the actual crack with a solid red color.

The middle image shows that the U-Net model is quite effective in detecting the central, continuous parts of the crack, which are highlighted in bright red. There are areas where the model confidence fades (shown by the yellow to blue gradient), indicating either weaker detection or actual variations in the crack visibility or width.

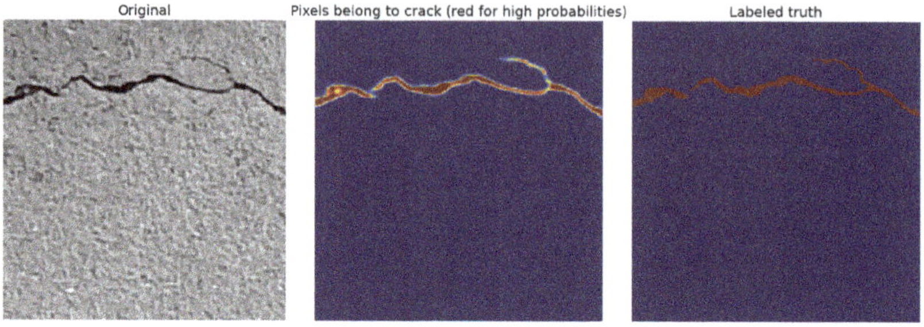

Figure 11. Example of semantic segmentation results for the trained U-Net model.

In Table 5 and in Figure 12, the performance of the three networks is quantitatively compared using four metrics.

Table 5. Comparison of the performance of the three models on test data using evaluation metrics.

Model	Precision [%]	Recall [%]	F1 [%]	mIoU [%]
U-Net	86.23	86.94	86.58	90.64
Attention U-Net	85.76	86.10	85.93	90.57
TransUNet	86.18	88.08	87.12	86.03

In this comparison, Attention U-Net has the highest precision (84.89%), U-Net follows closely with 84.45% and TransUNet has the lowest precision at 83.40%. This indicates that Attention U-Net performs slightly better in reducing false positives compared to the other models, while TransUNet sacrifices some precision, potentially to improve other metrics. In the case of the recall metric, TransUNet stands out with the highest recall (90.92%), outperforming the other models by a significant margin. Attention U-Net and U-Net achieve recall values of 87.71% and 87.16%, respectively.This shows that TransUNet excels at identifying true positives, making it more effective when missing fewer relevant predictions is critical. In this comparison, using the F1 score, TransUNet performs the best with an F1 score of 86.75%, indicating a good balance between precision and recall. Attention U-Net comes second with an F1 score of 85.82%, while U-Net has a slightly lower F1 score of 85.33%. This suggests that while the precision for TransUNet is slightly lower, its high recall helps to maintain a strong overall F1 score.

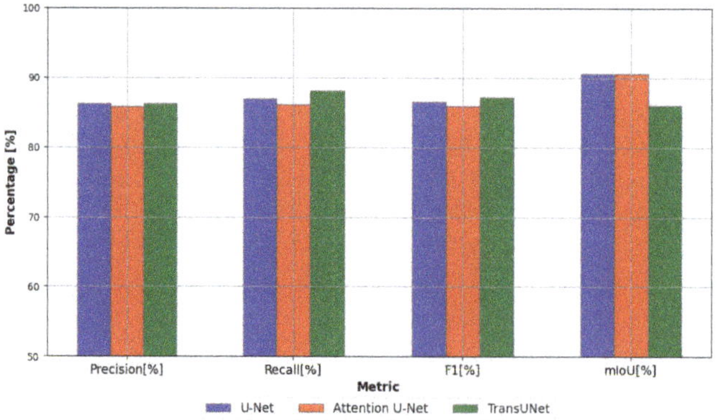

Figure 12. Comparison of model performance metrics.

In Figure 13, the performance of the three networks is quantitatively compared using confusion matrices.

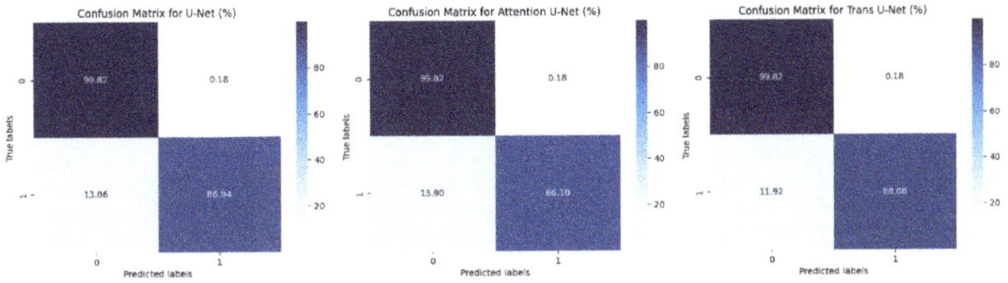

Figure 13. Comparison of confusion matrices.

In the case of the Mean Intersection Over Union (mIoU) metric, TransUNet achieves the highest mIoU (88.21%), indicating superior performance in terms of overall accuracy and overlap with ground truth segmentation. Attention U-Net (87.59%) and U-Net (87.35%) have similar performances, trailing just behind TransUNet.

In Figure 14, the performance of the three networks is quantitatively compared using binary IoU metrics.

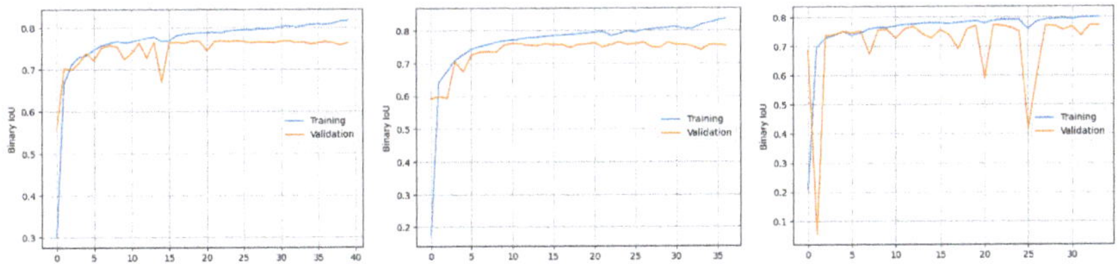

Figure 14. Comparison of binary IoU metrics.

In conclusion, TransUNet consistently performs the best in most metrics, especially in terms of recall, F1 score, and mIoU. This indicates that TransUNet is a well-balanced model, particularly strong in identifying true positives and providing accurate segmentations. Attention U-Net outperforms U-Net in every metric, particularly in precision, making it a solid improvement over the basic U-Net architecture. U-Net, while the original model, still performs competitively, but is slightly outclassed by its attention-based variant and TransUNet. These insights from Table 5 and the bar graph in Figure 12 indicate that TransUNet is probably the most suitable model for this specific semantic segmentation task, especially when recall and mIoU are prioritized.

In Figure 15 a comparison of the semantic segmentation of the cracks by the three models (U-Net, Attention U-Net, and TransUNet) is presented. In the first column, the selected images that contain cracks are presented. In the second column, the annotated ground-truth masks are displayed. The segmentation results for three networks are shown in columns 3–5, respectively. The segmentation results presented show that trained models are capable of predicting crack pixels with satisfactory accuracy, in a wide width range, from very thin cracks to very thick cracks. In the set of images presented, the continuousness of the cracks is also well maintained. Despite satisfactory predictions, there are few images in which the models fail to correctly classify some pixels as cracks. It is especially visible for the fourth image and the U-Net, where there are some pixels segmented as non-cracked pixels.

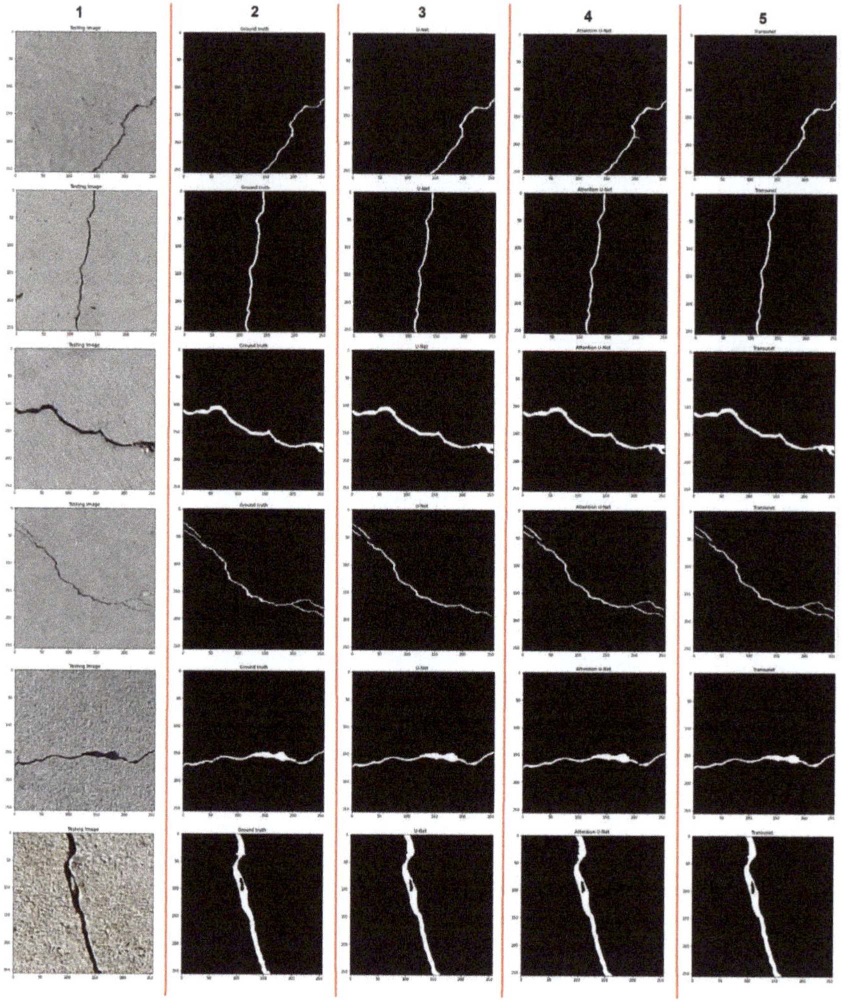

Figure 15. Examples of concrete cracks segmented by 3 models. Column 1: input images with cracks, column 2: annotated ground truth masks, column 3: predicted crack masks by U-Net, column 4: predicted crack masks by Attention U-Net, column 5: predicted crack masks by TransUNet, taken from [62].

There are also cases where the model performs better than the original mask. In Figure 16, it can be seen that the red and green marked elements do not appear in the input mask (the second image on the left); however, the result of the Attention U-Net and TransUNet model classified these areas as cracked. During the evaluation of the model, these areas were considered incorrectly predicted, but by manually analyzing the images, it can be determined that these areas are correct and can be considered cracked.

Figure 17 shows the contours predicted by each model that are more precise than the ground truth mask. Figure 18 shows the fillings predicted by each model. When analyzing the contours and fillings, it is possible to see the inaccuracy of the masks. This is possible because the masks come from the manual labeling of the crack images.

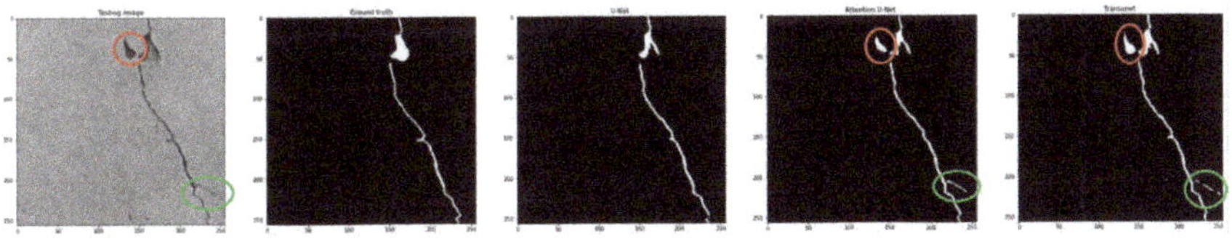

Figure 16. Another comparison of segmentation of concrete crack by 3 models. (**image 1**): input image with crack, (**image 2**): annotated ground truth mask (the red and green marked elements do not appear in the input mask), (**image 3**): predicted crack mask by U-Net, (**image 4**): predicted crack mask by Attention U-Net, (**image 5**): predicted crack mask by TransUNet, taken from [62].

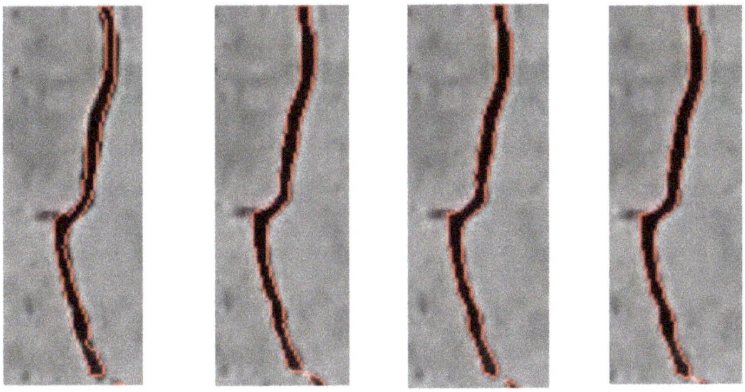

Figure 17. Ground truth mask contour (left) and contours predicted by each model (U-Net, Attention U-Net and TransUNet), taken from [62].

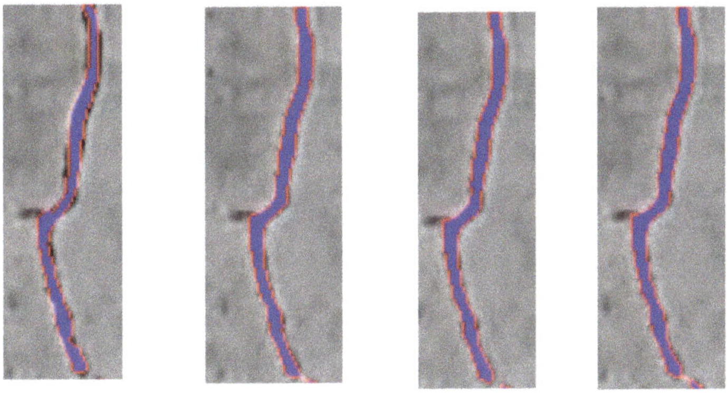

Figure 18. Ground truth mask filling (left) and fillings predicted by each model (U-Net, Attention U-Net and TransUNet), taken from [62].

Figure 19 shows the classification of pixels for each model in comparison to the ground-truth mask. The green color represents the common pixels for the ground truth mask and the segmentation results for each model. The red color illustrates pixels predicted by the models as cracked that are not in the ground-truth mask. The blue color represents pixels present in the ground truth mask that are absent in the predicted mask.

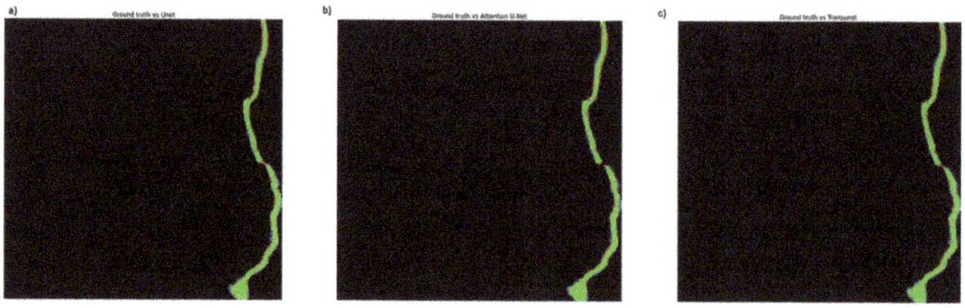

Figure 19. Classification of pixels for each model in comparison to the ground-truth mask: U-Net (**a**), Attention U-Net (**b**) and TransUNet (**c**), taken from [62].

TransUNet outperforms others models by combining the advantages of the Transformer architecture with the proven effectiveness of the U-Net framework. The Transformer component of TransUNet allows the model to capture global contextual information, which is crucial for understanding the broader area in which cracks are situated. TransUNet's ability to integrate features at different scales using both Transformer and convolutional layers results in richer feature representations. The combination of global and local processing in TransUNet makes it robust against variations in image quality, lighting conditions, and crack characteristics, increasing its utility in practical applications where such variability is common.

While TransUNet shows promising results, there is room for improvement, especially in terms of computational efficiency and real-time processing capabilities. Future work could focus on optimizing the model's architecture to reduce resource consumption without compromising performance. Furthermore, testing the model in a wider range of real-world scenarios, including outdoor environments with varying weather conditions, will be crucial to further validate its effectiveness and readiness for deployment in structural health monitoring applications.

While TransUNet demonstrates superior performance in crack segmentation, its deployment in real-world scenarios requires a critical assessment of its computational demands and scalability, especially in relation to larger datasets and real-time applications. The model's reliance on deep Transformer layers requires substantial memory and processing power, which can be a limiting factor in environments lacking specialized hardware. Due to its extensive network architecture, training TransUNet on large datasets involves longer durations, potentially making the model less feasible for projects with tight development cycles. As datasets grow in size and complexity, the computational resources required to train TransUNet can become prohibitively expensive. This is particularly challenging for organizations without access to high-performance computing facilities. For real-time crack detection and monitoring applications, the high inference time of TransUNet could be a bottleneck. Real-time processing demands quick turnaround times, which may not be achievable with the current model architecture without significant optimizations.

To address these challenges and improve the practical utility of TransUNet, developing simplified versions of TransUNet, which maintain a balance between performance and computational efficiency, could help reduce the model resource requirements. Leveraging more efficient computational hardware such as FPGAs (Field-Programmable Gate Arrays) or specialized AI accelerators can improve the model's performance in real-time scenarios. Implementing a hybrid approach that uses lightweight models for initial crack detection in real-time and reserves TransUNet for detailed analysis in batch processing scenarios can optimize resource usage while maintaining high accuracy levels. Future research should focus on addressing these computational challenges to improve the adaptability and efficiency of TransUNet.

5. Assessment of Precast Crane Beams Using DIC Method and U-Net Based Model

5.1. Overview

The effectiveness of the presented approach for the assessment of the precast crane beams was verified by analyzing the images taken during the laboratory evaluation of the precast, post-tensioned crane runway beams. The beams were manufactured between 1962 and 1963 and were removed from the industrial building in which they were located. The girder with an I section had a height of 80 cm and was composed of two pre-cast segments and had a total span of 600 cm. The tests were carried out on the girders at the Cracow University of Technology (CUT) Research Laboratory for Building Materials and Structures. The CivEng Vision system, developed at CUT, was used for collection, recording and analyzing the images [54,66]. Figure 20 shows the test stand prepared for the three-point bending test presenting also the numbering of acquisition devices [4].

Figure 20. Experimental setup arranged for monitoring post-tensioned, precast crane runway beam during the three-point bending test employing vision system described in [54].

5.2. Results for the Three-Point Bending Test

Figure 21 shows two images of the lateral face of the beam taken during the course of the bending test with the vision system, presenting also the numbering of developing cracks.

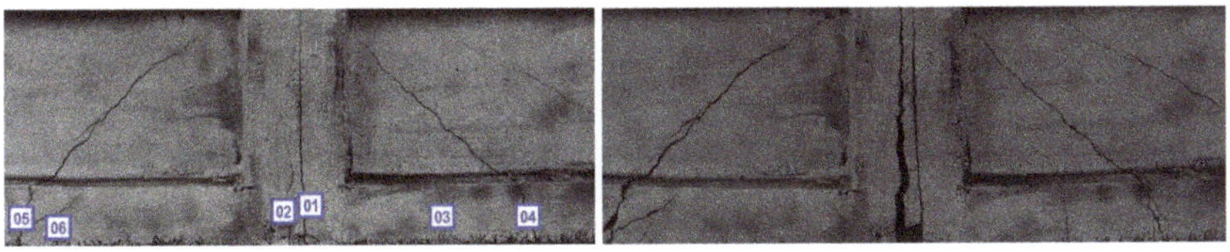

Figure 21. Two images of side surface of beam, captured using vision system, throughout three-point bending test, presenting also numbering of developing cracks (on the left).

Subsequently, the images were subjected to processing using the DIC method, and the deformation fields and width of the cracks were computed. Figure 22 illustrates the changes in the deformation fields along the X and Y axes as the beam undergoes loading.

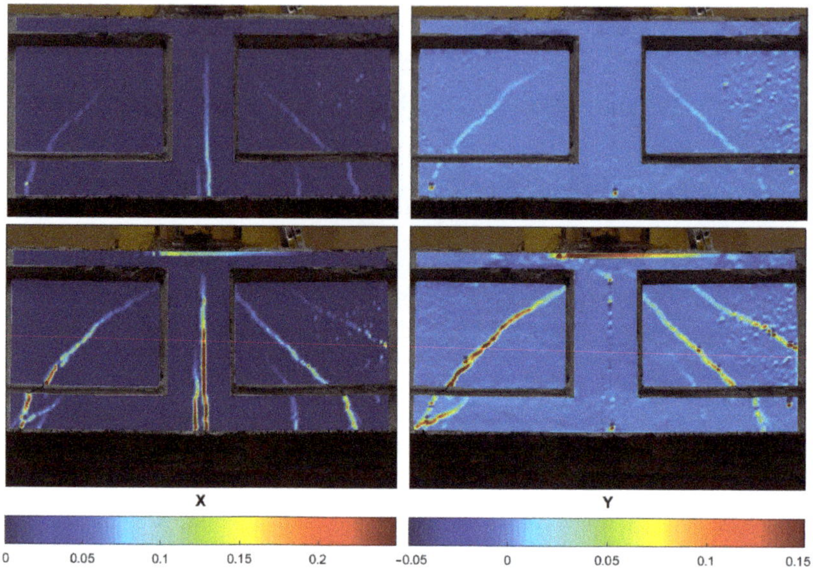

Figure 22. Deformation fields on beam surface in the X and Y directions during loading phase as measured by DIC, from [54].

As shown in Figure 23, the central vertical crack (01) developed precisely in the center of the beam. It formed at the beginning of the loading. The load then was decreased to zero and the crack was nearly fully closed. It reappeared in the course of the subsequent stage of the experiment. The images also were examined by the fine-tuned U-Net model to autonomously identify and segment the cracks present on the beam surface. Using a model fine-tuned with the SDNET dataset, image segmentation was performed. Figure 24 presents the result of the crack pixel density evaluation for 256 × 256 pixel frames.

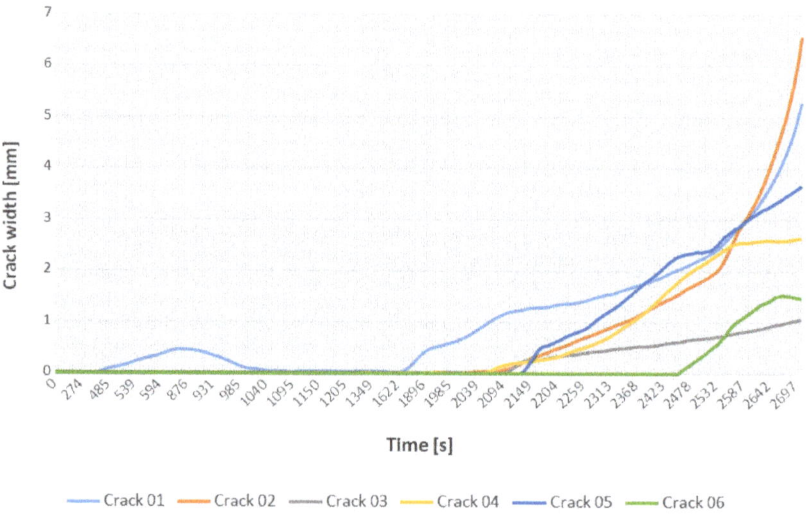

Figure 23. The cracks width development during the loading phase.

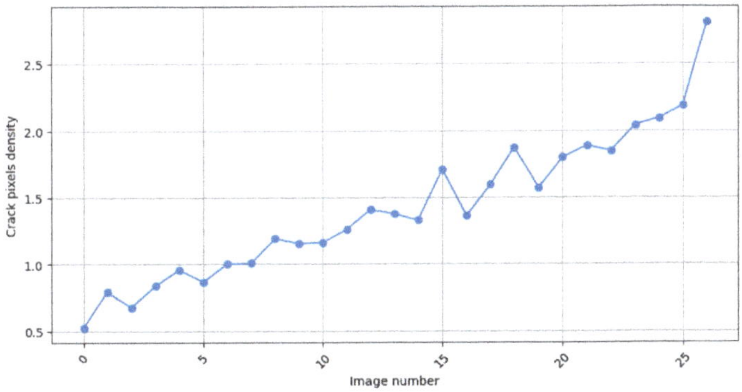

Figure 24. The cracks pixel density evaluation for 256 × 256 pixel frames.

5.3. *Results for the Shearing Test*

The left-hand side of Figure 25 shows an image of the beam lateral surface, captured using the vision system, in the process of shear test. The images were also processed using the DIC method and the strain distributions and width of the cracks were calculated. Figure 26 presents the evolution of strain fields in X and Y axis during the load imposition of the beam.

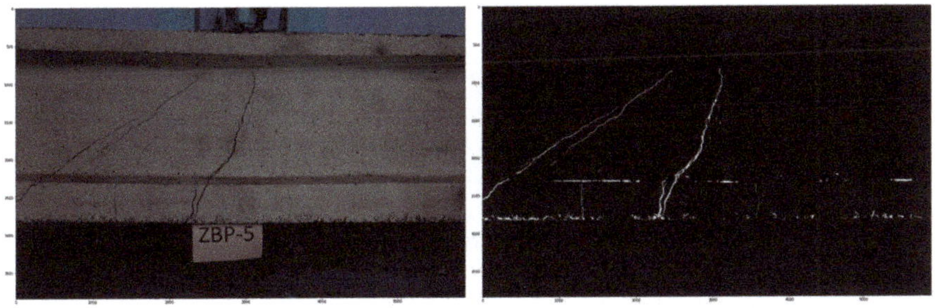

Figure 25. (**Left**): Image taken during laboratory experiments of the beam under shear loading, from [54]. (**Right**): Result of surface cracks segmentation after removing noise, from [62].

Ultimately, the images were examined by the trained U-Net model to autonomously identify and segment the cracks on the outer layer of the beams. The right-hand side of Figure 25 presents the result of image segmentation.

While this study focuses primarily on the assessment of pre-cast concrete crane beams using advanced imaging and segmentation models, the underlying methodologies, particularly the integration of U-Net variants with digital image correlation (DIC), hold significant potential for broader applications across various structural components and materials. The methodologies can be adapted to monitor the health of bridge decks and supports, where early detection of cracks and structural weaknesses is crucial to prevent failures. In urban settings, ensuring the integrity of the facades of high-rise buildings is essential for safety. The precision of U-Net-based models in detecting tiny cracks can help in routine inspection and maintenance of these structures.

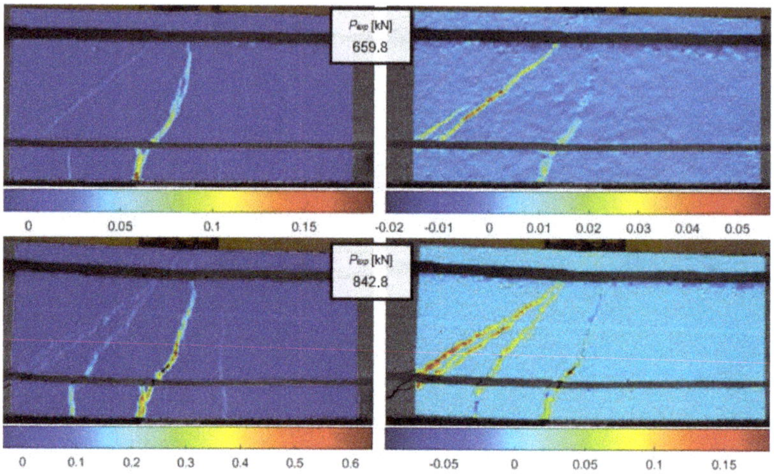

Figure 26. Strain full fields in X and Y axis on surface of beam under shear loading measured using DIC method, taken from [67].

6. Conclusions

This study has successfully demonstrated the integration of U-Net based deep learning models with digital image correlation (DIC) for the detailed and accurate assessment of concrete precast crane beams. The research not only proves the efficacy of these technologies in detecting and analyzing cracks but also underscores their potential to improve infrastructure monitoring and maintenance practices.

This study presented a system designed for automatic evaluation of crack evolution visible on the outer layer of a concrete test sample beam under laboratory test conditions. The methodology integrates the digital image correlation (DIC) technique to track the progression of crack width with a trained U-Net convolutional neural network for automated identification and segmentation of cracks.

An analysis and comparison of U-Net model-based architectures used in image segmentation were performed, a dataset was created that can be used in tasks related to the topic, and a method for measuring cracks in concrete elements was proposed. Using the described approach allows for the automation and acceleration of the inspection process of construction objects. Knowing the exact location of the device that takes the photo, it is possible to determine the actual size of the crack.

The applied computer vision system was assessed through static tests conducted in the CUT laboratory to establish a solution capable of performing this task. The images were acquired using DSLR cameras linked to an automatic actuator. The presented system demonstrated high precision in the evaluation of surface cracks. It can automatically detect the number of cracks and determine their locations. This process is repeatable for static analyses of other concrete elements. This study aims to present a computer vision system designed to quickly assess the visible crack width on the surface of concrete test specimens, as well as to automatically detect and segment cracks for monitoring purposes.

While the presented models demonstrate high efficacy in detecting and segmenting cracks, there are several limitations. The performance of the models is highly dependent on the quality and diversity of the training data. In scenarios where crack characteristics significantly differ from those in the training set, the accuracy of the model might be degraded. The models were primarily tested under controlled laboratory conditions. Real-world applicability might see reduced effectiveness due to varying environmental factors and operational conditions that were not simulated during the training phases. Advanced architectures, especially TransUNet, require significant computational resources, which could limit their deployment in real-time applications or on portable devices that are commonly used in field inspections.

The versatility of the applied models allows for adaptation to various types of infrastructure beyond crane beams, including bridges, roads, and building facades, suggesting a wide-reaching impact in the field of civil engineering. Automating structural evaluations reduces the need for extensive manual labor and mitigates the risks associated with manual inspections, leading to significant cost savings and safety improvements. The quantitative data provided by these models supports more informed decision-making in maintenance planning. This shift towards data-driven practices helps optimize resource allocation and scheduling in maintenance operations, ultimately extending the useful life of infrastructure assets.

Looking forward, the methodologies presented here set the groundwork for further research on integrating these technologies with other non-destructive testing methods to create even more comprehensive monitoring systems. Furthermore, expanding the training data sets to include a wider range of defects and environmental conditions will improve the accuracy and generalizability of the models, making them more effective tools in the global effort to ensure infrastructure safety and resilience.

The successful application of U-Net variants and digital image correlation (DIC) in the detection and analysis of cracks in concrete precast crane beams offers promising avenues for further research and development. To expand the scope and enhance the effectiveness of this methodology, several next steps can be proposed. Integrating additional data sources such as thermal imaging, acoustic emissions, and radar scans could enrich the input of the models and provide a more comprehensive view of structural health.This comprehensive technique is capable of detecting both visible fractures and underlying flaws, as well as material deterioration, which conventional imaging techniques may miss. Extending the crack detection methodologies to incorporate 3D imaging data, such as from LiDAR or structured light systems, would allow for a more detailed analysis of crack depth and volume, which are crucial for assessing the severity of damage. Developing 3D convolutional neural network models specifically tailored for this type of data could significantly improve the accuracy and utility of crack detection in complex structural components. Using transfer learning to adapt existing models to different types of materials and environmental conditions without the need for extensive retraining. This approach can accelerate the deployment of models in new areas and reduce computational costs.

These future directions seek to broaden the scope of present methods and inspire innovation in the area of structural health monitoring. By exploring these avenues, researchers can continue to push the limits of what is possible in infrastructure maintenance, ensuring that these critical systems remain safe and functional for the communities they serve.

Funding: This research obtained no third-party funding.

Data Availability Statement: The original contributions presented in this study are included in the article. Further inquiries can be directed to the corresponding author.

Acknowledgments: We acknowledge assistance given by Marcin Tekieli by providing results of experiments from his Doctoral thesis and Jakub Niemiec by providing results of experiments from his Master thesis.

Conflicts of Interest: The author states no conflict of interest.

References

1. di Marzo, M.; Tomassi, A.; Placidi, L. A Methodology for Structural Damage Detection Adding Masses. *Res. Nondestruct. Eval.* **2024**, *35*, 172–196. [CrossRef]
2. Aalami, B.O. *Post-Tensioning in Building Construction*; CRC Press: Boca Raton, FL, USA, 2023.
3. Berkowski, P.; Kosior-Kazberuk, M. Material and structural destruction of concrete elements in the industrial environment. *Procedia Eng.* **2017**, *172*, 96–103. [CrossRef]
4. Derkowski, W.; Walczak, R. Problem of condition assessment of precast, post-tensioned concrete crane beams in an extended period of use. In Proceedings of the Concrete—Innovations in Materials, Design and Structures: Proceedings of the *fib* Symposium 2019, Kraków, Poland, 27–29 May 2019; Derkowski, W., Ed.; International Federation for Structural Concrete: Lausanne, Switzerland, 2019; pp. 1507–1514.
5. Piątek, B.; Howiacki, T.; Kulpa, M.; Siwowski, T.; Sieńko, R.; Bednarski, Ł. Strain, crack, stress and shape diagnostics of new and existing post-tensioned structures through distributed fibre optic sensors. *Measurement* **2023**, *221*, 113480. [CrossRef]
6. Brownjohn, J.M. Structural health monitoring of civil infrastructure. *Philos. Trans. R. Soc. Math. Phys. Eng. Sci.* **2007**, *365*, 589–622. [CrossRef] [PubMed]
7. Spencer, B.F.J.; Hoskere, V.; Narazaki, Y. Advances in computer vision-based civil infrastructure inspection and monitoring. *Engineering* **2019**, *5*. [CrossRef]
8. Frangopol, D.M.; Soliman, M. Life-cycle of structural systems: Recent achievements and future directions. In *Structures and Infrastructure Systems*; Routledge: London, UK, 2019; pp. 46–65.
9. Koch, C.; Georgieva, K.; Kasireddy, V.; Akinci, B.; Fieguth, P. A review on computer vision based defect detection and condition assessment of concrete and asphalt civil infrastructure. *Adv. Eng. Inform.* **2015**, *29*, 196–210. [CrossRef]
10. Mohan, A.; Poobal, S. Crack detection using image processing: A critical review and analysis. *Alex. Eng. J.* **2018**, *57*, 787–798. [CrossRef]
11. Yuan, Q.; Shi, Y.; Li, M. A Review of Computer Vision-Based Crack Detection Methods in Civil Infrastructure: Progress and Challenges. *Remote. Sens.* **2024**, *16*, 2910. [CrossRef]
12. Garcia-Garcia, A.; Orts-Escolano, S.; Oprea, S.; Villena-Martinez, V.; Garcia-Rodriguez, J. A review on deep learning techniques applied to semantic segmentation. *arXiv* **2017**, arXiv:1704.06857.
13. Munawar, H.S.; Hammad, A.W.; Haddad, A.; Soares, C.A.P.; Waller, S.T. Image-based crack detection methods: A review. *Infrastructures* **2021**, *6*, 115. [CrossRef]
14. Azouz, Z.; Honarvar Shakibaei Asli, B.; Khan, M. Evolution of Crack Analysis in Structures Using Image Processing Technique: A Review. *Electronics* **2023**, *12*, 3862. [CrossRef]
15. Dorafshan, S.; Thomas, R.; Maguire, M. Comparison of deep convolutional neural networks and edge detectors for image-based crack detection in concrete. *Constr. Build. Mater.* **2018**, *186*, 1031–1045. [CrossRef]
16. Islam, M.M.; Kim, J.M. Vision-based autonomous crack detection of concrete structures using a fully convolutional encoder–decoder network. *Sensors* **2019**, *19*, 4251. [CrossRef]
17. Mousavi, M.; Bakhshi, A. Optimized U-shape convolutional neural network with a novel training strategy for segmentation of concrete cracks. *Struct. Health Monit.* **2023**, *22*, 1211–1224. [CrossRef]
18. Fan, Z.; Li, C.; Chen, Y.; Wei, J.; Loprencipe, G.; Chen, X.; Di Mascio, P. Automatic crack detection on road pavements using encoder-decoder architecture. *Materials* **2020**, *13*, 2960. [CrossRef] [PubMed]
19. Panella, F.; Lipani, A.; Boehm, J. Semantic segmentation of cracks: Data challenges and architecture. *Autom. Constr.* **2022**, *135*, 104110. [CrossRef]
20. Ai, D.; Jiang, G.; Lam, S.K.; He, P.; Li, C. Computer vision framework for crack detection of civil infrastructure—A review. *Eng. Appl. Artif. Intell.* **2023**, *117*, 105478. [CrossRef]
21. Dung, C.; Anh, L. Autonomous concrete crack detection using deep fully convolutional neural network. *Autom. Constr.* **2019**, *99*, 52–58. [CrossRef]
22. Yang, X.; Li, H.; Yu, Y.; Luo, X.; Huang, T.; Yang, X. Automatic pixel-level crack detection and measurement using fully convolutional network. *Comput.-Aided Civ. Infrastruct. Eng.* **2018**, *33*, 1090–1109. [CrossRef]
23. Pu, R.; Ren, G.; Li, H.; Jiang, W.; Zhang, J.; Qin, H. Autonomous Concrete Crack Semantic Segmentation Using Deep Fully Convolutional Encoder–Decoder Network in Concrete Structures Inspection. *Buildings* **2022**, *12*, 2019. [CrossRef]
24. Shen, Y.; Yu, Z.; Li, C.; Zhao, C.; Sun, Z. Automated Detection for Concrete Surface Cracks Based on Deeplabv3+ BDF. *Buildings* **2023**, *13*, 118. [CrossRef]
25. Razveeva, I.; Kozhakin, A.; Beskopylny, A.N.; Stel'makh, S.A.; Shcherban', E.M.; Artamonov, S.; Pembek, A.; Dingrodiya, H. Analysis of Geometric Characteristics of Cracks and Delamination in Aerated Concrete Products Using Convolutional Neural Networks. *Buildings* **2023**, *13*, 3014. [CrossRef]

26. Guo, P.; Meng, X.; Meng, W.; Bao, Y. Monitoring and automatic characterization of cracks in strain-hardening cementitious composite (SHCC) through intelligent interpretation of photos. *Compos. Part Eng.* **2022**, *242*, 110096. [CrossRef]
27. Xu, G.; Yue, Q.; Liu, X.; Chen, H. Investigation on the effect of data quality and quantity of concrete cracks on the performance of deep learning-based image segmentation. *Expert Syst. Appl.* **2023**, *237*, 121686. [CrossRef]
28. Forest, F.; Porta, H.; Tuia, D.; Fink, O. From Classification to Segmentation with Explainable AI: A Study on Crack Detection and Growth Monitoring. *arXiv* **2023**, arXiv:2309.11267. [CrossRef]
29. Ahmadi, M.; Lonbar, A.G.; Sharifi, A.; Beris, A.T.; Nouri, M.; Javidi, A.S. Application of segment anything model for civil infrastructure defect assessment. *arXiv* **2023**, arXiv:2304.12600.
30. König, J.; Jenkins, M.; Mannion, M.; Barrie, P.; Morison, G. What's Cracking? A Review and Analysis of Deep Learning Methods for Structural Crack Segmentation, Detection and Quantification. *arXiv* **2022**, arXiv:2202.03714.
31. Feng, C.; Liu, M.Y.; Kao, C.C.; Lee, T.Y. Deep active learning for civil infrastructure defect detection and classification. In *Computing in Civil Engineering*; Mitsubishi Electric Research Laboratories: Cambridge, MA, USA, 2017; pp. 298–306.
32. Sari, Y.; Prakoso, P.B.; Baskara, A.R. Road crack detection using support vector machine (SVM) and OTSU algorithm. In Proceedings of the 2019 6th International Conference on Electric Vehicular Technology (ICEVT), Bali, Indonesia, 18–21 November 2019; IEEE: New York, NY, USA, 2019; pp. 349–354.
33. Zhang, L.; Yang, F.; Zhang, Y.D.; Zhu, Y.J. Road crack detection using deep convolutional neural network. In Proceedings of the 2016 IEEE International Conference on Image Processing (ICIP), Phoenix, AZ, USA, 25–28 September 2016; IEEE: New York, NY, USA, 2016; pp. 3708–3712.
34. Zhou, S.; Canchila, C.; Song, W. Deep learning-based crack segmentation for civil infrastructure: Data types, architectures, and benchmarked performance. *Autom. Constr.* **2023**, *146*, 104678. [CrossRef]
35. Deng, J.; Singh, A.; Zhou, Y.; Lu, Y.; Lee, V.C.S. Review on computer vision-based crack detection and quantification methodologies for civil structures. *Constr. Build. Mater.* **2022**, *356*, 129238. [CrossRef]
36. Ronneberger, O.; Fischer, P.; Brox, T. U-net: Convolutional networks for biomedical image segmentation. In Proceedings of the International Conference on Medical Image Computing and Computer-Assisted Intervention, Munich, Germany, 5–9 October 2015; Springer: Berlin/Heidelberg, Germany, 2015; pp. 234–241.
37. Long, J.; Shelhamer, E.; Darrell, T. Fully convolutional networks for semantic segmentation. In Proceedings of the IEEE conference on Computer Vision and Pattern Recognition, Boston, MA, USA, 7–12 June 2015; pp. 3431–3440.
38. Badrinarayanan, V.; Kendall, A.; Cipolla, R. Segnet: A deep convolutional encoder-decoder architecture for image segmentation. *IEEE Trans. Pattern Anal. Mach. Intell.* **2017**, *39*, 2481–2495. [CrossRef]
39. Chen, L.C.; Papandreou, G.; Kokkinos, I.; Murphy, K.; Yuille, A.L. Deeplab: Semantic image segmentation with deep convolutional nets, atrous convolution, and fully connected crfs. *IEEE Trans. Pattern Anal. Mach. Intell.* **2017**, *40*, 834–848. [CrossRef] [PubMed]
40. Rao, A.S.; Nguyen, T.; Le, S.T.; Palaniswami, M.; Ngo, T. Attention recurrent residual U-Net for predicting pixel-level crack widths in concrete surfaces. *Struct. Health Monit.* **2022**, *21*, 2732–2749. [CrossRef]
41. Hang, J.; Wu, Y.; Li, Y.; Lai, T.; Zhang, J.; Li, Y. A deep learning semantic segmentation network with attention mechanism for concrete crack detection. *Struct. Health Monit.* **2023**, *22*, 3006–3026. [CrossRef]
42. Shamsabadi, E.A.; Xu, C.; Rao, A.S.; Nguyen, T.; Ngo, T.; Dias-da Costa, D. Vision transformer-based autonomous crack detection on asphalt and concrete surfaces. *Autom. Constr.* **2022**, *140*, 104316. [CrossRef]
43. Wang, W.; Su, C. Automatic concrete crack segmentation model based on transformer. *Autom. Constr.* **2022**, *139*, 104275. [CrossRef]
44. Xiang, C.; Guo, J.; Cao, R.; Deng, L. A crack-segmentation algorithm fusing transformers and convolutional neural networks for complex detection scenarios. *Autom. Constr.* **2023**, *152*, 104894. [CrossRef]
45. Shi, Z.; Jin, N.; Chen, D.; Ai, D. A comparison study of semantic segmentation networks for crack detection in construction materials. *Constr. Build. Mater.* **2024**, *414*, 134950. [CrossRef]
46. Li, H.; Wang, W.; Wang, M.; Li, L.; Vimlund, V. A review of deep learning methods for pixel-level crack detection. *J. Traffic Transp. Eng. (Engl. Ed.)* **2022**, *9*, 945–968. [CrossRef]
47. Słoński, M. A comparison of deep convolutional neural networks for image-based detection of concrete surface cracks. *Comput. Assist. Methods Eng. Sci.* **2019**, *26*, 105–112.
48. Sutton, M.; Orteu, J.J.; Shreier, H. *Image Correlation for Shape, Motion and Deformation Measurements*; Springer: Berlin/Heidelberg, Germany, 2009.
49. Pan, B. Digital image correlation for surface deformation measurement: Historical developments, recent advances and future goals. *Meas. Sci. Technol.* **2018**, *29*, 082001. [CrossRef]
50. Zhao, J.; Sang, Y.; Duan, F. The state of the art of two-dimensional digital image correlation computational method. *Eng. Rep.* **2019**, *1*, e12038. [CrossRef]

51. Peters, W.; Ranson, W. Digital imaging techniques in experimental stress analysis. *Opt. Eng.* **1982**, *21*, 427–431. [CrossRef]
52. Sutton, M.; Wolters, W.; Peters, W.; Ranson, W.; McNeill, S. Determination of displacements using an improved digital correlation method. *Image Vis. Comput.* **1983**, *1*, 133–139. [CrossRef]
53. Olufsen, S.N.; Andersen, M.E.; Fagerholt, E. μDIC: An open-source toolkit for digital image correlation. *SoftwareX* **2020**, *11*, 100391. [CrossRef]
54. Tekieli, M. A Vision-Based Measurement System for the Analysis of Structural Element Deformation Fields. Ph.D. Thesis, Cracow University of Technology, Cracow, Poland, 2019.
55. Planche, B.; Andres, E. *Hands-On Computer Vision with TensorFlow 2: Leverage Deep Learning to Create Powerful Image Processing Apps with TensorFlow 2.0 and Keras*; Packt Publishing Ltd.: Birmingham, UK, 2019.
56. Bahdanau, D.; Cho, K.; Bengio, Y. Neural machine translation by jointly learning to align and translate. *arXiv* **2014**, arXiv:1409.0473.
57. Oktay, O.; Schlemper, J.; Folgoc, L.L.; Lee, M.; Heinrich, M.; Misawa, K.; Mori, K.; McDonagh, S.; Hammerla, N.Y.; Kainz, B.; et al. Attention U-Net: Learning where to look for the pancreas. *arXiv* **2018**, arXiv:1804.03999.
58. Vaswani, A. Attention is all you need. *arXiv* **2017**, arXiv:1706.03762.
59. Chen, J.; Lu, Y.; Yu, Q.; Luo, X.; Adeli, E.; Wang, Y.; Lu, L.; Yuille, A.L.; Zhou, Y. Transunet: Transformers make strong encoders for medical image segmentation. *arXiv* **2021**, arXiv:2102.04306.
60. Dorafshan, S.; Thomas, R.; Maguire, M. SDNET2018: An annotated image dataset for non-contact concrete crack detection using deep convolutional neural networks. *Data Brief* **2018**, *21*, 1664–1668. [CrossRef]
61. Team, S. Segments.ai—The Training Data Platform for Computer Vision Engineers. 2022. Available online: https://segments.ai (accessed on 22 January 2024).
62. Niemiec, J. Deep Neural Networks in Structural Health Monitoring. Master's Thesis, Cracow University of Technology, Cracow, Poland, 2022. (In Polish)
63. Sudre, C.H.; Li, W.; Vercauteren, T.; Ourselin, S.; Jorge Cardoso, M. Generalised dice overlap as a deep learning loss function for highly unbalanced segmentations. In Proceedings of the Deep Learning in Medical Image Analysis and Multimodal Learning for Clinical Decision Support: Third International Workshop, DLMIA 2017, and 7th International Workshop, ML-CDS 2017, Held in Conjunction with MICCAI 2017, Québec City, QC, Canada, 14 September 2017; Proceedings 3; Springer: Berlin/Heidelberg, Germany, 2017; pp. 240–248.
64. Sha, Y. Keras-Unet-Collection. 2021. Available online: https://github.com/yingkaisha/keras-unet-collection (accessed on 14 April 2024). [CrossRef]
65. Google. Google Colaboratory. 2024. Available online: https://colab.research.google.com/ (accessed on 24 May 2024).
66. Tekieli, M.; Słoński, M. Digital image correlation and Bayesian filtering in inverse analysis of structures. In *Recent Advances in Civil Engineering: Computational Methods*; Cecot, W., Ed.; Publishing House of Cracow University of Technology: Kraków, Poland, 2015; pp. 111–124.
67. Walczak, R. Shear Capacity of Concrete Crane Beams in an Extended Period of Service in Uncertain Conditions of Prestressing Cables Anchorage. Ph.D. Thesis, Cracow University of Technology, Faculty of Civil Engineering, Cracow, Poland, 2022. (In Polish)

materials

MDPI

Article

Analysis of the Thermal Properties of Soft Silica Limestone Walls of Traditional Buildings in Central Poland

Aleksandra Gorączko *, Paula Szczepaniak and Marcin Gorączko

Faculty of Civil and Environmental Engineering and Architecture, Bydgoszcz University of Science and Technology, Kaliskiego 7, 85-796 Bydgoszcz, Poland; paula.szczepaniak@pbs.edu.pl (P.S.); gorgon@pbs.edu.pl (M.G.)

* Correspondence: agora@pbs.edu.pl

Abstract: The challenge of thermally upgrading traditional stone masonry buildings is addressed through the analysis of a representative example typical of regional rural architecture in central Poland, constructed using soft silica limestone and clay mortar. These buildings, which form an important part of the local cultural heritage, are increasingly becoming the subject of interdisciplinary research and conservation initiatives. This study presents a detailed characterization of the materials and architectural features specific to this building typology. Thermal transmittance calculations were performed and analyzed, with the use of THERM 7.6.1.0 software enabling precise modeling of the wall's heterogeneous structure. The physical and thermal properties of natural materials—particularly soft silica limestone and clay—were taken into account. The analysis included evaluation of the heat transfer coefficient, temperature distribution, and heat flux density for a reference wall model, as well as for variants with both internal and external insulation layers. The study explores thermal comfort and energy performance within the broader context of preserving and reusing historic rural buildings. Furthermore, the findings are discussed in relation to current European energy efficiency regulations and heritage protection frameworks. The scientific value of this work lies in its context-specific, material-sensitive methodology and in providing practical insight into balancing energy retrofitting with architectural conservation.

Keywords: stone walls; insulation; soft limestone; opoka; regional architecture

1. Introduction

The pursuit of ensuring thermal comfort and reducing energy consumption for heating and cooling have recently led to a growing interest in the thermal modernization of vernacular, often historic, residential buildings constructed using outdated technical standards [1]. Improving thermal comfort through retrofitting is often essential to maintaining these buildings as habitable spaces. Such actions are supported or even required by various international regulations that emphasize energy efficiency and adaptation to current and anticipated climate changes [2].

The challenge of thermal modernization applies, among others, to stone masonry buildings. Stone was historically a popular construction material due to its local availability, durability, and aesthetic qualities, such as color, texture, and finish. Stone buildings are found across different climatic zones worldwide, and their extensive use within specific regions often led to the development of unique local craftsmanship and architectural styles. Traces of these traditions can still be seen today. Such structures are studied from historical, architectural, and sociological perspectives. Some studies address traditional masonry techniques and the principles of preserving authenticity during restoration work [3,4]. Others focus on the physical and mechanical properties of natural stone typical for specific regions, which are critical for understanding both structural and thermal behavior [5,6]. A separate group of contributions highlights the cultural and environmental value of stone architecture and the importance of documenting and maintaining regional identity through material and architectural expression [7–9]. Broader comparisons with contemporary construction approaches demonstrate the energy-relevant qualities of vernacular architecture in various climates [10]. Further research [11] investigates how principles of vernacular architecture can be applied in contemporary rural development, highlighting the balance between cultural continuity and functional modernization Vernacular stone architecture is also increasingly recognized for its environmental performance and is frequently protected as an important component of cultural heritage [12,13].

However, traditional stone buildings rarely meet modern energy efficiency standards, which makes a detailed analysis of insulation methods necessary for each case. The need for thermal modernization depends largely on the climatic zone and the type of stone and construction used. In some milder climates, hygrothermal analyses show that traditional solutions provide sufficient insulation and can be preserved in their original state or even replicated in modern construction [14,15]. Advances in computational methods, which consider the thermal inertia of massive structures [16,17], also support these findings. Still, most traditionally constructed buildings do not meet today's standards, leading to numerous studies exploring retrofitting possibilities [18,19].

A common thermal modernization approach involves applying external insulation, which often covers the original stone façade. This creates a conflict between achieving technical standards and preserving the building's original appearance. Additionally, regional and national regulations, especially in countries with a large number of historic buildings, often prohibit energy-efficiency measures that alter a building's visual identity, proportions, or character [20]. Special provisions regarding the need to ensure the protection of cultural values during the thermal modernization of traditional buildings are also included in European-level documents [21,22]. In such cases, internal insulation often becomes the only viable solution to improve the energy performance. However, this method poses significant challenges, as it alters the hygrothermal behavior of the walls and changes boundary conditions [23–26], particularly in colder climates [26,27]. One example of such a change is the shift of the temperature gradient within the wall, which causes the dew point to move closer to the interior. As a result, the internal face of the original wall becomes colder and may remain within a temperature and humidity range that promotes moisture accumulation. This can lead to interstitial condensation at the insulation–wall interface, thereby increasing the risk of material degradation or mold growth. The use of vapor-permeable insulation materials may help reduce this risk by allowing gradual moisture transfer and drying, but it does not eliminate the need for careful hygrothermal assessment and proper indoor humidity control.

Analyzing the hygrothermal performance of stone masonry in vernacular and historic buildings presents many methodological difficulties. Case studies and meta-analyses [28–30] especially highlight challenges such as obtaining reliable material parameters, including density, composition, moisture content, and thermal properties. These issues are more pronounced for natural stone than for brick masonry. Moreover, accurately modeling wall structures is complex due to variations in thickness, material composition, and often heterogeneous structures. Additionally older buildings may also exhibit damage, material loss, or moisture issues. Standard calculations often assume monolithic walls, but the proportions of stone, mortar, and voids significantly impact results. Direct measurements frequently show discrepancies compared to analytical calculations [28,31,32].

Addressing the thermal performance of traditional stone masonry buildings presents a considerable scientific challenge due to the high regional variability in materials, construction techniques, and wall typologies. This diversity often necessitates localized, case-specific analyses, limiting the applicability of general models or assumptions. The present study contributes to bridging this knowledge gap by focusing on a representative example from central Poland, where soft silica limestone and clay were historically used in vernacular construction. These materials, particularly the lightweight and porous limestone, are difficult to characterize due to their heterogeneity and limited availability in modern laboratory datasets. By applying detailed hygrothermal simulations to this under-researched masonry system, the study expands knowledge of its thermal behavior under realistic boundary conditions.

The scientific and novel value of this work lies in its integrative and context-specific approach—combining in situ material characterization, simulation, and a comparative evaluation of retrofit strategies—to derive findings relevant for conservation, sustainability, and energy efficiency in heritage buildings. In particular, this study provides a practical comparison of internal and external insulation solutions, considering the architectural and physical limitations of small-scale traditional structures. The ultimate objective is to assess the thermal behavior of traditional stone masonry under realistic conditions and to evaluate the effectiveness and applicability of targeted energy retrofit measures, with implications for sustainability, building physics, and heritage conservation

2. Materials and Methods

2.1. Characterization of Vernacular Buildings

This study concerns traditional stone masonry structures in central Poland, built using soft silica limestone sourced from local quarries (Rożniatów, Poddębice, Poland). These buildings, numbering in the tens of thousands, were constructed in rural areas and small towns of this region from the late 19th century until the 1970s (Figure 1).

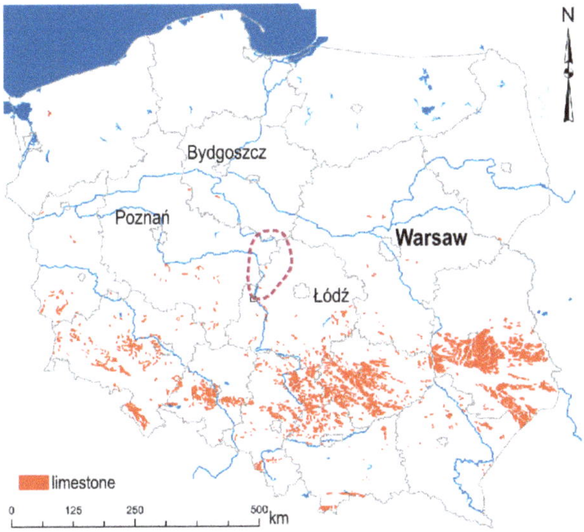

Figure 1. Map of Poland with limestone deposits and the distribution of described buildings (marked in the circle) in central Poland.

During that period, a distinctive building tradition emerged in the vicinity of the quarries, shaped by the technology of stone wall construction [33]. These structures are mostly small, single-story residential buildings with simple architectural forms, although more prominent and representative buildings, such as churches and manor houses, can also be found (Figure 2).

Figure 2. Examples of traditional buildings made of soft silica limestone in the analyzed area: (**a**–**c**) residential buildings; (**d**) farmstead.

Due to the widespread use of this construction technique, these buildings formed a characteristic architectural element of the region for many years. However, as a result of material degradation and incompatibility with modern technical and functional standards, this unique architecture is gradually disappearing. At the same time, it is increasingly recognized as an important cultural heritage and a source of inspiration [34]. Consequently, there is growing interest in a detailed assessment of their properties, including thermal performance.

2.2. Regional Method of Constructing Walls from Soft Silica Limestone

Traditional walls made of soft silica limestone were typically supported by a shallowly embedded and slightly raised foundation wall constructed from split fieldstones and leveled with a layer of ceramic brick. The stone material delivered from the quarry was manually processed on-site. The raw material was easy to work with, and individual stone elements could be shaped with varying degrees of precision depending on their placement within the wall. It was very rare for the walls to be intentionally designed for external plastering. Instead, the stone façade was meant to remain exposed, which is why the stone elements were generally processed with great care. The most meticulously worked elements, shaped by splitting and trimming, were those intended for the facing surface of the wall (Figure 3).

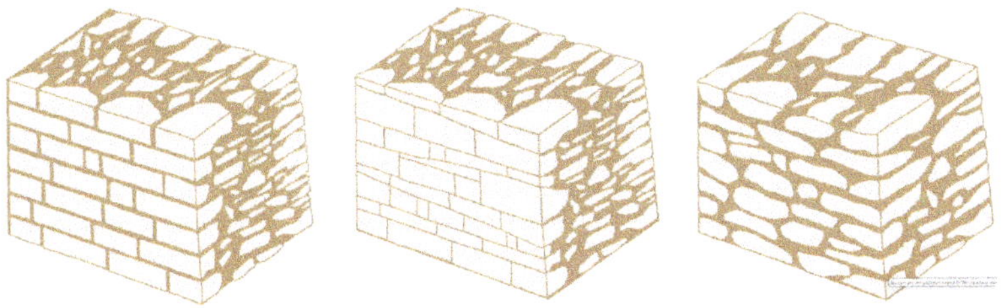

(**a**) (**b**) (**c**)

Figure 3. Typical layouts of traditional masonry walls made of soft silica limestone, representing the main types of masonry characteristic of the analyzed region: (**a**) regular wall (**b**) layered wall (**c**) wild wall.

The characteristic masonry texture is regular masonry (Figure 3a), a specific type of layered masonry (Figure 3b). Other solutions, such as mosaic or wild masonry (Figure 3c), were also used, though less frequently [35]. The layered and regular stone masonry, where all layers are horizontal and their height corresponds to that of a brick, represents the most recognizable architectural feature of buildings in this region. Ceramic bricks were used into the stone masonry at key structural points such as building corners, window and door frames, and cornice. The stone elements were bonded with clay mortar, made from natural clay extracted from the ground near the construction site. The joint width ranged from a few millimeters to 1 cm. The careful processing of stone elements primarily concerned the representative façades, whereas on the interior side, stone elements were laid in layers but not as rigorously as on the exterior.

The natural slaty structure of the rock in the deposit allowed stone elements to be incorporated into the masonry even without any processing. The goal was not to achieve a perfectly smooth wall surface; on the contrary, irregularities were deliberately left to enhance the adhesion of the internal clay plaster, which was applied in layers several centimeters thick. In cross-section, an infill layer was present between the external and internal layers of the wall (Figure 3). This core consisted of homogenized and plasticized natural clay mixed with smaller stone fragments and spalls, often waste material from processing the masonry elements. These fragments were arranged randomly or in a layered manner. The stability of the wall was ensured by the use of through-stones—long stone elements that alternately connected the outer and inner stone layers with the clay-stone infill.

According to historical building codes and guidelines, the typical and recommended thickness of soft silica limestone walls in residential buildings ranged from 50 to 60 cm [36]. Thicker walls were used for large-scale structures, such as barns and churches, whereas in farm buildings, the wall thickness was typically 40–50 cm.

2.3. *Physical and Thermal Properties of Traditional Materials*

2.3.1. Stone Properties

Silica limestone from Rożniatów is a sedimentary rock with a characteristic light creamy or light yellowish hue, composed mostly of calcium carbonate and organogenic silica [37]. The rock is known by various regional names such as opoka (Poland and Lithuania), opuka (Czech Republic), Pläner (Germany) or gaize (France). The composition is predominantly calcite, representing accounting for approximately 70%. Laboratory tests have confirmed that the mechanical properties of the limestone are entirely sufficient for constructing residential and farm buildings. Its strength ranges from 13 to 16 MPa [34]. The values of the basic physical properties determined for the rocks from the primary extraction site (Rożniatów) are provided in Table 1.

Table 1. Properties of soft silica limestone from Rożniatów—own research cited from [37,38].

Parameter	Unit	Mean Value	Min.–Max	Method
Apparent density (ρ_b)	[kg/m^3]	1420	1280–1550	EN 1936:2010 [39]
True density (ρ_r)	[kg/m^3]	2570	2520–2620	
Open porosity (n_e)	[%]	43.9	41.4–47.3	
Total porosity (n_t)	[%]	45.3	-	
Water absorption (A_b) at atmospheric pressure	[%]	27.6	21.5-30.7	
Porosity (n_c)	[%]	34.23	-	mercury intrusion porosimetry [38]

Soft silica limestone from Rożniatów is a rock with a relatively low apparent density and high total (n_t = 45.3%) and open porosity (n_e = 42.9%) [37]. Connected porosity, measured using mercury intrusion, yielded a value (n_c = 34.23%), which is higher than water absorption at atmospheric pressure (A_b = 27.6%) but lower than the total porosity and even lower than the open porosity. The most probable cause is the impenetrability of the finest pores to mercury intrusion (<5 nm). Additionally, large pores could not be measured (>200 μm) [38]. Thus, soft silica limestone from Rożniatów has a lower density than typical ceramic brick but a significantly higher open porosity.

In the current standard PN-EN ISO 10456:2009 [40], which provides thermal conductivity (λ) values, data is available only for very soft limestones with a density of 1600 kg/m^3. There is a lack of information for lighter and more porous rocks, such as tested soft silica limestone, which has a density of approximately 1420 kg/m^3. Therefore, archival literature data [41] was used, according to which the thermal conductivity of Rożniatów limestone is λ = 0.53 kcal/(m·h·°C), equivalent to 0.62 W/(m·K). Similar values for highly porous rocks are provided in the archival Polish standard related to thermal protection issues [42] (Table 2). This standard differentiates limestones properties based on their bulk density and moisture conditions. This classification aligns with studies showing that the thermal conductivity of porous materials increases significantly with rising moisture content [43,44].

Table 2. Thermal conductivity (λ) values of porous limestones in the Polish standard [42].

Material	Bulk Density	Thermal Conductivity (λ) [W/(m·K)]	
	[kg/m^3]	Moderately Humid Conditions [1]	Humid Conditions [2]
Porous limestone	1700	0.93	1.16
	1400	0.64	0.76

[1] Moderately humid conditions—rooms with air humidity up to 75% during winter. [2] Humid conditions—rooms with air humidity above 75% during winter.

2.3.2. Clay Mortar Properties

The clay material used as a masonry mortar, wall infill, and internal plaster is also of local origin. It is primarily composed of glacial clays, often (according to local interviews) lightened with the addition of sand and, in the case of internal plasters, sometimes with straw or sawdust. The granulometric composition [45] indicates a predominant content of sandy fractions and approximately 30% silt and clay fraction. And occasionally, a small amount of calcium carbonate (up to 5%) was present. The bulk density of clay mortars, according to our own research, is 1850 kg/m^3.

The thermal conductivity value for clay soils according to the standard [40] is λ = 1.5 W/(m·K). Significantly lower values of λ = 0.7 W/(m·K) for sandy clays with a density of 1800 kg/m^3 are provided by the standard [42], for both moderately humid and humid conditions. Meanwhile, in a more recent publication [46], the range of values for clay materials is given from 0.4 W/(m·K) (for lightweight clay plasters with fillers, e.g., microfibers) to 0.8 W/(m·K). Finally, the archival Polish building standard [47] was used, which provides the conductivity coefficient for clay masses used in construction from raw clay, depending on the bulk density (Table 3).

Table 3. Thermal conductivity (λ) values of clay materials in the Polish standard [47].

Material	Bulk Density	Thermal Conductivity (λ)	
	[kg/m^3]	[Kcal/(m·h·°C)]	[W/(m·K)]
Heavy clay	1700–1800	0.70–0.80	0.81–0.93
Medium clay	1600–1700	0.40–0.70	0.46–0.81
Lightweight clay	1500–1600	0.20–0.40	0.23–0.46

2.4. Assumptions of the Model and Calculation Variants

The reliability of numerical analyses using modern tools depends on the correct adoption of appropriate input data. In the previous subsections, the internal structure of the wall was presented, along with explanations of the basis for selecting thermal conductivity coefficients for the wall's constituent materials. The lack of data on moisture-related parameters (diffusion resistance coefficient) for soft silica limestone led to the decision to perform only heat flow analyses. However, the undeniable influence of changes in the moisture content of masonry materials was accounted for only indirectly. When defining the calculation cases, likely moisture-change processes and the associated changes in thermal conductivity values were assumed.

The analyses were performed using THERM 7.6.1.0 software, which allows the calculation of two-dimensional steady-state heat flow with a non-orthogonal mesh grid. This enables the modeling of the complex wall structure, distinguishing between its components—stone masonry elements, mortar, and infill.

A geometric model of the wall was created, representing a typical masonry of the analyzed structures. The model assumes a wall thickness of 50 cm and dimensions reflecting the average real elements in the stone masonry while maintaining the layered structure and the proportions between the stone material and clay (Figure 4).

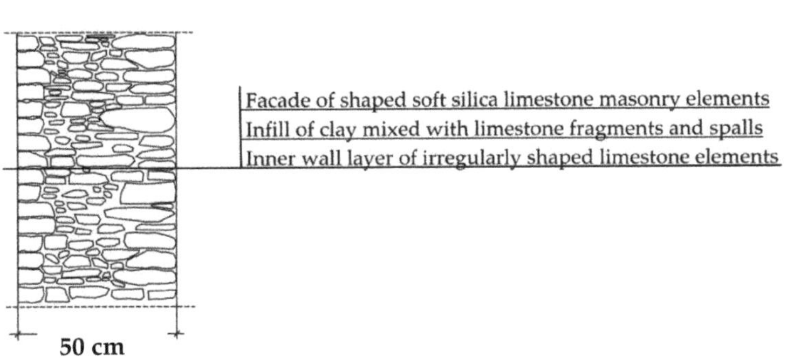

Figure 4. Basic geometric model of the wall for calculations in Therm software.

Heat transfer coefficient (U) calculations were carried out for six wall variants: three without insulation (A–C) and three with additional thermal insulation (D1, D2, E):

- Variant A: the basic model—a layered stone wall according to Figure 4, without plaster coatings. Possible moisture penetration of the external wall layer from atmospheric precipitation was assumed. For the internal layers, thermal parameters for stone and mortar in moderately humid conditions were adopted. It was assumed that the high open porosity of the stone and clay mortar allows for relatively free moisture movement and periodic drying of the walls. In the calculations, the surface resistance on the external and internal sides was taken according to standard [48], with values of $R_{se} = 0.04\ (m^2{\cdot}K)/W$ and $R_{si} = 1.13\ (m^2{\cdot}K)/W$, respectively;
- Variant B: the version takes into account the commonly occurring internal clay plaster, typically 3.0 cm thick. Given the low vapor diffusion resistance reported for clay plasters (μ_{H_2O} = 6–10) [40,46], it was assumed that moisture flow was not significantly disrupted. The parameter values were adopted as those for the clay mortar used in the wall construction, as no significant differences in granulometric composition and density were identified [45];
- Variant C: in this variant, calculations were made considering the external cement plaster (1.5 cm thick). This is a situation commonly encountered in practice, typically applied when the outer wall layer begins to show signs of damage due to natural material wear over time or water exposure, leading to excessive water accumulation on parts of the façade. The thermal conductivity coefficient was adopted according to [37] $\lambda = 1.0\ W/(m{\cdot}K)$. The cement plaster layer, on one hand, limits the periodic soaking of the external stone layer. Nevertheless, considering its higher diffusion resistance [45], it blocks moisture flow (especially in the case of insufficient ventilation) and may cause moisture accumulation in the partition. This was taken into account by assuming an increased value of λ for the stone and clay mortar elements, as in humid conditions;
- Variant D: the case with external insulation layer of polystyrene—the most commonly used variant in practice for thermal modernization. The calculations were performed for both the wall in a damp condition (variant D1) and in a moderately damp condition (variant D2), taking into account the phenomenon of gradual drying of the damp insulated wall over time [49];
- Variant E: the version with internal insulation, which allow the preservation of the original texture of the wall using high vapor-permeable panels, recommended for use in the renovation of historic buildings.

All parameters adopted for the calculations of the model, including geometry and boundary conditions, are presented in Table 4 (variants A–C) and Table 5 (variants D–E).

For the calculation of temperature distribution in the partition, the external temperature was assumed to be the long-term average based on the minimum daily temperatures for the winter months (December to February), which is approximately $\theta_e = -4\ °C$ according to data from the Koło meteorological station [50], and the internal temperature $\theta_i = +20\ °C$.

Table 4. Comprehensive list of parameters assumed for thermal calculations in the THERM software for variants A–C (without insulation).

Variant	Parameter						
	External Surface Resistance (R_{se})	Thermal Conductivity (λ) [1]					Internal Surface Resistance (R_{si})
		External Cement Plaster	Façade Limestone Elements	Clay Mortar and Clay Infill	Internal Limestone Elements	Internal Clay Plaster	
	$(m^2 \cdot K)/W$	$W/(m \cdot K)$	$W/(m \cdot K)$	$W/(m \cdot K)$	$W/(m \cdot K)$	$W/(m \cdot K)$	$(m^2 \cdot K)/W$
A 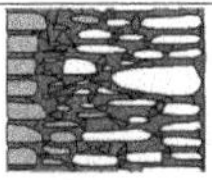	0.04	-	0.76	0.81	0.64	-	0.13
B 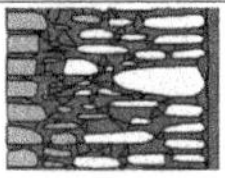		-	0.76	0.81	0.64	0.81	
C 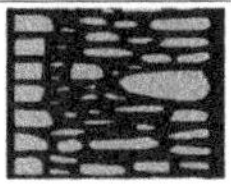		1.00	0.76	0.93	0.76	0.93	

[1] Gray scale on the model proportional to the thermal conductivity (λ) value.

Table 5. Comprehensive list of parameters assumed for thermal calculations in THERM software for variants D–E (with insulation).

Variant	Parameter								
	R_{se}	Thermal Conductivity (λ) [1]							R_{si}
		Polystyren	External Cement Plaster	Façade Lime-stone Elements	Clay Mortar and Clay Infill	Internal Lime-stone Elements	Internal Clay Plaster	Internal Porous Insulation	
	$(m^2 \cdot K)/W$	$W/(m \cdot K)$	$W/(m \cdot K)$	$W/(m \cdot K)$	$W/(m \cdot K)$	$W/(m \cdot K)$	$W/(m \cdot K)$	$W/(m \cdot K)$	$(m^2 \cdot K)/W$
D1 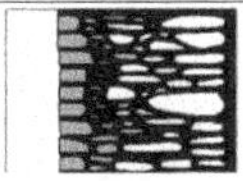	0.04	0.36	-	0.76	0.93	0.76	-	-	0.13
D2 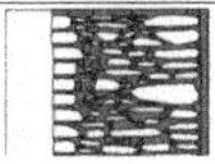		0.36	-	0.64	0.81	0.64	0.81	-	
E 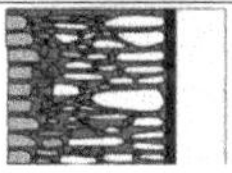		-	-	0.76	0.81	0.64	0.81	0.36	

[1] Gray scale on the model proportional to the thermal conductivity (λ) value.

3. Results

The results obtained from the THERM software for variants A-C without wall insulation are summarized in Table 6, while Table 7 presents the results for variants D-E, which include insulation.

The least favorable U-value of 1.19 $W/(m^2 \cdot K)$ was obtained for the wall in variant A without internal and external plaster. The application of a 3 cm internal clay plaster resulted in a slight decrease in the thermal transmittance coefficient to U = 1.14 $W/(m^2 \cdot K)$ (variant B). However, the additional application of an external cement plaster (variant C), assuming a reduction in the wall's diffusivity leading to increased moisture content, had an adverse effect, causing the U-value to rise to 1.24 $W/(m^2 \cdot K)$. For each variant, the equivalent thermal conductivity coefficients (λ_{equiv}) corresponding to a homogeneous wall were calculated, ranging between 0.57 and 0.62 W/(m·K).

The calculated surface temperatures on the interior side range from 13.5 °C for the damp wall with external plaster (variant C) to 13.7 °C for the wall with internal clay plaster (variant B).

Table 6. Results of calculations in the THERM software for variants A–C (without insulation).

Variant		Thermal Parameters	Isotherms	Temperature Profile	Heat Distribution	Flux Rate
A	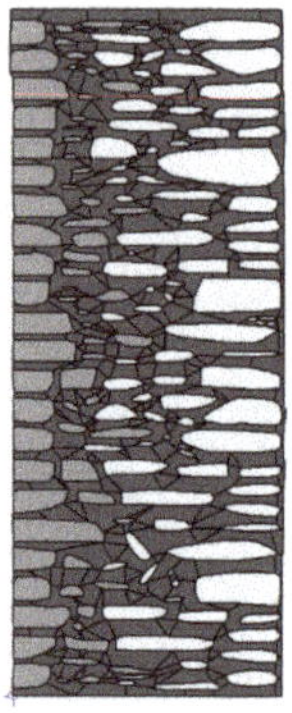	R = 0.84 $(m^2·K)/W$ λ_{equiv} = 0.59 W/(m·K) U =1.19 $W/(m^2·K)$ θ_{min} = +13.6 °C f_{Rsi} = 0.73	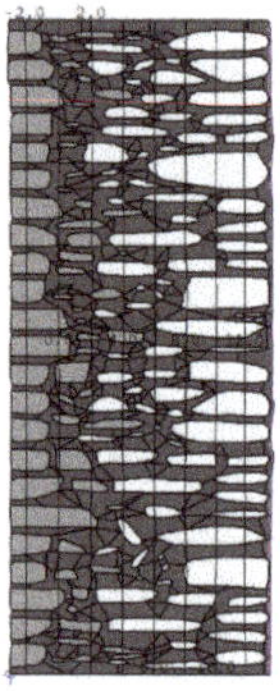	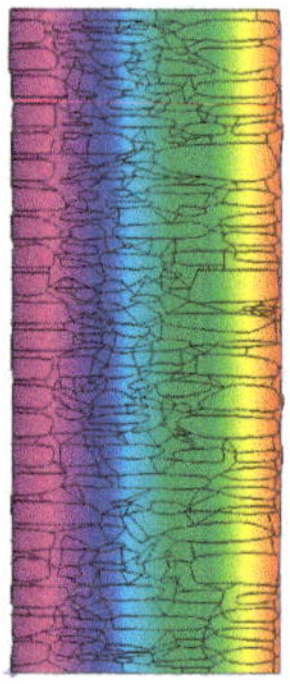	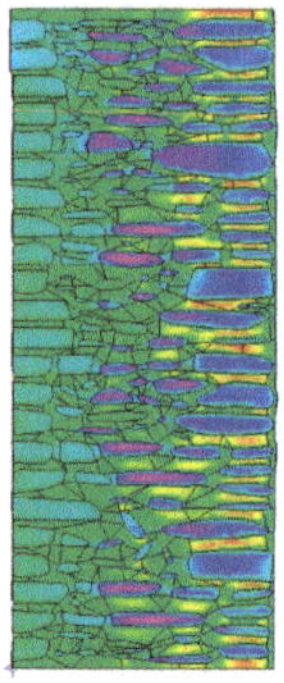	q = 24.03 ÷ 31.04 W/m^2
B	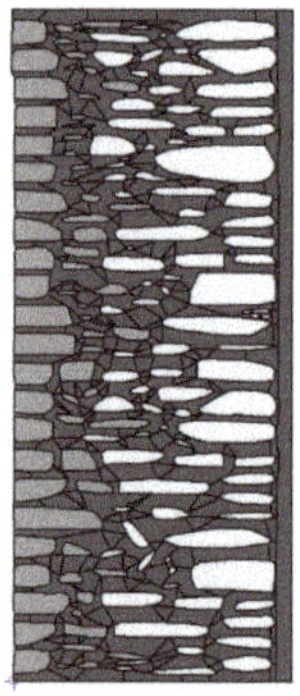	R = 0.88 $(m^2·K)/W$ λ_{equiv} = 0.57 W/(m·K) U = 1.14 $W/(m^2·K)$ θ_{min} = +13.7 °C f_{Rsi} = 0.74	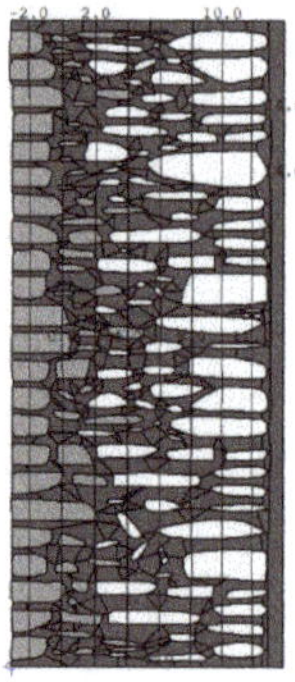	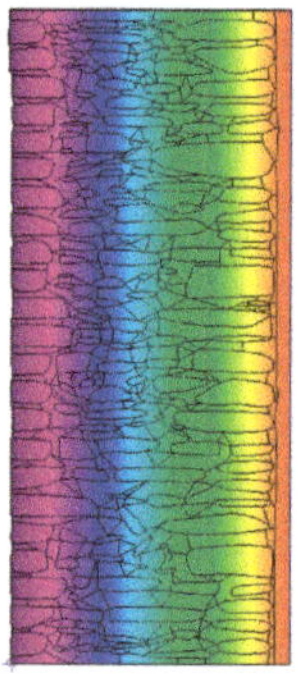	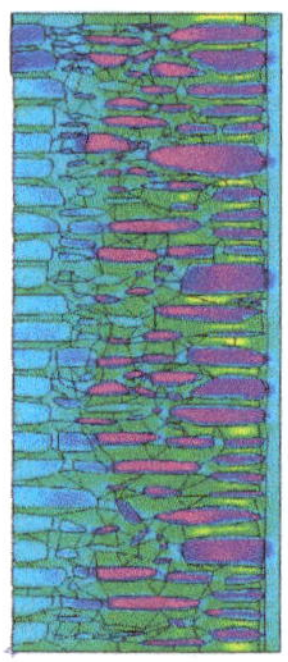	q = 23.03 ÷ 32.46 W/m^2
C	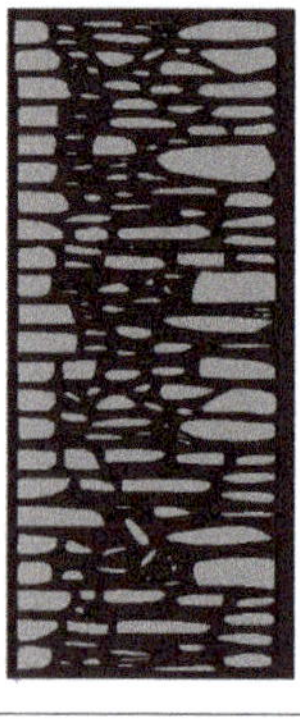	R = 0.81 $(m^2·K)/W$ λ_{equiv} = 0.62 W/(m·K) U = 1.24 $W/(m^2·K)$ θ_{min} = +13.5 °C f_{Rsi} = 0.73		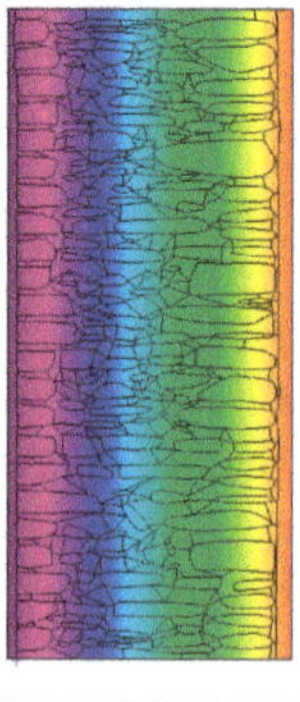	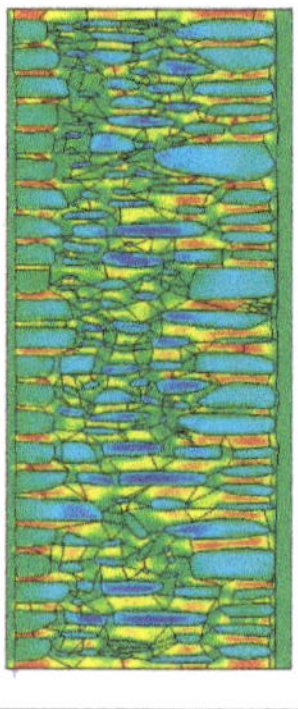	q = 25.45 ÷ 35.02 W/m^2

Table 7. Results of thermal calculations in THERM software for variants D–E (with insulation).

Variant		Thermal Parameters	Isotherms	Temperature Profile	Heat Distribution	Flux Rate
D1	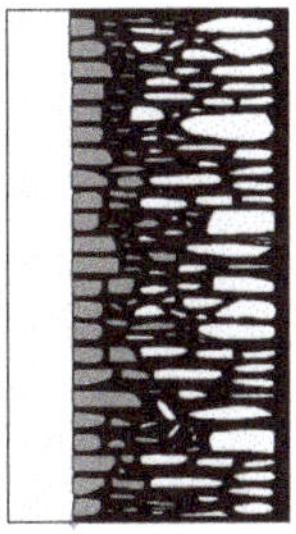	$R = 4.84$ (m^2·K)/W $U = 0.21$ W/(m^2·K) $\theta_{min} = +18.8$ °C $f_{Rsi} = 0.95$	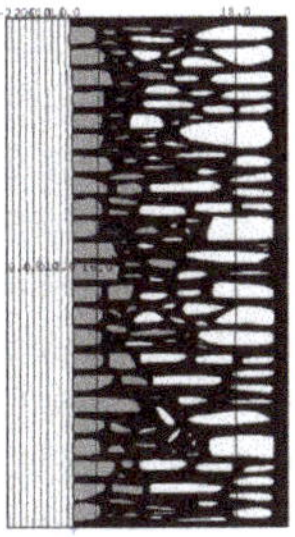	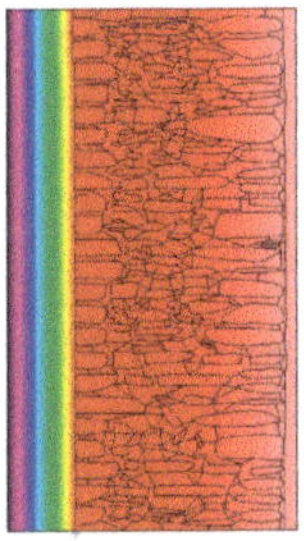	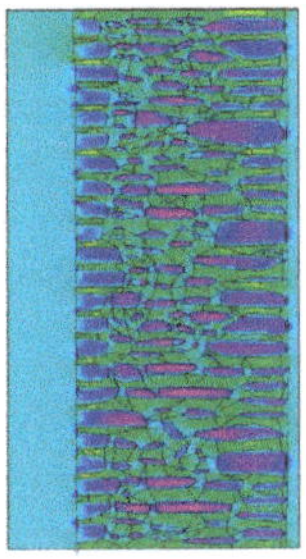	$q = 3.67 \div 6.87$ W/m^2
D2	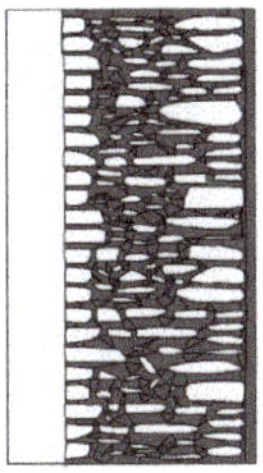	$R = 4.90$ (m^2·K)/W $U = 0.20$ W/(m^2·K) $\theta_{min} = +18.8$ °C $f_{Rsi} = 0.95$	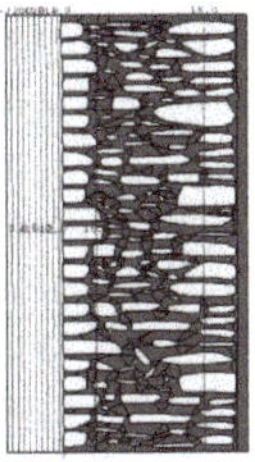	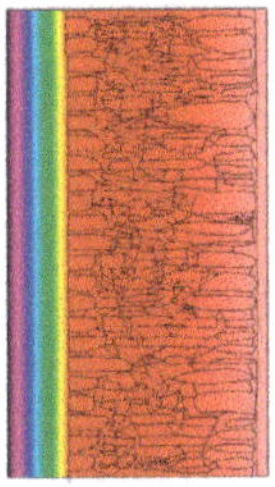	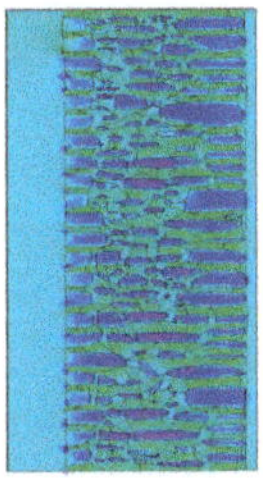	$q = 3.68 \div 6.59$ W/m^2
E	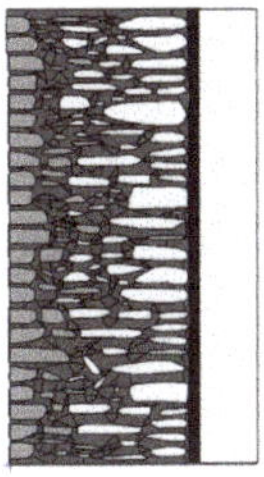	$R = 4.90$ (m^2·K)/W $U = 0.20$ W/(m^2·K) $\theta_{min} = +18.8$ °C $f_{Rsi} = 0.95$	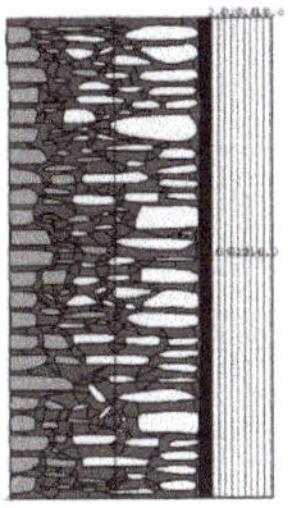	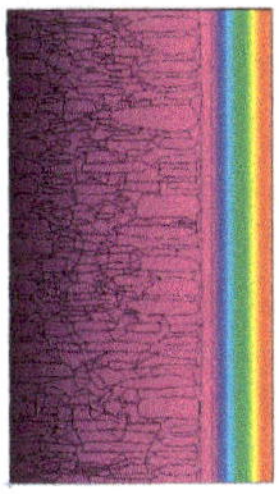	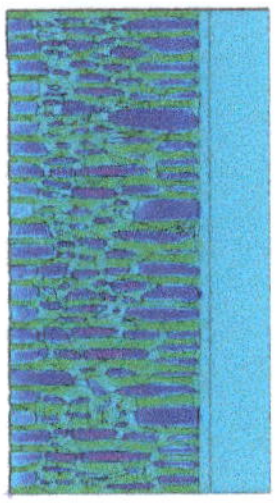	$q = 4.08 \div 6.09$ W/m^2
				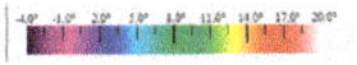		

The f_{Rsi} coefficient values range from $f_{Rsi} = 0.73$ to 0.74. This value is above the minimum hygiene criterion (0.72) according to [51]. This means that under normal usage conditions (i.e., residential spaces with standard humidity), the risk of condensation is low and may only occur in high-humidity areas (e.g., bathrooms, kitchens), especially during colder periods.

Due to the relatively similar thermal conductivity (λ) values of the individual wall components, i.e., soft silica limestone and clay mortar, which bind the elements and fill the inner part of the wall, the isotherms within the wall are very evenly distributed. According to the calculations, only the external layer of the wall, up to a thickness of approximately 10 cm, is exposed to low temperatures and cyclic freezing.

The heat flow rate slightly decreases after taking into account the internal clay plaster. However, the external cement plaster with low diffusivity, which, as assumed, can cause condensation of water vapor and moisture accumulation in stone and clay materials, results in an increase in the heat flow rate through the wall.

The results presented in Table 7 show that in the cases of external and internal insulation (Variant D2 and E), the calculated thermal transmittance (U-value) for the partition is U = 0.20 for 15 cm of insulation with a thermal conductivity of $\lambda = 0.36$. A slight difference appears in Variant D1, where a temporary increase in moisture content in the partition material after insulation was assumed, causing this parameter to rise to U = 0.21. The temperature distribution chart indicates that almost the entire thickness of the stone and clay wall is exposed to cyclic freezing.

4. Discussion

Calculations based on the actual wall structure and thermal transmittance coefficients (U) for variants A–C, which lack wall insulation, showed that the heat transfer coefficient was sufficient to meet the standards in place at the time of construction, i.e., in the 1960s, and remained compliant with Polish regulations until the 1980s [52]. However, it falls far short of current requirements specified in contemporary technical and building standards [51]. To meet these standards, a minimum insulation thickness of 15 cm is required to achieve the currently mandated U-value of 0.20, assuming the use of an insulation material with a thermal conductivity of $\lambda = 0.36$ W/(m·K).

In this context, the aspect of the significant thermal inertia of walls made of soft silica limestone should be addressed. Although Directive 2018/844 [21] does not specify recommended values for thermal inertia, it emphasizes its importance for energy efficiency. Thermal capacity does not directly affect the amount of heat loss through a partition under steady-state heat flow conditions. However, a high thermal mass provides the building with substantial thermal inertia, making the indoor environment resistant to short-term temperature fluctuations, which positively impacts the perception of thermal comfort throughout the year. It also plays a crucial role in maintaining the thermal stability of the indoor environment and can significantly reduce energy expenditures for heating and cooling the building.

The insulation variants considered in the calculations include the commonly used external insulation solution. However, this approach is highly destructive from the perspective of architectural heritage preservation, as it compromises the characteristic texture and color of the soft silica limestone masonry, which is typical for the region. The recommendations for the energy retrofit of traditional buildings in Directive 2018/844 [21] emphasize that the selection of energy efficiency measures should take into account the cultural value of the building, and that modernization efforts must not compromise the historical and architectural significance of the structure. Furthermore, when adapting buildings to contemporary energy standards, the directive recommends preserving original materials and construction techniques. This implies that for well-preserved stone facades, external insulation—despite being the most effective solution from a thermal modernization standpoint—should not be considered an acceptable option.

It is indicated that existing thermal insulation methods—if proper ventilation is ensured, preferably using diffusion-open insulation systems [21]—can protect the building from negative effects such as water vapor condensation. Nevertheless, a significant practical drawback in the case of the discussed buildings is the already limited internal volume, particularly the usable room height. Another concern with internal insulation is the unpredictable consequences of exposing the entire wall thickness to increased cyclic freezing, especially given the likely increase in wall moisture content.

In light of the above findings, the issue of improving the thermal insulation of traditional buildings made of soft silica limestone while preserving their architectural value is undoubtedly very complex, which places investors in a difficult decision-making situation.

Directive 2023/1791 [22] emphasizes that EU Member States should support the education of their citizens and spread knowledge about the validity of renovation solutions, as well as provide transparent advisory systems in the field of investments aimed at improving the energy efficiency of buildings, especially in the context of the required improvement of buildings' energy efficiency classes [2] one-stop shops for energy efficiency in buildings.

In Poland, for cases such as the one described, it would be advisable to implement a new tool supporting the energy renovation process in EU countries, namely 'one-stop shops for energy efficiency in buildings' [22] at the regional level. This would enable a more individualized assessment and selection of thermal insulation improvement options, tailored to the technical condition and cultural value of each building. The ongoing accelerated process of degradation of rural architectural heritage is also being observed in other regions of the country [53], leading to the development of interesting proposals for a comprehensive model of energy renovation management for traditional rural houses that takes national conditions into account.

5. Conclusions

This study presents a thermal analysis of traditional masonry constructed from soft silica limestone, characteristic of vernacular architecture in central Poland. Addressing a gap in current research, the work simulates the thermal behavior of this under-researched natural masonry material—soft silica limestone with clay mortar—considering both dry and moisture-affected thermal conductivity values. The results show that while uninsulated walls do not meet modern thermal performance standards under Polish climate conditions, their substantial thermal mass contributes significantly to indoor thermal stability. The study compares retrofit options and highlights the conflict between improving energy efficiency and preserving architectural heritage. Although external insulation provides superior thermal performance, it often compromises the appearance of historic façades. Internal insulation is visually less intrusive but constrained by limited interior space and the risk of moisture accumulation. Future research should focus on quantifying vapor permeability parameters and conducting long-term in situ monitoring to better evaluate retrofit outcomes.

These findings highlight the importance of retrofit strategies that are responsive to the specific physical behavior of traditional materials while respecting the cultural significance of heritage buildings. The results contribute to broader interdisciplinary discourse on enhancing thermal performance without compromising regional architectural identity.

Author Contributions: Conceptualization, A.G. and M.G.; methodology, A.G. and P.S.; software, P.S.; validation, A.G., M.G. and P.S.; investigation, A.G., M.G. and P.S.; resources, M.G.; writing—original draft preparation, A.G.; writing—review and editing, A.G. and M.G.; visualization, P.S. All authors have read and agreed to the published version of the manuscript.

Funding: This research received no external funding.

Institutional Review Board Statement: Not applicable.

Informed Consent Statement: Not applicable.

Data Availability Statement: The original contributions presented in this study are included in the article. Further inquiries can be directed to the corresponding author.

Conflicts of Interest: The authors declare no conflicts of interest.

References

1. Costa-Carrapiço, I.; González, J.N.; Raslan, R.; Sánchez-Guevara, C. Understanding the challenges of determining thermal comfort in vernacular dwellings: A meta-analysis. *J. Cult. Herit.* **2022**, *58*, 57. [CrossRef]
2. European Council 'Fit for 55': EU Strengthens Emission Reduction Targets for Member States. 2023. Available online: https://www.consilium.europa.eu/en/press/press-releases/2022/11/08/fit-for-55-eu-strengthens-emission-reduction-targets-for-member-states/ (accessed on 23 March 2025).
3. Margani, G. Costruzione e Recupero Dell'Opera Muraria. "Regola Dell'Arte" e Interventi di Consolidamento. 2009. Available online: https://www.researchgate.net/publication/259889584/ (accessed on 23 March 2025).
4. Kapetanović, A. Restoration of a Traditional Montenegrin Stone House. The Site in the Village of Gačevići. 2006. Available online: https://www.hdm.lth.se/fileadmin/hdm/alumni/papers/CMHB_2007/Montenegro_A_Kapetanovic__Montenegrin_Stone_House.pdf (accessed on 23 March 2025).
5. Mustafaraj, E.; Luga, E.; Corradi, M.; Borri, A.; Muceku, Y.; Zharkalli, A. Physical-Mechanical Properties of Stone Masonry of Gjirokastër, Albania. *Materials* **2021**, *14*, 1127. [CrossRef] [PubMed]
6. Careddu, N.; Grillo, S.M. "Trachytes" from Sardinia: Geoheritage and Current Use. *Sustainability* **2019**, *11*, 3706. [CrossRef]
7. Tomanović, D.; Rajković, I.; Grbić, M.; Aleksić, J.; Gadžić, N.; Lukić, J.; Tomanović, T. Houses Based on Natural Stone; A Case Study—The Bay of Kotor (Montenegro). *Sustainability* **2019**, *11*, 3866. [CrossRef]
8. Pereira, D.; Baltuille, J.M. Documenting natural stone to preserve our cultural and architectonic heritage. *Mater. Construcción* **2014**, *64*, ed002.
9. Hou, X.; Cheng, B.; Yang, J. A quantitative study on the exterior wall texture of stone-built dwellings in traditional villages in China: A case study of the Xisuo village in the Jiarong Tibetan area. *J. Build. Eng.* **2021**, *42*, 102357. [CrossRef]
10. Serghides, D.K. The wisdom of Mediterranean traditional architecture versus contemporary architecture–the energy challenge. *Open Constr. Build. Technol. J.* **2010**, *4*, 29–38. [CrossRef]
11. Zhou, M.; Bonenberg, W.; Wei, X.; Qi, L. The Study on Adaptation of Vernacular Architecture in Modern Rural Architecture Design During the Process of Rural Revitalization. In Proceedings of the AHFE 2020 Virtual Conference on Human Factors in Architecture, Sustainable Urban Planning and Infrastructure, San Diego, CA, USA, 16–20 July 2020; pp. 185–191.
12. Stovel, H. *Risk Preparedness: A Management Manual for World Cultural Heritage*; ICCROM: Rome, Italy, 1998; p. 145. Available online: https://www.iccrom.org/sites/default/files/ICCROM_17_RiskPreparedness_en.pdf (accessed on 23 March 2025).
13. *Standard PrEN 16682*; European Committee for Standardization (CEN), Conservation of Cultural Heritage. Guide to the Measurements of Moisture Content in Materials Constituting Movable and Immovable Cultural Heritage. CEN: Bruxelles, Belgium, 2013.
14. Morel, J.C.; Mesbaha, A.; Oggerob, M.; Walkerc, P. Building houses with local materials: Means to drastically reduce the environmental impact of construction. *Build. Environ.* **2001**, *36*, 1119. [CrossRef]
15. Mariani, S.; Rosso, F.; Ferrero, M. Building in Historical Areas: Identity Values and Energy Performance of Innovative Massive Stone Envelopes with Reference to Traditional Building Solutions. *Buildings* **2018**, *8*, 17. [CrossRef]
16. Stefanizzi, P.; Fato, I.; Di Turi, S. Energy and environmental performance of trullo stone building. An experimental and numerical survey. *Int. J. Heat Technol.* **2016**, *34*, S396. [CrossRef]
17. Akander, J.; Bakhtiari, H.; Ghadirzadeh, A.; Mattsson, M.; Hayati, A. Development of an AI Model Utilizing Buildings' Thermal Mass to Optimize Heating Energy and Indoor Temperature in a Historical Building Located in a Cold Climate. *Buildings* **2024**, *14*, 1985. [CrossRef]
18. Ekici, B.B.; Gulten, A.A.; Aksoy, U.T. A study on the optimum insulation thicknesses of various types of external walls with respect to different materials, fuels and climate zones in Turkey. *Appl. Energy* **2012**, *92*, 211. [CrossRef]
19. Georgiou, L.; Panteli, C.; Kylili, A.; Fokaides, P.A. Thermal performance of brick and stone masonry: Cumulative heat flux dataset for main orientations and under diverse seasonal conditions. *Data Brief* **2020**, *33*, 106599. [CrossRef] [PubMed]
20. Galatioto, A.; Ciulla, G.; Ricciu, R. An overview of energy retrofit actions feasibility on Italian historical buildings. *Energy* **2017**, *137*, 991. [CrossRef]
21. Directive (EU) 2018/844 of the European Parliament and of the Council of 30 May 2018 amending Directive 2010/31/EU on the energy performance of buildings and Directive 2012/27/EU on energy efficiency, Brussels, European Parliament. *Off. J. EU* **2018**, *61*, L156/75.
22. Directive (EU) 2023/1791 of the European Parliament and of the Council of 13 September 2023 on energy efficiency and amending Regulation (EU) 2023/955, Brussels, European Parliament. *Off. J. EU* **2023**, *66*, L231/1.
23. Blumberga, A.; Freimanis, R.; Biseniece, E.; Kamenders, A. Hygrothermal Performance Evaluation of Internally Insulated Historic Stone Building in a Cold Climate. *Energies* **2023**, *16*, 866. [CrossRef]

24. Blumberga, A.; de Place Hansen, E.J. Written Guidelines for Decision Making Concerning the Possible Use of Internal Insulation in Historic Buildings. Robust Internal Thermal Insulation of Historic Buildings, Project no: 637268; Horizon 2020. 2020. Available online: https://orbit.dtu.dk/files/238206107/Written_guidelines_for_decision_making_concerning_the_possible.pdf (accessed on 23 March 2025).
25. Straube, J.; Ueno, K.; Schumacher, C. *Internal Insulation of Masonry Walls: Final Measure Guideline*; Prepared for U.S. Department of Energy; Office of Energy Efficiency and Renewable Energy: Somerville, MA, USA, 2012.
26. Buda, G.; Taylor, B.; Bennadji, A. The nature of mass masonry granite walling and the potential for retrofit internal wall insulation strategies. *J. Build. Surv. Apprais. Valuat.* **2013**, *2*, 36. [CrossRef]
27. Orlik-Kożdoń, B.; Krause, P.; Steidl, T. Wpływ wybranego rozwiązania materiałowego ocieplenia od wewnątrz na procesy cieplno-wilgotnościowe w przegrodzie. *Fiz. Budowli W Teor. I Prakt.* **2015**, *7*, 55.
28. Lucchi, E. Thermal transmittance of historical stone masonries: A comparison among standard, calculated and measured data. *Energy Build.* **2017**, *151*, 393. [CrossRef]
29. Ascione, F.; Ceroni, F.; De Masi, R.F.; de'Rossi, F.; Pecce, M.R. Historical buildings: Multidisciplinary approach to structural/energy diagnosis and performance assessment. *Appl. Energy* **2017**, *185*, 1517. [CrossRef]
30. Cesaratto, P.G.; De Carli, M.; Marinetti, S. Effect of different parameters on the in situ thermal conductance evaluation. *Energy Build.* **2011**, *43*, 1792. [CrossRef]
31. Li, F.G.N.; Smith, A.Z.P.; Biddulph, P.; Hamilton, I.G.; Lowe, R.; Mavrogianni, A.; Oikonomou, E.; Raslan, R.; Stamp, S.; Stone, A.; et al. Solid-wall U-values: Heat flux measurements compared with standard assumptions. *Build. Res. Inf.* **2014**, *43*, 238. [CrossRef]
32. Baker, J.P. *U-Values and Traditional Buildings: In Situ Measurements and Their Comparisons to Calculated Values*; Historic Scotland Technical Paper 10; Glasgow Caledonian University: Glasgow, UK, 2011.
33. Gorączko, M.; Gorączko, A. Vernacular architecture and traditional rural landscape in new socio-economic realities—A case study from Central Poland. *Bull. Geography. Socio-Econ. Ser.* **2015**, *30*, 43. [CrossRef]
34. Gorączko, M.; Gorączko, A. Heritage, reinterpretation or imitation?—Contemporary inspirations from traditional soft limestone architecture in the Uniejów region. *Inżynieria I Bud.* **2024**, *LXXX*, 196. [CrossRef]
35. Gorączko, A.; Gorączko, M. Application of geometric parameters in texture analysis of traditional stone walls. *Mater. Bud.* **2024**, *627*, 157. [CrossRef]
36. Wyganowski, Z.; Próchnicki, F. *Budynki z Kamienia*; Arkady: Warszawa, Poland, 1958.
37. Gorączko, A.; Gorączko, M. Wykorzystanie lokalnego kamienia budowlanego w świetle przepisów. *Mater. Bud.* **2016**, *7*, 61. [CrossRef]
38. Gorączko, A.; Gorączko, M. Results and Conclusions of Laboratory Characteristics of Soft Limestone from Central Poland as a Construction Material. *IOP Conf. Ser. Mater. Sci. Eng.* **2019**, *471*, 1. [CrossRef]
39. *EN 1936:2010*; Natural Stone Test Methods—Determination of Real Density, Apparent Density, and Total and Open Porosity. Polish Committee for Standardization: Warsaw, Poland, 2010.
40. *PN-EN ISO 10456:2009*; Building Materials and Products—Thermal Properties—Determination of Declared and Design Thermal Values. Polish Committee for Standardization: Warsaw, Poland, 2009.
41. Różalski, J. Właściwości budowlane kamienia. In *Budownictwo Wiejskie*; Politechnika Warszawska: Warsaw, Poland, 1958; Volume 12, pp. 21–26.
42. *PN-B-02020:1991*; Ochrona Cieplna Budynków—Wymagania i Obliczenia. PKN. University of Zielona Góra: Zielona Góra, Poland, 1991.
43. Llorente-Alvarez, A.; Camino-Olea, M.S.; Cabeza-Prieto, A.; Saez-Perez, M.P.; Rodríguez-Esteban, M.A. The thermal conductivity of the masonry of handmade brick Cultural Heritage with respect to density and humidity. *J. Cult. Herit.* **2022**, *53*, 212. [CrossRef]
44. Suchorab, Z.; Barnat-Hunek, D. The analysis of heat conductivity coefficient of the aerated concrete building barriers depending on moisture changes. *Bud. I Archit.* **2011**, *8*, 107. [CrossRef]
45. Gorączko, A.; Gorączko, M. Clay mortars in traditional masonry buildings of soft limestone—Research on composition and structure. *Mater. Bud.* **2024**, *623*, 24. [CrossRef]
46. Rokiel, M. Właściwości, parametry i zastosowanie tynków glinianych. *Izolacje* **2017**, *1*, 48.
47. *BN-62/6738-02*; Materiały Budowlane. Materiały Wiążące, Spoiwa, Betony. Budownictwo z Gliny. Masy Gliniane. Politechnika Lubelska: Lubelska, Poland, 1962.
48. *PN-EN ISO 6946:2017-10*; Building Components and Building Elements—Thermal Resistance and Thermal Transmittance—Calculation Methods. PKN. Intertek Inform: London, UK, 2017.
49. Bobociński, A.; Pogorzelski, J. Analiza procesu wysychania ścian z zewnętrznym systemem izolacji cieplnej. *Pr. Inst. Tech. Bud.* **2001**, *30*, 29.

50. Institute of Meteorology and Water Management-National Research Institute. Available online: https://danepubliczne.imgw.pl/ (accessed on 25 December 2024).
51. Regulation of the Minister of Infrastructure of 12 April 2002 on the Technical Conditions to Be Met by Buildings and Their Location. 2002. Available online: https://isap.sejm.gov.pl/isap.nsf/DocDetails.xsp?id=wdu20020750690 (accessed on 23 March 2025).
52. *PN-74/B-03404;* Ochrona Cieplna Budynków. Wymagania i Obliczenia. Politechnika Lubelska: Lubelska, Poland, 1974.
53. Szarejko, M. Zarządzanie renowacją energetyczną tradycyjnych wiejskich domów mieszkalnych. *Bud. I Archit.* **2024**, *23*, 173. [CrossRef]

materials

MDPI

Article

Detection and Location of Defects in Externally Bonded FRP Concrete Structures—Comparison of Selected Methods

Paweł Tworzewski [1,*], Kamil Bacharz [2], Wiktor Wciślik [2], Michał Teodorczyk [3,*], Sylwia Wciślik [4] and Justyna Tworzewska [1]

1 Department of Building Structures, Kielce University of Technology, al. Tysiąclecia Państwa Polskiego 7, 25-314 Kielce, Poland; j.tworzewska@tu.kielce.pl
2 Department of Strength of Materials and Structures Diagnostics, Kielce University of Technology, al. Tysiąclecia Państwa Polskiego 7, 25-314 Kielce, Poland; kbacharz@tu.kielce.pl (K.B.); wwcislik@tu.kielce.pl (W.W.)
3 Building Structures, Geotechnics and Concrete Department, Instytut Techniki Budowlanej, ul. Filtrowa 1, 00-611 Warsaw, Poland
4 Faculty of Environmental Engineering, Geomatics and Renewable Energy, Kielce University of Technology, al. Tysiąclecia Państwa Polskiego 7, 25-314 Kielce, Poland; sylwiazw@tu.kielce.pl
* Correspondence: ptworzewski@tu.kielce.pl (P.T.); m.teodorczyk@itb.pl (M.T.); Tel.: +48-41-342-4579 (P.T.); +48-22-57-96-117 (M.T.)

Abstract: This paper compares three nondestructive methods used to detect and locate defects such as delaminations or voids in externally bonded fiber reinforced polymer (FRP) concrete structures: infrared thermography, ground-penetrating radar, and measurement of acoustic wave velocity. One of the main goals was to check whether it was possible to distinguish overlapping defects. For this purpose, eight concrete samples were made with a bonded carbon fiber reinforced polymer (CFRP) strip with the following dimensions 100 × 100 × 500 mm. Two samples had no defects, four had single defects varying in location (at the edge of the strip or in the centre) simulating delamination or voids in the concrete cover, and the remaining samples had overlapping defects. Both infrared thermography and acoustic wave velocity measurement methods allow the detection of defects/voids in the adhesive layer and a concrete defect (void in the concrete cover). However, ground penetration failed to detect defects in the adhesive layer. Only infrared thermography allows for the differentiation of overlapping defects. On the basis of the conducted research, the methodology, differences, advantages, and limitations of each method were described, along with recommendations based on the authors' experience.

Keywords: nondestructive testing (NDT); Externally Bonded FRP (EBR); structural health monitoring (SHM); ground penetrating radar (GPR); infrared thermography (IRT); acoustic wave velocity (WVMM)

1. Introduction

Carbon fibre-reinforced polymers (CFRP) are widely used in various industries, including aviation, transportation, and construction, because of their exceptional characteristics such as low weight, high mechanical strength, and resistance to corrosion. In the CFRP construction sector, the application is steadily increasing, with projected usage of 6.2 kilotons by 2025 [1]. In particular, approximately 80–90% of CFRP employed in this field is dedicated to strengthening and rehabilitating existing structures that have suffered deterioration. It is important to emphasise that the number of buildings requiring strengthening is constantly growing. This is due to the natural ageing of building materials, increased service loads, and changes in the standards requirements. New buildings also require strengthening due to changing investor requirements, often already in the construction stage [2].

Materials such as FRP plates, strips, or sheets are commonly used to strengthen RC structures and are among the most effective solutions. However, this solution still has certain limitations related to fully utilising the potential of FRP materials. This is, of course, the issue of debonding, which limits the effectiveness of these solutions and is the most common cause of failure. The causes and factors that initiate debonding are varied, but the main ones include [3–6]:

- Debonding caused by shear cracks or flexural cracks, i.e., defects in the reinforced concrete structure;
- Debonding related to insufficient anchoring of the FRP material;
- Debonding caused by unevenness or other defects in the concrete surface;
- Delamination in FRP;
- Debonding caused by improper preparation of the concrete surface before applying the FRP material or by improper application.

It should be emphasised that the above-mentioned factors mentioned above are of different significance depending on the strengthening system used. Systems made with externally bonded reinforcement (EBR) are most susceptible to these factors. Near-surface mounted reinforcement (NSM), due to its different application method, excludes or limits the impact of some of the factors mentioned above, which does not eliminate the possibility of debonding, but only limits it [7,8].

The surface to which the FRP EBR is bonded should be level, as concave surfaces can result in FRP debonding. Requirements for the depth of surface variation with respect to a straight base length depend on the type of FRP EBR used. Example requirements [9] are presented in Table 1.

Table 1. Requirements for concave surfaces of the concrete element [9].

Type of FRP EBR		Permissible Unevenness [mm]	
		For 2000 mm Base	For 300 mm Base
prefabricated	thickness > 1 mm	10	4
	thickness < 1 mm	6	2
not prefabricated (made in situ)		4	2

All of the requirements listed above are intended to eliminate or limit potential defects that could lead to premature debonding of the FRP material and failure to achieve the required load-bearing capacity.

Detection and localization of potential FRP debonding initiation point is crucial in the diagnostics and monitoring of objects [3,10,11]. These mainly concern concrete defects (cracks and voids), bond defects (debonding in the adhesive, debonding at the interfaces between concrete and adhesive or adhesive and FRP), and defects in the FRP composite itself, whether prefab type or wet lay-up type.

In this paper, the authors verify the possibility of using selected methods, i.e., infrared thermography, ground-penetrating radar, and measurement of acoustic wave velocity, to locate and determine various types of defects such as voids in the adhesive layer and voids in the concrete cover for externally bound FRP Concrete Structures.

This work not only tested the ability to detect various types of defects, as has already been presented in other works, but also examined whether it was possible to distinguish overlapping defects. In this case, it was a defect/void in the adhesive layer and a concrete defect (void in the concrete cover). It should be noted that such a configuration occurs quite often in the case of FRP deponding. Typical triggers for this phenomenon are cracks or other defects, such as voids in the concrete cover, concave surfaces that lead to the debonding of the FRP material around it.

1.1. Infrared Thermography (IRT)—Description of the Method

Infrared thermography (IRT) is a fairly well-known diagnostic method. It is typically associated with investigating the impact of climatic parameters on a building, but its application is much broader [12–14]. The method's operating principle is quite simple, as it involves measuring the temperature distribution across the recorded surface. Observed differences indicate uneven heat diffusion, and such anomalies in a structural element or material may indicate defects such as voids, cracks, or delamination. The main advantages of this method include ease of use, the ability to conduct real-time inspections, and non-contact measurements. In the case of concrete or reinforced concrete structures, infrared thermography is used to detect cracks [15–17]. However, such tests have shown certain limitations, including sensitivity to external conditions such as wind, rain, sunlight, etc. To reduce such interference, active thermography (Figure 1) is recommended, which involves applying an independent heat source, a thermal effect, to the area examined. This measurement involves recording the response of the observed material to thermal effects generated for this purpose (e.g., halogen lamps) using a thermal imaging camera. Due to the small size of the defects, another limitation arises: the small examination area. The resolution of these cameras is not high, so achieving high accuracy requires a reduced distance from the detected surface. The research presented in [16] showed that in the case of concrete structures, the method allows the detection of internal cracks (only at a shallow depth). However, thinner cracks with a width of less than 0.5 mm have been shown to only be observed with an additional stimulus, that is, water.

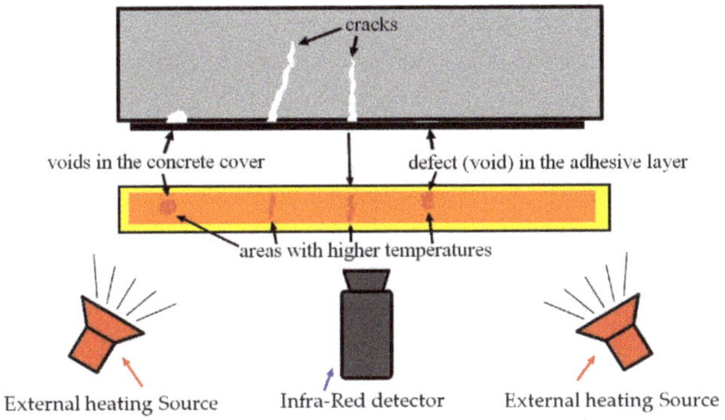

Figure 1. Scheme of using thermography to detect defects in building structures.

This method has also been used in the diagnosis of FRP composite materials themselves. An example is the use of long pulse thermography (LPT) to detect both depth and impact defects in GFRP materials [18]. These studies also used image processing techniques to classify composite defects and automatically highlight defective areas. Active thermography can also be used in this type of research, as demonstrated in [19].

Currently, this method is increasingly used to locate defects in concrete elements reinforced with FRP materials [3,20]. Infrared thermography allows in this case to detect invisible cracks, hidden under the glued FRP material as described in [21,22], as well as other types of voids [23]. In this type of research, it should be noted that we are dealing with three materials with significantly different thermal diffusivities: CFRP strip, resin, and concrete. CFRP conducts heat very well, and, for example, in the studies presented in [24], the thermal diffusivity of the tested material along the CFRP fibres was in the range of 240–280.2 $\times$ 10^{-6} m^2/s. This also contributes to the rapid temperature drop in the defect area when the heat source is disconnected. It should also be remembered that this material is anisotropic, meaning that in the second direction, this value is over 100 times lower. Epoxy resin, on the other hand, has the lowest thermal diffusivity, which means that it acts as an insulator; an example range of values obtained in the studies presented in [25] was 0.7–1.1 $\times$ 10^{-7} m^2/s. Concrete has moderate diffusivity and conducts heat better than resin, but still relatively slowly. In [26], the authors, based on their research, present thermal properties of concretes of various strength classes. The obtained thermal diffusivity values ranged from 0.69 to 1.34 $\times$ 10^{-6} m^2/s. The thermal conductivity of both resin and concrete is significantly greater than that of the air filling the defects, so in the area where they occur, we can observe material's temperature changes—a hot spot occurs, a higher temperature area. This means that as the volume of the void increases, the area and temperature recorded by the camera will also increase.

1.2. Ground-Penetration Radar (GPR)—Description of the Method

GPR is a geophysical method that involves emitting an electromagnetic wave into the interior of the medium being examined and then recording the reflected wave. The evaluation of the internal structure of the medium is based on the analysis of the properties of the reflected wave, primarily the amplitude and propagation time. Generally, wave reflection from an object within the medium results in a local increase in amplitude.

Although the first applications of GPR involved assessing the structure of near-surface geological layers [27], in the late 20th century, with the widespread introduction of high-frequency antennas and advanced data analysis methods, GPR was introduced in other areas, including construction. It is particularly often used to inventory the internal structure of concrete structures [28–30], as well as to assess the moisture content [31,32] and the degree of corrosion [33–35]. The GPR technique can also be used to diagnose structures made entirely of FRP composites [36].

In recent years, GPR has been increasingly used for the non-destructive evaluation of externally reinforced concrete structures with composite elements [3,37]. Most authors claim that high-frequency GPR antennas (1–2 GHz) are effective for detecting subsurface voids in concrete [38–40].

Riad et al. [41,42], found that GPR with a 2.6 GHz antenna accurately detected voids and discontinuities at the FRP strip–concrete interface. Furthermore, the authors observed the influence of other bonding parameters (FRP type, number of epoxy layers, epoxy type, surface roughness, etc.) on the reflected wave amplitude. In contrast, the authors of [43] noted that the GPR was useful only in detecting voids, while other defects (for example, delaminations resulting from improper surface preparation before bonding) were not visible.

Ortiz et al. [43] pointed out the practical limitations of using GPR to study CFRP external reinforcements. Carbon fibres, as a conductive material, constitute a barrier to radio waves, making detection of subsurface defects impossible. This limitation does not apply to insulating composites. For example, Li and Liu [44] demonstrated that a 10 GHz microwave system is effective in mapping the structure of concrete reinforced with GFRP mats, including detecting voids and interface delaminations. Similar conclusions were drawn in [45].

However, some authors [38] emphasise that GPR is not a fully reliable tool and should be combined with other NDT methods, such as infrared thermography (IRT).

1.3. Wave Velocity Measurement Method (WVMM)—Description of the Method

Nondestructive diagnostic methods are increasingly popular due to the fact that they do not generate structural damage during testing. A group of methods is based on the analysis of acoustic waves at various frequencies and their behaviour within the tested element. Examples include ultrasonic methods [46–48]. The second group includes methods using acoustic emission [49], which include, among others, the IADP method [50], based on the recording of acoustic waves generated by active destructive processes occurring automatically, for example, from shrinkage [51–53] or under load [54,55] within the element. This method is also used to analyse various materials, including, in addition to concrete [51–53] and steel [56,57], composite materials [58] and even soil [59,60]. The diagnostic method used in this study, the WVMM wave velocity measurement method, is also based on the analysis of acoustic elastic waves. In this case, unlike the IADP method, an acoustic signal in the form of an elastic wave is generated externally. This can be used to detect and confirm the location of defects. In this case, the reference Hsu-Nielsen method (Hsu-Nielsen-Source) was used as the wave source, as described in the European [55] and American [61] standards.

The Hsu-Nielsen method involves the generation of an elastic wave as a result of fracture of a graphite stylus with a diameter of 0.5 mm and a hardness of 2H, the protruding part of which should be 3.0 ± 0.5 mm long. During fracture, the stylus generates an impulse with an amplitude of approximately 1.6 N [62]. Furthermore, to achieve the most repeatable fracture positions, a Teflon ring is applied to the tip of the mechanical pencil, which, together with a guide, allows the pencil to be positioned at the same 30 ° angle during the test.

The HSU–Nielsen source has its advantages:

- A broad and flat frequency band ranging from 40 to 400 kHz, similar to the acoustic emission resulting from concrete cracking [63].
- The generated signal is strong, allowing this source to be used for wave measurement at various distances, including measuring the attenuation of longitudinal wave velocity [63,64].
- Simplicity of construction, universality, and ease of use both in the laboratory on samples and in the field on existing structures [63,64].
- Allows for determining the reference wave velocity in concrete [64].
- Allows for checking the sensor connection to the concrete and its correct operation; easy to use [64].

The HSU–Nielsen source also has the following disadvantages:

- Two sensors are required to measure the longitudinal wave velocity, as the source is not synchronized with the measuring equipment, which complicates analysis [63].
- A broken graphite guide may reflect off the sensor or the surface of the element, causing additional light and erroneous results [63].
- The person breaking the guide should maintain a constant pressure, but in practice this is difficult, and each break of the graphite is not always identical [63].
- Material costs increase with the number of tests; pencil lead must be purchased only from a recommended company with the appropriate parameters [63,64].
- Not suitable for tests that require high precision [65].

2. Materials

2.1. Materials Used to Prepare Specimens and Their Properties

The six identical beams were made, each 0.5 m long and with a square cross section of 0.10 m × 0.10 m. The elements do not have steel reinforcement. All beams were cast at a similar time) in several-day intervals because of the precast form, allowing for the production of 4 elements/beams at one time. The concrete samples strength test was carried out in parallel to the beam tests. Concrete average cylinder compressive strength f_c = 49.4 MPa hence the designated concrete class is C45/55. The concrete recipe is given in Table 2.

Table 2. Concrete recipes for 1 m^3.

Material	Values
Cement 42.5	360 kg
Sand	726 kg
Basalt aggregate 2/8	563 kg
Basalt aggregate 8/16	738 kg
Water	97 L
Plasticizer—Adva Flow 440 (0.5% cement mass)	1.80 kg
Aerator—Darex (0.1% cement mass)	0.36 kg

As strengthening, a CFRP strip manufactured by Sika, designated Sika® CarboDur® S 512 and Sikadur-330 resin were used (basic material properties are given in Tables 3 and 4).

Table 3. CFRP strip properties [66].

Sika® CarboDur® S 512 (Manufacturer's Catalog Data)	
Properties	**Values**
Width	50 mm
Thickness	1.2 mm
Cross-sectional area	60 mm^2
Average tensile strength of the FRP	3100 MPa
Fiber volume content	>68%

Table 4. Resin properties [67].

Sikadur®-330 (Manufacturer's Catalog Data)	
Properties	**Values**
Tensile modulus	30 MPa
Tensile strength	4500 MPa

2.2. *Preparation of the Specimens*

The beams were divided into four groups, three of which contained defects (Figure 2):

- N1 and N2—beams without defects;
- D1 and D2—beams with a bond defect in the adhesive layer;
- C1 and C2—beams with a concrete defect in the form of drilled voids in the concrete cover;
- M1 and M2—beams with overlapping defects of both types.

The preparation of the strengthened specimens was the same for each beam. It required several steps:

1. Grinding and cleaning the concrete surface of the samples;
2. Preparation of defects—voids in the concrete cover in samples marked with symbols by drilling holes in accordance with the drawing using a hammer drill with a drill bit of 10 mm diameter;
3. Cleaning the surface of the CFRP strip;
4. Applying a thin layer of resin to the prepared surface and spreading it with a putty knife to fill all pores, avoiding defects. Removal of excess resin with a putty knife;
5. Attaching a plastic strip to the place of simulated defects (15 mm wide);

6. Applying the resin to the strip using a previously prepared trough-type tool with dimensions consistent with the manufacturer's requirements;
7. Adherence of the CFRP strip to the prepared concrete surface and pressing it to remove excess resin;
8. Sliding the plastic strip out from under the CFRP strip, creating a void in the resin layer between the strip and the concrete surface;
9. Removing the excess resin using a putty knife.

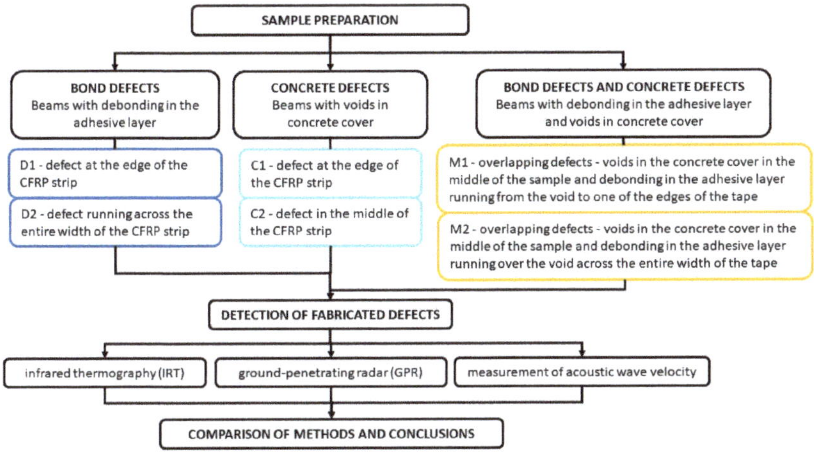

Figure 2. Flow chart of the research methodology.

The type and location of defects in the individual beams are shown in Figure 3. Photos of the subsequent stages of the concrete beam are shown in Figures 4 and 5.

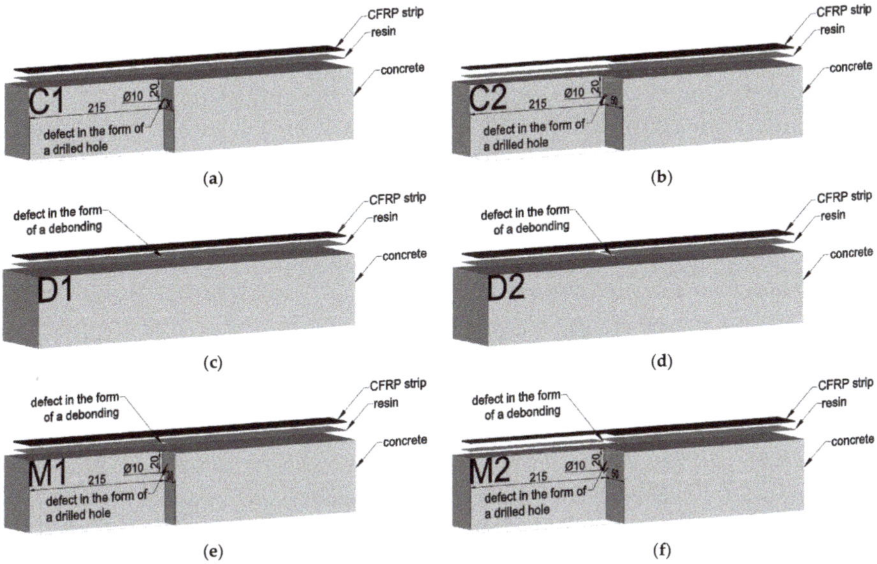

Figure 3. Location of defects in beams (**a**,**b**) defects in the form of a hole under the composite–concrete defects, (**c**,**d**) defect in the form of debonding–bond defect in the adhesive layer, (**e**,**f**), samples with both types of damage—units are given in [mm].

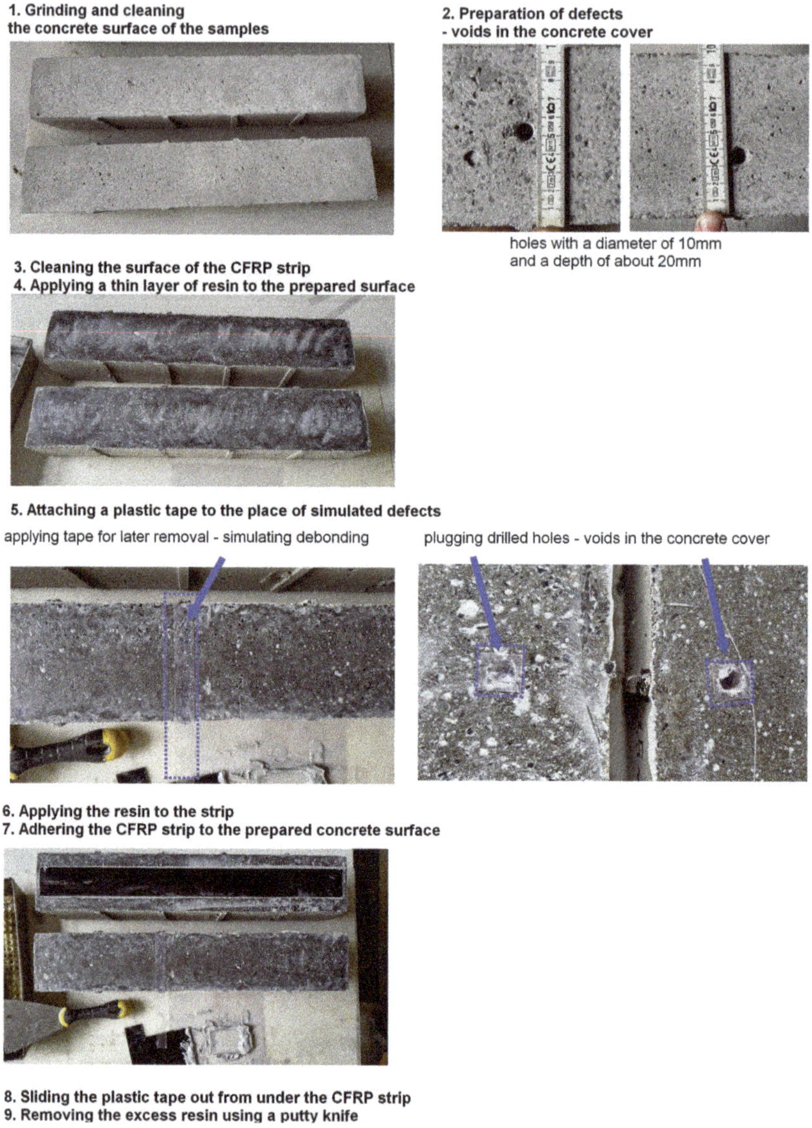

Figure 4. Preparation of the strengthened specimens.

Proper surface preparation of the structural element before bonding FRP materials is a critical step in ensuring the effectiveness of the reinforcement. When executed correctly, it can significantly delay debonding and improve the ultimate strength of the system. Key preparatory actions include removing any loose concrete, levelling the surface to eliminate irregularities, and thoroughly cleaning both the concrete substrate and the FRP material to remove contaminants. It is equally important that the epoxy resin remain clean and uncontaminated. It is also possible to apply additional treatments to the prepared surface, which is described in more detail in [68]. These treatments include transverse, diagonal, and longitudinal grooving. They aim to increase the contact surface area and further help to postpone the onset of debonding.

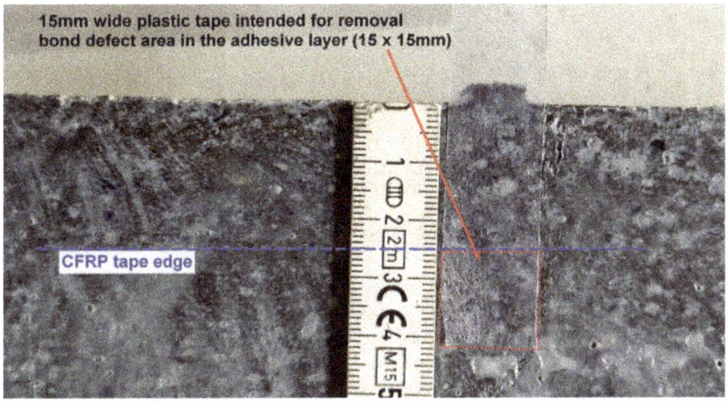

Figure 5. Dimensions of the bond defect area in the adhesive layer for sample D1.

In the case of strengthening real structures, concrete surface preparation should be preceded by verification of the concrete substrate's soundness. Depending on the requirements, the minimum concrete tensile strength should be greater than 1.5 N/mm^2 (PN-EN 1542 [69]) or 1.4 N/mm^2 (ACI 440.3 L.1 [70]). Rebar corrosion should be stopped to avoid damage to the concrete cover due to extensive rust. The depth of carbonation and chloride content (concentrations greater than 0.3% by weight of cement are assumed dangerous) should also be verified. According to the requirements described in [9], horizontal cracks, cracks wider than 0.2 mm, and holes in the concrete cover should be filled or sealed with injection. This aims to reduce the risk of reinforcing steel corrosion, solve leakage problems, and avoid weak bond strength.

Before testing, the thickness of bond defects/voids in the adhesive layer was measured for samples D1, D2, M1, and M2 using a feeler gauge. The depth of concrete defect/drilled voids in the concrete cover for samples C1, C2, M1 and M2 was measured using a digital caliper. The results are summarised in a Table 5, and a sample measurement is shown in Figure 6.

Table 5. Size and volume of the individual materials and defects in the prepared samples.

	Width [mm]	Length [mm]	Thickness [mm]	Volume [mm^3]
Concrete	100	500	100	5,000,000
CFRP strip	50	490	1.2	29,400

		Bond defect in the adhesive layer				Concrete defect—drilled voids			
		Width [mm]	Length [mm]	Average thick-ness [mm]	Volume 1 [mm^3]	Hole diameter [mm]	Depth [mm]	Volume 2 [mm^3]	Total volume [mm^3]
Defects	D1	15	15	0.3	67.5	-	-	-	67.5
	D2	15	50	0.35	262.5	-	-	-	262.5
	C1	-	-	-	-	10	18.9	1483.7	1483.7
	C2	-	-	-	-	10	19	1491.5	1491.5
	M1	15	33	0.65	321.8	10	21.3	1672.05	1993.8
	M2	15	50	0.525	393.8	10	22.2	1742.7	2136.5

(a) (b)

Figure 6. Example measurement of defect size: (**a**) thickness of the defect in the adhesive layer; (**b**) concrete defect in the form of drilled voids in the concrete cover.

3. Methods and Results

3.1. Infrared Thermography (IRT)

To detect defects, active thermography was used; i.e., an external heat source was used to heat the CFRP strip in the area captured by the thermal imaging camera.

The surface of the specimens was heated using two types of radiation sources: a halogen lamp (KFB 93325, 2 × 1000 W, Kaiser Fototechnik GmbH & Co. KG, Buchen, Germany) and an infrared radiant heater (Einhell QHT 1500, 1500 W, Einhell Germany AG, Landau/Isar, Germany) (Figure 7). Both lamps were placed at a distance of about 1 m from the surfaces of the specimens. The heating duration was approx. 5 min. It allowed the CFRP strip to reach a temperature of approximately 37–38 °C. A radiant heater was placed above the samples perpendicular to the surface with the CFRP strip applied, so that the strip, not the concrete sample, was heated. The short heating time was intended to limit the amount of heat supplied to increase the concrete temperature. The scale effects associated with the small size of the concrete sample were negligible, as smaller samples heat up faster but have a higher surface area relative to their volume, which means that heat dissipates more efficiently and the temperatures obtained for the concrete should be similar. In the case of CFRP strip, which has the highest thermal conductivity, the scale effect may be much more significant, but this requires confirmation on full-size samples. Before testing, the CFRP strip surface was cleaned using isopropyl alcohol. This is important to remember, as any contamination, such as dust or grease, can cause uneven surface heating. The temperature increase in the central part of the CFRP strip during heating was recorded by an additional thermal camera (UNI-T Uti260B, Uni-Trend Technology Co., Ltd. Dongguan, China) every minute. The average results are presented in Figure 8.

The ambient temperature during the test was 22 ± 1 °C and the emissivity of the concrete surface was set to 0.95 on the infrared camera (Testo 890, Testo SE & Co. KGaA, Black Forest, Germany). Air humidity was 31.5%, measured using a Benetech GM1365 humidity and Temperature Recorder.

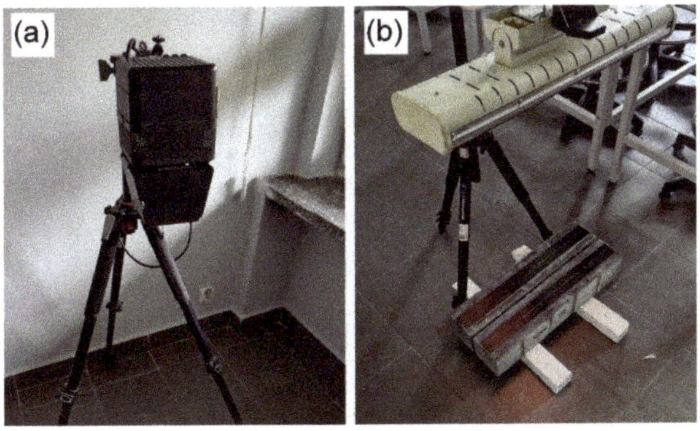

Figure 7. Heat sources used in the test: (**a**) halogen lamp; (**b**) radiant heater.

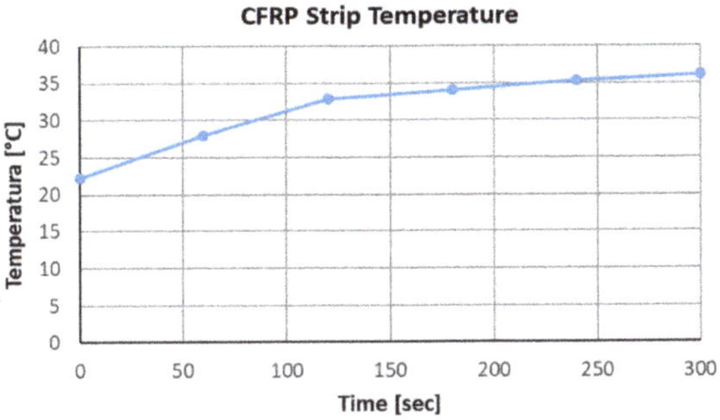

Figure 8. Averaged results for the temperature increase in the central part of the CFRP strip during heating.

Preliminary tests showed that the halogen lamp heated the surface unevenly, making defect localization difficult. In contrast, the infrared radiant heater provided a more uniform and controllable thermal excitation, and was therefore used in subsequent measurements. In addition to power, the heating efficiency depends on the spectral characteristics of the source, the distance and angle of incidence, and the surface properties of the specimen. Although the halogen lamp provides higher luminous flux, its emission in the near-infrared range results in less effective and less uniform heating compared to the infrared radiant heater.

A Testo 890 thermal imaging camera with an infrared resolution of 640 × 480 pixels (Figure 6) and other technical parameters specified in Table 6 were used for the measurements. It was placed on a tripod approximately 120 cm above the samples (Figure 9).

Table 6. Technical specifications of the infrared camera (Testo 890).

Parameter	Specification
Model	Testo 890
Detector type	Uncooled microbolometer
Infrared resolution	640 × 480 pixels (SuperResolution: 1280 × 960 pxs)
Spectral range	7.5–14 μm
Thermal sensitivity (NETD)	<40 mK at 30 °C

Table 6. *Cont.*

Parameter	Specification
Temperature measurement range	−30 °C to +650 °C (optional up to +1200 °C)
Accuracy	±2 °C or ±2% of reading
Field of view (standard lens)	30° × 23°
Minimum focus distance	0.1 m
Frame rate	33 Hz
Focus	Manual or motorised
Emissivity adjustment	0.01 to 1.00
Visual image camera	3.1 MP digital camera (integrated)

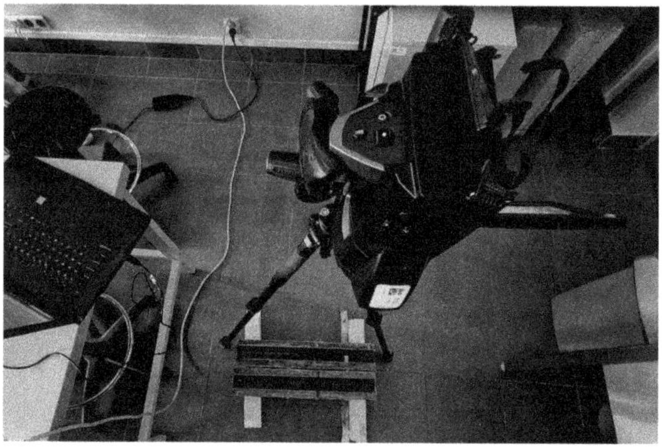

Figure 9. A thermal imaging camera placed above the test samples.

The measurement, that is, the image was recorded immediately after the heat source was removed. The defects were visible in the image for approximately 10 s, after which the temperature difference disappeared very rapidly. The results obtained, along with the defect indications, are presented in Figure 10. Images obtained from a thermal imaging camera allow for the location of defects, but only for a short time after the heat source is removed. All defects were detected, regardless of type and volume. The temperature within the defects is approximately 2 °C higher. This is clearly visible in the detailed analysis presented for sample M1, where cross sections were made as shown in Figure 11. The thickness of the defect in the adhesive layer in this case is 0.65 mm. The temperature readings are presented in the graphs shown in Figure 8. This analysis also allows us to observe a small temperature difference between overlapping defects, approximately 0.2–0.3 °C. A similar effect is visible in the case of sample M2, where varying defect thicknesses in the adhesive layer in the adhesive layer (from 0.4 to 0.65 mm) are observable. Higher temperatures—about 1 °C (from 34.3 °C to 35.3 °C, section P3 in Figure 12)—can be observed on the side with the larger defect thickness. However, these are very small differences that may be difficult to observe in a real structure due to contamination or environmental influences. The defect in the concrete cover is not visible in this sample. Re-examination of samples M1 and M2 carried out to verify the repeatability of the obtained results and a broader analysis of the possibility of locating overlapping defects gave similar results (Figure 12). Again, in the case of sample M1, slightly higher temperature values were obtained in the hole area of the drilled in the concrete cover. The maximum recorded temperatures within the defects for both samples were similar: 36 °C (sample M1) and

35.5 °C (sample M2). Both occurred in locations where the thickness of the defect in the adhesive layer was similar, at 0.65 mm. In sample C2, the temperature in the bond defect area was higher than in the concrete defect area of sample D2. On sample D1 (Figure 10), a lighter area can be seen to the left of the defect; this is where the strip was contaminated with resin during sample preparation.

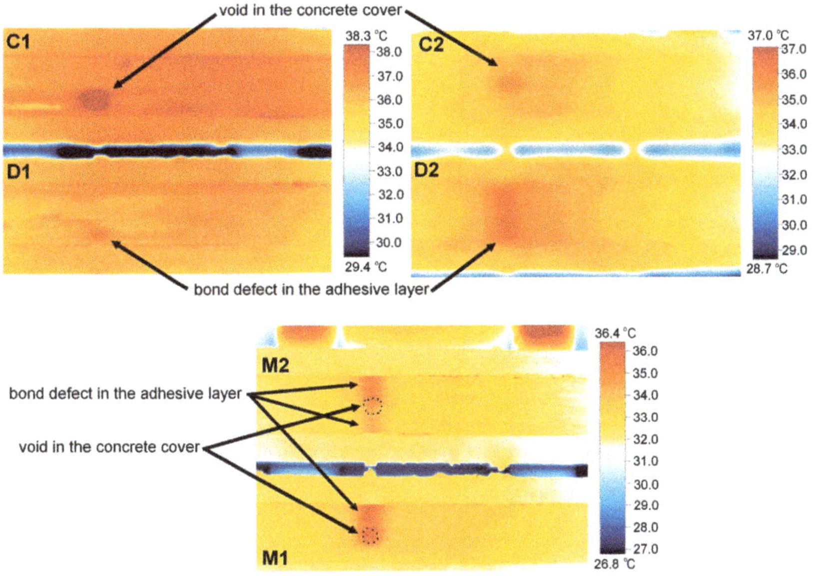

Figure 10. Infrared data obtained using the IRSoft software for the tested sample.

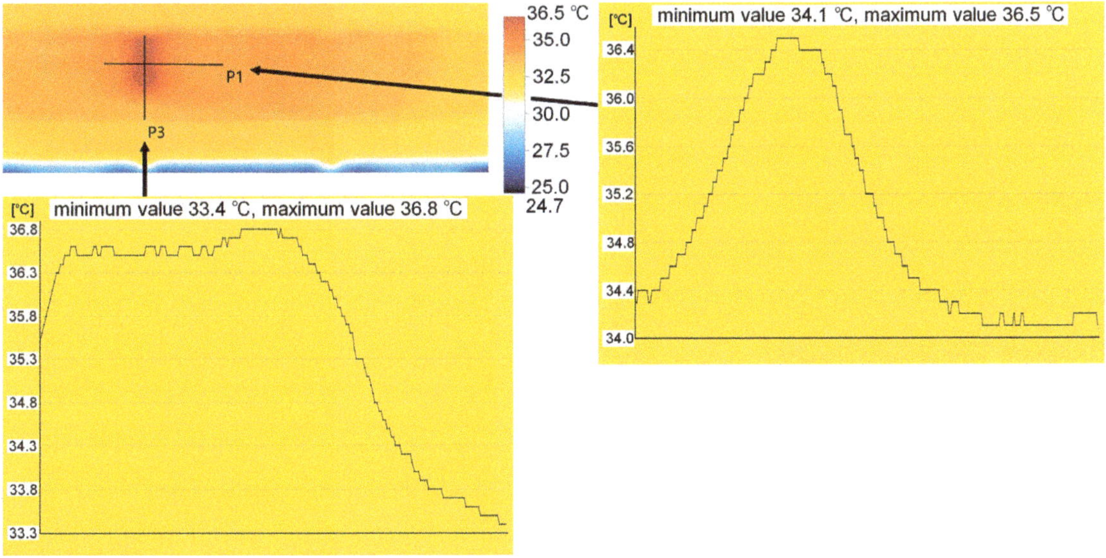

Figure 11. Temperature distribution along selected sections for sample M1.

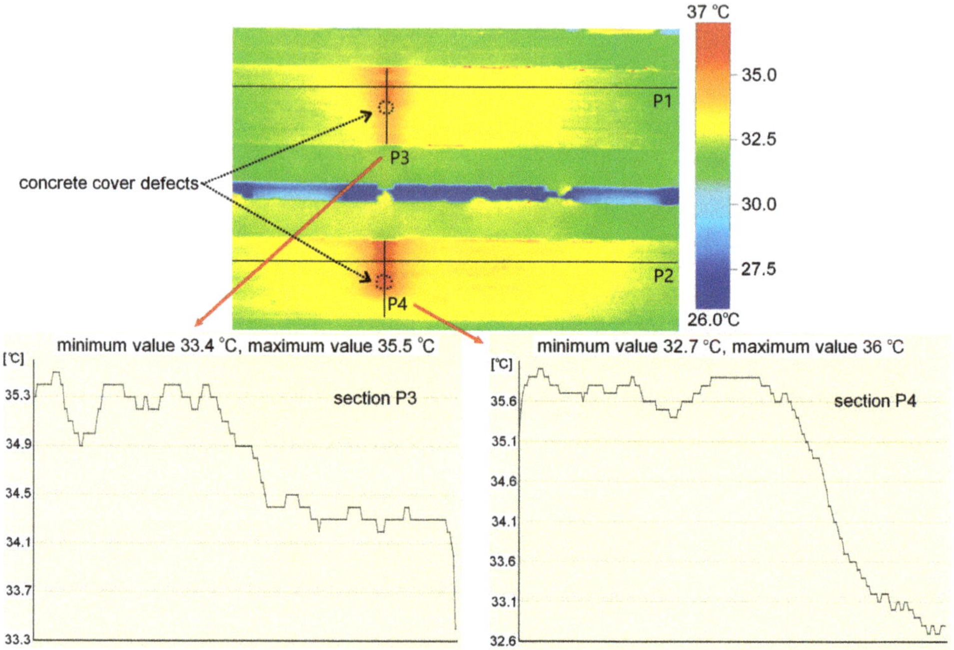

Figure 12. Results of temperature distribution measurements on the surface of samples M1 and M2—repeated testing.

The main factor influencing the increase in visible temperature in the area of defects is a significant change in the thermal conductivity of the material properties under the CFRP strip, meaning that instead of resin and concrete, air is filling the defects. The change in depth and, therefore, the volume of air in the defect, contributes to very small deviations in the recorded temperature, which may well result from other factors, such as surface contamination or concrete moisture.

All infrared data were analysed using the IRSoft v5.2 software (Testo SE & Co. KGaA, Germany). The software allowed the extraction of temperature profiles, contrast enhancement, and emissivity correction. The thermograms shown in Figures 10–12 were exported directly from IRSoft.

3.2. Ground-Penetrating Radar (GPR)

Regardless of the tests described above, an attempt was made to detect defects in the strip-sample surface connection using ground-penetrating radar (GPR). IDS Aladdin GPR system with a 2 GHz antenna was used. The choice of this type of antenna was justified by its wide practical application and compromise between measurement accuracy (antenna resolution) and depth range. Commercially available antennas with higher frequencies (up to 6 GHz) are characterised by higher resolution and accuracy, but their short range (on the order of several centimetres) makes their use ineffective in engineering applications.

Each experimental study involved passing the antenna along the test sample, directly over the defect. The studies were carried out in two variants: scanning the surface of the composite strip (Figure 13) and scanning the opposite (unreinforced) surface.

Figure 13. Scanning the sample using the IDS Aladdin georadar and a 2 GHz centre frequency antenna.

To ensure the good quality and repeatability of the results, the scanning was repeated five times in two directions.

The results in the form of radargrams (Figure 14) were analysed. The radargram can be interpreted as a "vertical cross section" of the tested object, placed along the antenna's path. The upper horizontal axis of the radargram represents the antenna's path across the tested surface. The vertical axis, in turn, represents the time of bidirectional wave propagation (from the transmitting antenna to the medium and back). Knowing the wave propagation velocity in the tested medium, one can convert the time scale to a depth scale, allowing one to determine the depth of the target object. In the case of concrete, the wave propagation velocity is approximately 10–13 cm/ns.

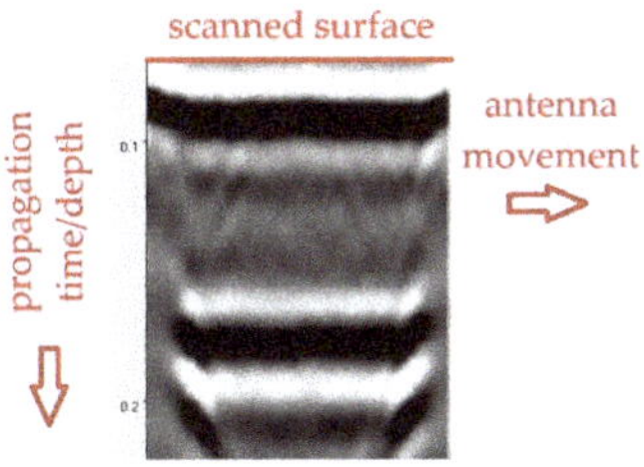

Figure 14. Interpretation of GPR results.

The greyscale in Figure 14 presents the amplitude of the wave reflected from individual areas of the medium. Negative amplitudes are marked in black, whereas positive amplitudes are marked in light colours. Discontinuities and objects within the medium are visible as local peaks in wave amplitude.

Examples of processed radargrams are presented in Figure 15. In each case, horizontal bands are visible, representing wave reflection from the top and bottom surfaces of the sample. Inverted hyperbolas can be observed at the beginning and end of each scan, indicating wave reflection from the edges of the sample.

Figure 15a illustrates an example radargram recorded on a defect-free sample (N2), In this case the antenna was moved across the unreinforced surface of the sample and the reinforced surface was located underneath. As mentioned above, the horizontal bands in Figure 15a indicate wave reflection from the sample surfaces. No signals were observed between the bands (in the sample volume). Therefore, no defects were detected, either in the concrete volume or in the reinforced area.

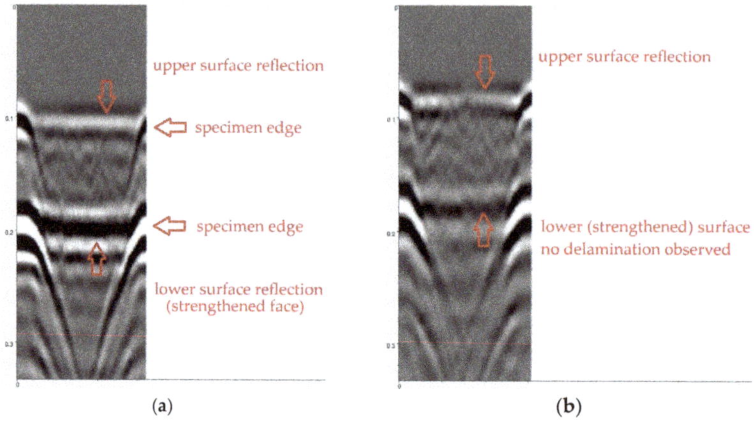

Figure 15. Example radargrams recorded during sample scanning: (**a**) sample without defects (N2), (**b**) sample with CFRP strip debonding (D1).

Figure 15b presents an example scan result for a sample with an adhesive layer defect (debonding, sample D1). As before, the scan was conducted on the unreinforced surface, so the surface with the strip and adhesive bond is visible in the lower part of the radargram. However, none of the measurements recorded a signal that could indicate the presence of a defect in the adhesive layer. This is due to the small size of the defect (approximately 2 mm thick) relative to the antenna resolution.

The vertical resolution of the antenna (the ability to distinguish objects located at different depths) can be estimated as 0.25 of the wavelength λ [71]. The value of λ is in turn, defined as the quotient of the wave speed to the centre frequency. Assuming a typical wave speed for concrete v = 10 cm/ns = 10^8 m/s and the central frequency of the antenna used f = 2 GHz = 2·10^9 Hz, one can obtain:

$$\lambda = \frac{v}{f} = \frac{10^8}{2{\cdot}10^9} = 0.05\text{m} = 5\text{cm}, \tag{1}$$

The vertical resolution R can be calculated as follows [72]:

$$R = \frac{\lambda}{4} = \frac{5}{4} = 1.25\text{cm}, \tag{2}$$

As can be seen, the resolution of the antenna used is much greater than the size of the defect in the adhesive layer. In general, it is assumed that GPR allows for the detection of objects measuring at least the antenna resolution determined by Equation (2). As discussed in [72], the detection of smaller objects is possible in principle but requires high-quality input data as well as complex processing techniques. In practical terms, it can therefore be assumed that the limiting size of a void that can be detected using a 2 GHz antenna is approximately 1–1.5 cm.

Furthermore, the dielectric properties of concrete are crucial for estimating resolution. These properties change, for example, depending on the humidity or the presence of chlorides. These factors directly influence the dielectric constant ε_r, which is a measure of its ability to store energy in an electric field. A change in ε_r (resulting, for example, from concrete moisture) directly translates into a change in wave velocity, according to the formula:

$$v = \sqrt{\frac{c^2}{\varepsilon_r}}, \tag{3}$$

where v—mean wave velocity in the tested medium, c—speed of light in vacuum, ε_r—dielectric constant of concrete.

Since increasing the moisture content of concrete leads to an increase in the dielectric constant (ε_r for concrete is between 9 and 12, while for water it is 81), according to Equation (3), the wave velocity decreases, which in turn results in a reduction in wavelength (Equation (1)) and improved resolution (Equation (2)). In practice, this means that smaller objects can be detected but at the cost of increased wave attenuation and reduced range.

Generally, however, objects smaller than the theoretically determined antenna resolution are not visible as a strong, single echo, but rather as a small wave disturbance that can easily be confused with background noise.

Taking the above into consideration, delamination can affect the characteristics of the reflected wave (namely, reducing the wave amplitude). This problem has been described in some detail for the case of reinforced concrete slabs in bridges [73]. In this paper, on this assumption, an attempt was made to evaluate the effect of a bond defect on the amplitude of the wave reflected from the CFRP strip. Figure 16 presents the results obtained for the case where scanning was performed on an unreinforced surface and the wave propagated through the thickness of the sample. The wave reflection from the strip is visible on the right side of the graph. In debonding, the peak amplitude was slightly lower (3.26 V) than in defect-free areas (3.42 V).

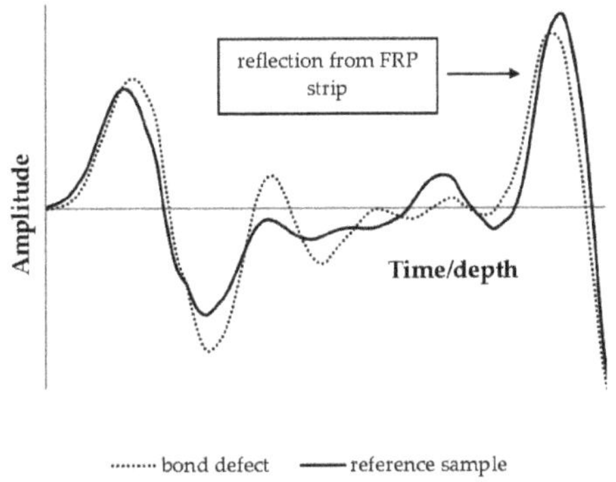

Figure 16. Influence of the bond defect on the amplitude of wave reflected from the CFRP strip.

Although visible and repeatable in all tested samples, this effect is not strong enough to be of practical use. Therefore, the 2 GHz GPR antenna is not a useful tool for assessing adhesive layer defects in joints.

Figure 17 presents the results of scanning C1 sample (with a void in the concrete under the strip). Scanning was performed on the unreinforced surface.

Figure 17a illustrates a raw radargram (without filters or processing). As before, wave reflections from the sample surface and edges are clearly visible. In addition to these signals, an echo in the form of an inverted hyperbola is visible in the right part of the radargram, originating from a void in concrete. However, due to environmental noise and multiple reflections in the radargram, this signal is poorly visible and can easily be missed during interpretation.

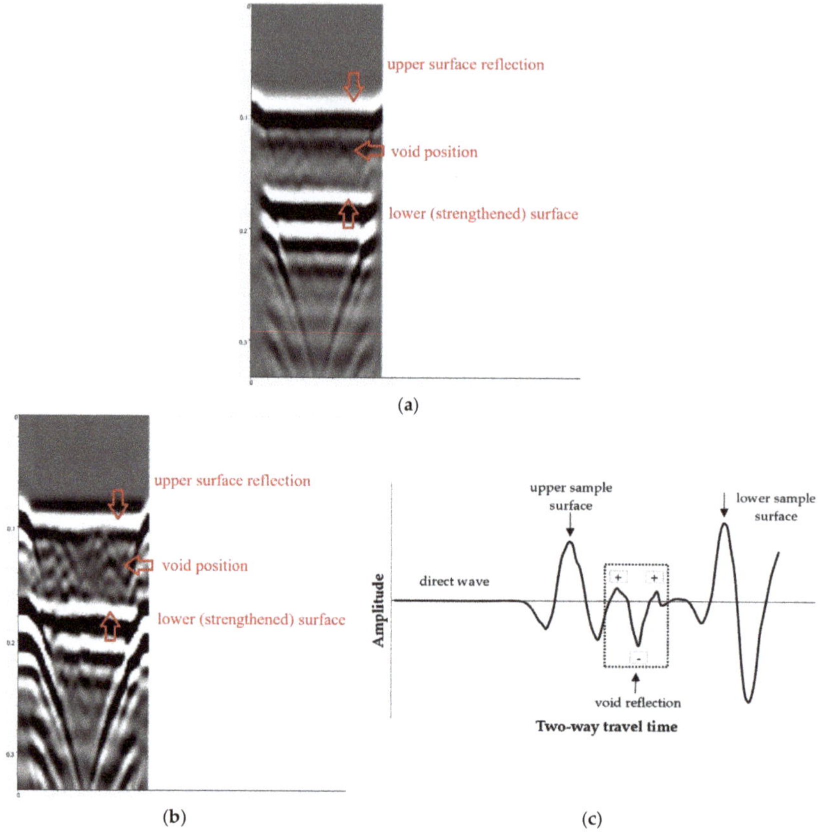

Figure 17. Radargrams recorded on sample with a void under the strip (C1): (**a**) raw data, (**b**) radargram form (**a**) after background removal and contrast enhancement, (**c**) A-scan through void section.

Several radargram processing procedures were used to highlight this echo, namely: background removal, bandpass filter, Hilbert envelope, and linear gain. The void was the most clearly visible in radargrams subjected to background removal, possibly combined with linear gain or contrast enhancement. Figure 17b illustrates an example radargram after applying background removal and increasing contrast. The hyperbola representing the wave reflection from the void is clearly discernible.

This conclusion is confirmed by analysis of the polarity of the reflected wave. As can be seen in Figure 17b, the reflection from the void actually includes 3 hyperbolas (looking from top to bottom, the hyperbola is white, the one below is black and then white again). This means that the signal includes 3 amplitudes in the order $+/-/+$. The negative (middle) amplitude has the highest value. In the case of the system used, this means that the wave was reflected, passing from a medium with a higher dielectric constant ε_r to a medium with a lower constant. Since the value of ε_r for concrete is 9–12 and for air it is 1, the result obtained indicates that the wave was most likely reflected from a void filled with air.

The behaviour of the wave after reflection can also be analysed in terms of the reflection coefficient Γ. It is defined as the ratio of the amplitude of the wave reflected from the object A_{ref} to the amplitude of the incident wave, A_{inc}. While A_{ref} is directly read from the radargram, accurate determination of A_{inc} is not possible in practice. A certain approximation of A_{inc}

can be the amplitude of the wave reflected from a metal plate, but such a measurement does not take into account a number of phenomena, such as wave attenuation and scattering in concrete, wave reflection from the sample surface, environmental noise, and others.

Therefore, alternatively, assuming the direction of wave incidence normal to the surface of the object, the reflection coefficientwas determined from the formula:

$$\Gamma = \frac{\sqrt{\varepsilon_{object}} - \sqrt{\varepsilon_{concrete}}}{\sqrt{\varepsilon_{object}} + \sqrt{\varepsilon_{concrete}}}, \tag{4}$$

where Γ—wave reflection coefficient, $\varepsilon_{\text{object}}$—dielectric constant of the object (void filled with air was assumed, therefore $\varepsilon_{\text{object}} = 1$), $\varepsilon_{\text{concrete}}$—dielectric constant of concrete.

The value of $\varepsilon_{\text{concrete}}$ was determined by transforming Equation (3) into the form:

$$\varepsilon_{concrete} = \frac{c^2}{v^2}, \tag{5}$$

The wave velocity v was calculated as the quotient of the doubled distance from the sample surface to the void (the wave path from antenna to the void and back) divided by the time of bidirectional wave propagation read from the radargram. The average wave velocity between the sample surface and the void was 9.21 $\frac{\text{cm}}{\text{ns}}$, and the value of the $\varepsilon_{\text{concrete}}$ from Equation (5) was 10.61. Substituting into Equation (4) $\varepsilon_{\text{object}} = 1$ and $\varepsilon_{\text{concrete}} = 10.61$, the reflection coefficient was obtained:

$$\Gamma = \frac{\sqrt{1} - \sqrt{10.61}}{\sqrt{1} + \sqrt{10.61}} = -0.53, \tag{6}$$

The (−) sign in this case indicates a change in the polarity of the reflected signal.

Since the depth of the void was 3 cm and the estimated resolution of the apparatus was 1.25 cm (Equation (2)), detecting the defect with the 2 GHz antenna was not problematic.

However, due to its small diameter (1 cm), the void should be regarded as a point defect. In this situation, the horizontal resolution of the GPR apparatus becomes particularly important, as it indicates the ability to distinguish objects located next to each other horizontally. The horizontal resolution R_{h} depends on the depth of the defect and is determined from the following relationship:

$$R_h = \sqrt{\frac{\lambda \cdot z}{2}}, \tag{7}$$

where R_{h}—horizontal antenna resolution, λ—wavelength according to Equation (1), z—depth of defect location.

In the analysed case, assuming $\lambda = 5$ cm (see Equation (1)), and $z = 7$ cm (distance from the surface of the scanned sample to the void), R_{h} can be calculated:

$$R_h = \sqrt{\frac{5 \cdot 7}{2}} = 4.18 \text{ cm}, \tag{8}$$

As can be seen, the diameter of the void is significantly smaller than the system's horizontal resolution. However, unlike vertical resolution, this does not mean that the defect cannot be detected. In this case, the void is visible as a point defect, and its small dimensions compared to the R_{h} mean that its shape and extent cannot be determined from the radar image.

Summarising the results of the scan on the unreinforced surface (Figures 15–17), it can be concluded that the GPR system was found to be an effective tool to detect voids in concrete located under the composite reinforcement, but it is impossible to diagnose defects and adhesive joint defects using this system. The test results for samples containing both types of defects (void and adhesive joint defect, samples M1 and M2) confirm these conclusions and will not be discussed in detail here.

Scanning the samples on the composite-reinforced surface did not reveal the internal structure of the concrete sample or any defects in the composite reinforcement. This is due to the fact that the CFRP fibre used in the reinforcing strips is an electrical conductor and therefore reflects electromagnetic waves. As a result, depending on the fiber content antenna frequency and the resin used the dielectric constant of CFRP ranges from 3 to 30 [74] or more. Therefore, the penetration of the wave into the sample is very difficult or even impossible. An example of the results of scanning sample C1 (with a void) from the composite reinforcement side is shown in Figure 18. As can be seen, the reflected signals are weaker than before, and unlike scanning the opposite surface (Figure 17), it was impossible to detect the void.

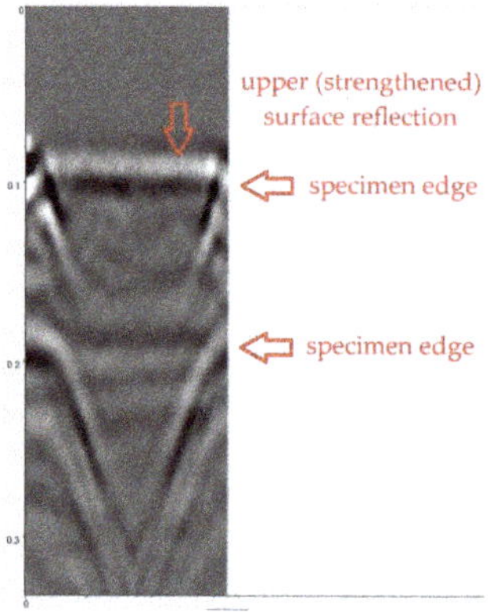

Figure 18. Radargram recorded on sample with a void (C1), scanned on the surface of the composite strip.

3.3. *Wave Velocity Measurement Method (WVMM)*

The study used the Pocket AE unit [75], which allows the connection of two sensors and an external input parameter that records the load or stress for correlation with the measurements of acoustic emission. The kit is portable and allows for conducting research both in the laboratory and in the field for up to six hours. The device has an 18-bit analog-to-digital converter, which allows for sampling rates of up to 10 MSPS (Mega Samples Per Second). An interesting feature of the device is the ability to excite the crystal in the acoustic emission sensor using repetitive pulse sequences (Auto Sensor Test). During the study, acoustic wave parameters such as counts, amplitude, energy, rise time, and waveform are recorded and can be analyzed in real-time as graphs. Two PK6I [68] resonant sensors with a built-in 26 dB preamplifier, with operating frequency range 35–65 kHz and peak sensitivity 106 dB (Figure 19a,c) were used in the study, which can both generate and receive acoustic emission waves.

The study (Figure 19b) aimed to measure the elastic wave velocity [76] in concrete and composite materials based on the path between the sensors and the difference in the wave's arrival time at the sensors. The study was preceded by preliminary work consisting of cleaning and grinding the surfaces to which the sensors were attached. Additionally, a silicone layer was applied to the surface to ensure adequate contact between the sensors and the elastic wave registration surface.

The tests were carried out on six samples marked with symbols C1 and C2, D1 and D2, and M1 and M2 (Figure 20b), characterized in Chapter 2.2 of this study. Elastic wave velocity measurements were performed for each sample using PK6I sensors located on opposite sample walls, as shown in (Figure 20a), at three measurement points designated on the element surface, enabling the estimation of elastic wave velocity in the assumed cross-sections (Figure 21):

- P1 indicates the measurement point located above the damage;
- P2 indicates the measurement point 10 cm from the damage;
- P3 indicates the measurement point located on the wall perpendicular to the plane with the composite adhered.

(a)

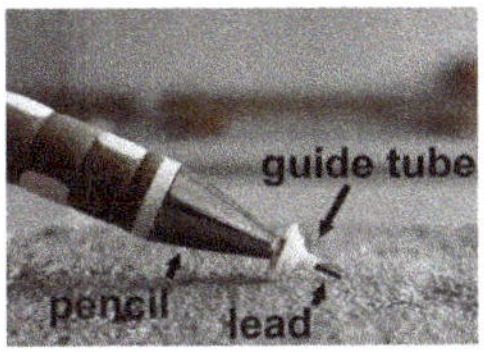

(b)

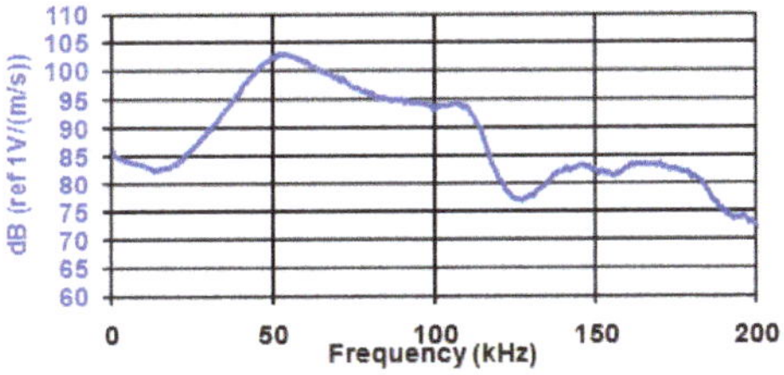

(c)

Figure 19. (**a**) PK6I resonant sensor: (**b**) example of a pencil with a guide prepared to generate a wave using the Hsu-Nielsen method (**c**) graph of the sensor sensitivity as a function of frequency [77].

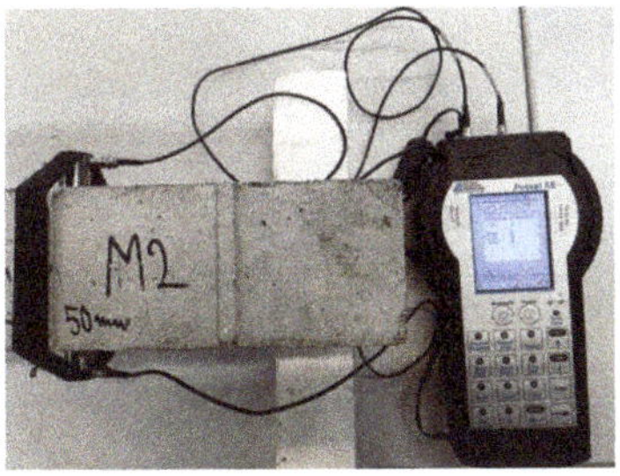

(a)

(b)

Figure 20. Specimen under the test: (**a**) apparatus during the test; (**b**) specimens prepared to test.

The measurement points indicated above (Figure 21) were located to, among other things, obtain the velocity of an elastic wave passing through an intact medium (composed of composite, resin, and concrete), i.e., the so-called baseline elastic wave velocity. They were also used to determine the velocity for a medium composed of the same layers, assuming that various defects were present at the measurement location. Therefore, as already mentioned, the first measurement point (P1) was positioned at the defect location so that the elastic wave passed through the prepared defects. The second point (P2) was approximately 10 cm from P1 so that the elastic wave did not encounter any defects along its path and, as a result, passed through the baseline composite–resin–concrete system, allowing for the baseline elastic wave velocity value to be obtained. The third point (P3) was set so that the sensors were directly adjacent to the concrete surface, hence they were placed on walls perpendicular to the surface on which the composite was attached, as a result of which a comparative value of the elastic wave velocity passing only through the concrete was obtained.

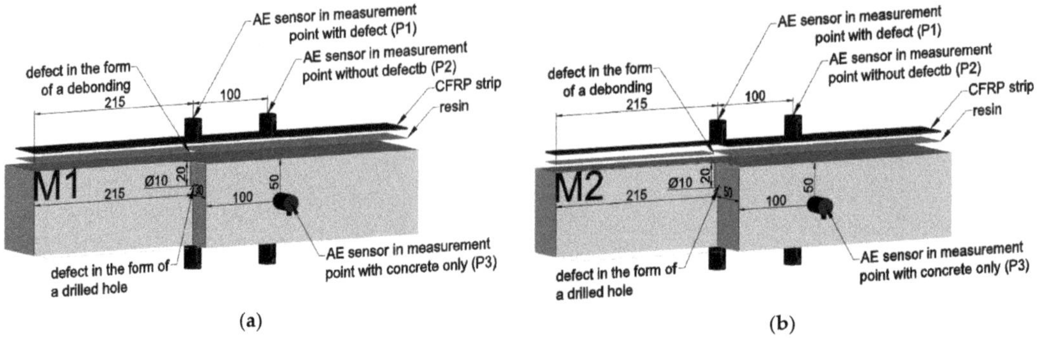

Figure 21. AE sensor arrangement diagram for the M1 and M2 samples; (**a**) M1 sample with a defect in the form of a drilled hole and a detachment under a part of the composite; (**b**) sample with a defect in the form of a drilled hole and a detachment along the entire width of the composite—units are given in [mm].

It should also be noted that, regardless of the measurement point, at least one of the sensors was always placed on the concrete, while the other was located on the opposite side. As a result, in the case of points P1 and P2, it was placed directly on the composite surface, and in point P3, it was placed on the concrete (Figure 21).

Furthermore, at each measurement point (P1–P3), the elastic wave excitation always took place in the immediate vicinity of sensor no. 1 by breaking a graphite stylus 10 times at an angle of 30°, as illustrated schematically in Figure 22. The average values of the elastic wave speed recorded at the measurement points was shown in Table 7.

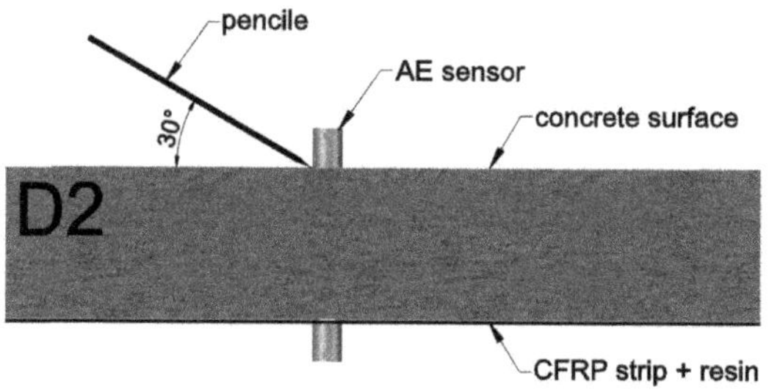

Figure 22. Schematic diagram of elastic wave excitation using sample D2 as an example.

Table 7. Summary of the average elastic wave velocity V_W and its difference recorded at the measurement points.

Test Samples	AE Wave Velocity Vw												Difference in AE Wave Velocity Between Position 2 and 1	
	P1				P2				P3					
	V_w [m/s]	s [m/s]	CV_w [%]	SE [m/s]	V_w [m/s]	s [m/s]	CV_w [%]	SE [m/s]	V_w [m/s]	s [m/s]	CV_w [%]	SE [m/s]	ΔV_w [m/s]	ΔV_w [%]
C1	2940	93.0	3.2	29.5	3410	41.5	1.2	13.1	3779	60.2	1.6	19.0	470	14%
C2	3403	90.6	2.7	30.5	3774	118.4	3.1	37.6	3744	208.9	5.6	66.1	371	10%
D1	3473	130.9	3.8	41.4	3540	178.9	5.1	56.6	3668	72.6	2.0	23.0	67	2%
D2	2941	59.8	2.0	18.9	3707	68.6	1.9	21.7	3668	56.3	1.5	17.8	766	21%
M1	2940	108.8	3.7	34.4	3601	142.3	4.0	45.0	3732	77.9	2.1	24.6	661	18%
M2	3004	95.4	3.2	30.2	3688	122.6	3.3	38.8	4150	354.1	8.5	112.0	684	19%

The time between the two sensors was determined using the standard threshold method. In all experiments, the threshold was set to 45 dB, which corresponds to 0.18 mV after conversion. Figures 23 and 24 present example waveforms recorded on the side opposite the pencil-lead break (Hsu–Nielsen test). The time of arrival (TOA) was defined as the moment of the first threshold crossing of the wave, as shown in Figures 23b and 24b. Figure 24a shows a waveform recorded in the M2 sample without defects. The peak amplitude rises quickly to about 35 mV and then rapidly decays to about 5 mV. Figure 24b shows the first wave peak, which drops to −1.5 mV. Figure 24a shows a waveform recorded in the M2 sample with defects. The peak amplitude rises quickly to about −8 mV and then slowly decays to about 2 mV. Figure 24b shows the first wave peak, which increases to 2 mV. Based on Figures 23 and 24, clear differences can be observed between the waveforms recorded in defect-free and defective samples. Therefore, waveform analysis can serve as an additional indicator of structural changes in the sample.

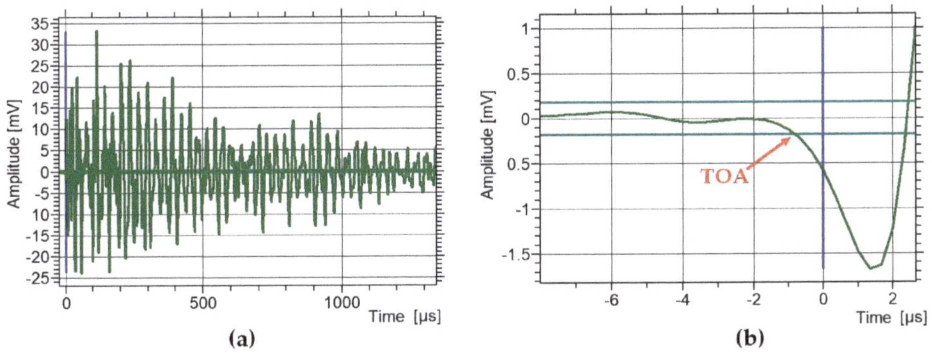

Figure 23. Waveform recorded in the M2 sample without defects: (**a**) full scale, (**b**) beginning of the wave with the red marked time of arrival (TOA).

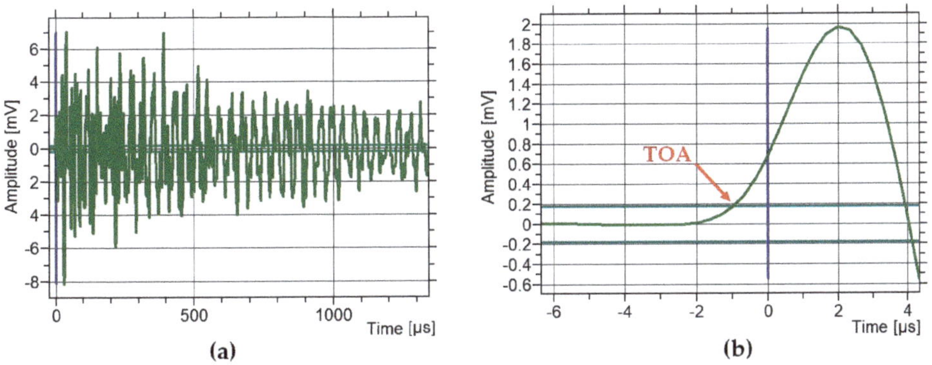

Figure 24. Waveform recorded in the M2 sample with defects: (**a**) full scale, (**b**) beginning of the wave with the red marked time of arrival (TOA).

The elastic wave speeds obtained in this way allowed for estimating the average values of the elastic wave speed (V_W) together with the basic statistical parameters in the form of standard deviation (s) coefficient of variation (CV_W) and standard error of the mean (SE) (Table 7). It should be mentioned here that the obtained results were very good repeatability, because in most cases the determined coefficient of variation was within the range of 1–4%, reaching a maximum value of no more than 9%. Also a low standard error of the mean (SE) indicates that the mean obtained from the analyzed sample corresponds with a high probability to the actual population mean. The mean wave velocity V_W values in concrete (P3) were consistent between 3650 and 3750 m/s, with the exception of sample M2, which the authors attribute to improved concrete compaction. This confirms that the observed decreases in mean wave velocity in P1 are primarily due to a defect in the CFRP/adhesive/concrete propagation path, rather than to the overall quality of the concrete. It was also noted that the larger and more extensive the defect, the greater the decrease in V_W velocity over the defect (P1) relative to the defect-free reference (P2):

- D1 (debonding at the edge of the CFRP strip): $\Delta V_W = 2\% \rightarrow$ barely measurable effect.
- C1, C2 (voids at the edge and in the middle of the CFRP strip): $\Delta Vw = 10–14\% \rightarrow$ moderate decrease.
- M1, M2 (overlapping defects): $\Delta V_W = 18–19\% \rightarrow$ significant speed drop.
- D2 (debonding running across the entire width of the CFRP strip): $\Delta V_W = 21\% \rightarrow$ greatest speed drop.

Additionally, in order to illustrate the observed relationships, the values of the average elastic wave velocity are presented in Figures 25 and 26. The type of defect itself did not determine the mean velocity V_W. The decrease was primarily determined by the size, extent, and location of the discontinuity relative to the wave propagation path. Both adhesion loss in the adhesive layer and voids in the concrete cover resulted in comparable decreases in mean velocity V_W. Therefore, the WVMM method is useful for detecting the presence of damage, but not for its typing.

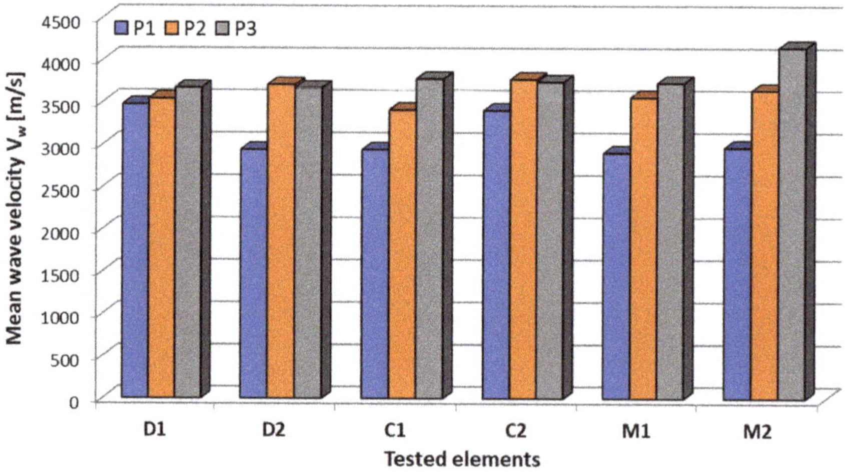

Figure 25. Comparison chart for the obtained average values of elastic wave velocity V_W for individual samples at each of the three measurement points P1 (measurement through the composite at the effect location), P2 (measurement through the composite at the defect-free location), P3 (measurement only through the concrete).

Moreover, based on the above results of the average wave velocity, it can also be stated that the phenomenon of anisotropy related to the presence of the composite does not occur in the context of the conducted research. This is confirmed by the comparison of the average velocity of the wave passing through the composite–resin–concrete system and the wave recorded after passing through the concrete medium itself, and comparing these two values, no significant differences were found.

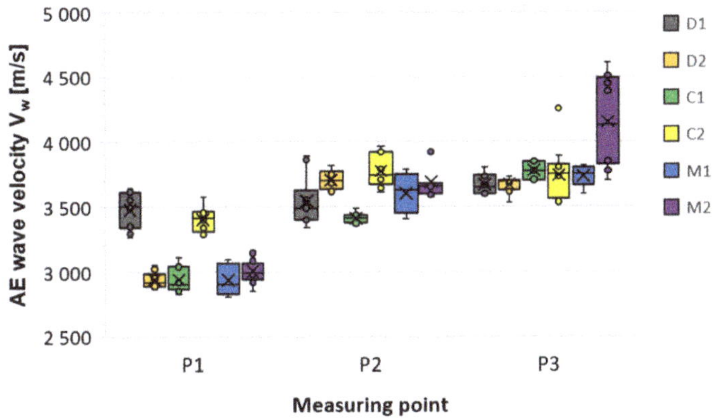

Figure 26. Comparison chart for the obtained average values and statistical parameters of elastic wave velocity V_W for individual samples at each of the three measurement points P1–P3.

4. Discussion

Infrared thermography (IRT) is an effective method for detecting and locating defects in externally bonded FRP concrete structures. This method allows for relatively quick and easy detection of defects and is also economical, but certain limitations must be kept in mind. The tested surface must be uniformly heated; the recording time after removing the heating source is relatively short; surface contamination significantly affects the temperature distribution. A rough estimation of defect size is possible, but assessing their depth is somewhat more difficult. During the study, it was observed that areas with a bond defect in the adhesive layer cooled faster after removing the heating source. In [24], the authors suggest that it is possible to detect the depth of the voids by heating and scanning the adjacent sides of the samples, but this is not always possible and usually takes time.

Moreover, the influence of the heating source on defect detection is also important. The choice of heating source was found to have a significant impact on the quality of thermal contrast and defect visibility. When using the halogen lamp, the surface heating was spatially non-uniform, which resulted in irregular temperature distributions and made the identification of subsurface defects more difficult. The uneven illumination led to local hot spots unrelated to actual defects.

In contrast, the infrared radiant heater produced a more homogeneous heat flux over the inspected area, enabling clearer thermal patterns corresponding to defect locations. This uniform excitation improved the repeatability of the measurements and allowed for more consistent evaluation of defect-induced temperature differences (ΔT). Consequently, the radiant heater was selected for the subsequent experiments as a more reliable excitation source for active thermography.

Summarizing the results of scanning the samples with a 2 GHz GPR antenna, it should be noted that the usefulness of GPR in diagnosing defects in composite reinforcements is limited, requiring the parallel use of other methods. Of the two types of defects, only the void under the reinforcing strip could be identified. Due to the antenna's low resolution compared to the defect size, the GPR method did not allow for the identification of defects in the adhesive bond. These conclusions are consistent with literature data [23,38–42].

Nevertheless, according to [42], delamination had effect on the amplitude of the wave reflected from the CFRP strip. The difference in amplitude between the defect-free and debonding areas was very small (about 5%), making this relationship difficult to use in practice, especially for on-site inspections.

GPR diagnostics of composite reinforcements using CFRP fibers also presents other practical limitations. Due to the strong reflection of the wave from the composite, scanning by moving the antenna along the reinforced surface is ineffective [44]. As presented above, scanning the opposite (unreinforced) surface may give better results. However, in engineering practice, access to this surface is often limited. Furthermore, the opposite surfaces are often also reinforced with composite strips (e.g., beam webs), which makes obtaining qualitative results difficult.

It should be mentioned, however, that this limitation does not apply to GFRP (glass fiber reinforced polymer) composite reinforcements, as this material is an insulator, therefore allowing electromagnetic waves to penetrate into the interior of the tested element.

GPR can therefore be used for joint diagnostics only as a supporting method, useful for detecting voids in concrete occurring under composite reinforcement.

In the case of the WVMM method, as previously demonstrated, it is possible to approximate the location of defects. It should be noted, however, that this method has certain limitations. Primarily, it is impossible to distinguish between a surface defect, such as a debonding (samples D1 and D2), and a deep defect, such as a concrete loss (samples C1 and C2). According to the presented results, in both cases, a decrease in elastic wave velocity was recorded relative to the undamaged location, while the difference in velocities for defected locations between samples does not allow for distinguishing the type of defect. Furthermore, in the case of very small defects, such as minor debonding, it is possible that a change in elastic wave velocity (sample C1) may not be detected. However, these limitations do not preclude the use of this approach as a tool for confirming the presence of defects, whether in significant locations or locations identified by other methods, as a control method. Other limitations of the WVMM method include the influence of factors on wave velocity that may interfere with defect identification, such as the maximum length of the wave velocity measurement section, water-cement ratio, aggregate size, and the repeatability of results. In [64], the influence of the water-cement ratio and the maximum aggregate size on acoustic wave velocity attenuation in concrete was investigated. It was observed that the wave velocity attenuation with distance was greater when the water-cement ratio decreased and when the aggregate size was larger. For a section length between sensors of 1000 mm, a 50% wave velocity attenuation was noted. In [78], the influence of aggregate grain size 2/16 and 8/16 on wave velocity was observed at a constant sensor spacing of 500 mm. Aggregate grain size significantly changes the longitudinal wave velocity in concrete without load. A smaller aggregate fraction of 2/16 mm causes a lower wave velocity, which may be the result of wave dispersion with a larger number of grains and air voids. In [65], it was found that wave excitation using the Hsu-Nielsen test is poorly repeatable; the relative standard deviation for the amplitude spectrum was 11% for the full frequency range. To increase the repeatability of the results, in [65], control of the length of the broken graphite was proposed. The relative standard deviation was reduced to about 70% when the graphite length was in the range of 2.8–2.9 mm.

The presented studies are preliminary. It is planned to expand the research to include full-size samples—beams strengthened with strips and/or sheets in various configurations and a greater variety of defects (including cracks) and other detection methods.

4.1. Uncertainty Analysis of Thermographic Measurements

Due to the precision and reliability of the data, a detailed uncertainty analysis was performed based on the specifications of the Testo 890 thermal imaging camera and the measurement geometry, following the guidelines of the Guide to the Expression of Uncertainty in Measurement (GUM) [79]. The uncertainty was evaluated for the two key aspects of the study: temperature measurement (affecting defect detection and contrast) and spatial resolution (affecting defect localisation).

4.1.1. Temperature Measurement Uncertainty (u_T)

The combined standard uncertainty of the temperature measurement (u_T) incorporates several independent components ($u_c(T)$) related to the instrument's accuracy, sensitivity (NETD), and the assumed emissivity value (ε).

A. Uncertainty Components (Type B):

- Instrument Accuracy (u_{acc}): The manufacturer's specification for the Testo 890 accuracy is ±2 °C. For measurements around 40 °C, the 2 °C limit governs. Assuming a rectangular distribution, the standard uncertainty is: $u_{acc} = \frac{2\ °C}{\sqrt{3}} \approx 1.15$ °C.
- Thermal Sensitivity (u_{NETD}): The Noise Equivalent Temperature Difference (NETD) defines the camera's thermal resolution: $u_{NETD} = 40$ mK $= 0.04$ °C.
- Emissivity Uncertainty (u_ε): The surface emissivity was set to 0.95. Assuming an uncertainty margin of $\Delta\varepsilon = \varepsilon \pm \varepsilon$Th (that is, range 0.93–0.97) due to surface heterogeneity and non-ideal black body behaviour, this variation affects the calculated temperature. Based on an approximate sensitivity analysis for $T_{target} = 40$ °C and $T_{ambient} = 22$ °C, this error is estimated to be ±0.5 °C

The temperature deviation due to emissivity uncertainty was estimated using the relation:

$$u_T = \left| \frac{\partial T}{\partial \varepsilon} \right| u(\varepsilon), \tag{9}$$

assuming constant apparent radiance.

$$u_\varepsilon = \frac{0.5\ °C}{\sqrt{3}} \approx 0.29\ °C.$$

B. Combined and Expanded Uncertainty

The combined standard uncertainty $u_c(T)$ is calculated by root-sum-of-squares (RSS) method:

$$u_c(T) = \sqrt{u_{acc}^2 + u_{NETD}^2 + u_\varepsilon^2} \tag{10}$$

$$u(\Delta T) = \sqrt{u_{TypeA}^2 + u_{TypeB}^2} \tag{11}$$

$$u_c(T) = \sqrt{(1.15)^2 + (0.04)^2 + (0.29)^2} \approx 1.19\ °C$$

Expanded uncertainty (U), used for reporting the result with a 95% confidence level (coverage factor k = 2), is:

$$U(\Delta T) = k \cdot u(\Delta T) \tag{12}$$

$$U(T) = 2 \cdot u_c(T) \approx 2.38\ °C$$

4.1.2. Spatial Resolution and Defect Localization Uncertainty u_L

The uncertainty in localising and sizing defects is determined by the camera's instantaneous field of View (IFOV) at the working distance. This uncertainty does not affect ΔT directly, but defines the spatial confidence of defect localization.

A. Instantaneous Field of View (IFOV) Calculation

The camera's standard lens provides a Field of View (FOV) of 30° × 23° with a resolution of 640 × 480 pixels. The angular IFOV is calculated from the horizontal FOV and resolution. Converting this to milliradians (mRad):

$$IFOV_{mRad} = 0.046875° \cdot \left(\frac{\pi}{180}\right) \cdot 1000 \approx 0.818 \text{ mRad}$$

B. Physical Pixel Size and Localization Uncertainty

The physical size of a single pixel u_{pixel} on the object plane at the measurement distance of 1 m defines the limit of the spatial resolution:

The uncertainty in determining the physical location of the defect edge $u_{lokalizacja}$ is assumed to be bounded by the size of one pixel due to the visual ambiguity of the thermal gradient (edge blur): $u_{location} = \pm u_{pixel} = \pm 0.82$ mm.

This value establishes the precision with which the boundaries and dimensions of the internal defects can be reported.

4.1.3. Statistical Uncertainty from Repeatability (Type A)

Although the primary sources of uncertainty are related to the instrument's specifications (Type B), the repeatability of the measurements provides an essential statistical contribution (Type A), especially for the crucial parameter: the thermal contrast ΔT between the defect area and the sound concrete surface.

To quantify repeatability, we consider the standard deviation of multiple measurements of the same physical quantity under the same conditions (repeatability conditions).

A. Quantification of Repeatability

If the thermal contrast ΔT was measured n times for a representative defect (e.g., the largest or most characteristic one), the standard uncertainty of Type A u_A is the experimental standard deviation of the mean:

$$u(\Delta T) = \frac{s}{\sqrt{n}}, \tag{13}$$

where s is the experimental standard deviation of the n repeated measurements of ΔT, n is the number of repetitions of the measurement.

If the thermal contrast ΔT was measured 5 times n = 5 and the standard deviation was

$$s = 0.35 \text{ °C};$$

$$u(\Delta T) = \frac{0.35 \text{ °C}}{\sqrt{5}} = 0.16 \text{ °C}.$$

B. Inclusion in Combined Uncertainty

To obtain the most accurate total uncertainty for thermal contrast ΔT, the standard uncertainties from the instrument u_{acc}, u_ε and the statistical uncertainty u_A must be combined. Since ΔT is a difference between two temperature readings T_{defect} and T_{sound}, and assuming the Type B uncertainties are correlated (one camera measures both points), we simplify the combined Type B uncertainty $u_B(\Delta T)$ to be equal to $u_C(T)$ calculated previously.

The final Combined Standard Uncertainty for Thermal Contrast $u_C(\Delta T)$ is:

$$u_C(\Delta T) = \sqrt{u_B^2 + u_A^2} \tag{14}$$

Using the previous result for the combined Type B uncertainty, $u_B \sim 1.19\ ^\circ\mathrm{C}$ and the example Type A uncertainty, $u_A \sim 0.16\ ^\circ\mathrm{C}$:

$$u_c(\Delta T) = \sqrt{(1.19\ ^\circ\mathrm{C})^2 + (0.16\ ^\circ\mathrm{C})^2} \approx 1.20\ ^\circ\mathrm{C} \quad (15)$$

The total uncertainty is only marginally increased, but the inclusion of the repeatability component scientifically validates the robustness of the experimental setup and methodology.

Although the uncertainty is low (<0.05 °C), the main source of variability in ΔT arises from nonuniform surface heating. This component, characterised by the standard deviation between repeated thermal contrasts, is included as u_{heat} in the combined uncertainty.

4.1.4. Summary of Uncertainty Analysis

In summary, the combined expanded uncertainty of the thermal contrast measurement using the Testo 890 camera is approximately 0.01–0.05 °C (k = 2), which is negligible compared to the thermal contrast values observed ($\Delta T \approx$ 1–3 °C). The dominant source of uncertainty originates from the variability of external heating and surface emissivity rather than the infrared camera itself. The repeatability analysis confirmed that the measurement method is robust and suitable for defect detection in FRP–concrete systems.

5. Conclusions

- Infrared thermography (IRT), or more precisely, active thermography, which involves applying an independent heat source, a thermal effect, to the examined area, allows the detection and location of defects in externally bonded FRP concrete structures. This applies to both bond defects (voids) in the adhesive layer and concrete defects in the form of voids in the concrete cover. In these areas, the temperature is not transferred to the concrete, so it is higher than in the remaining area. However, it should be remembered that, after the heat source, the surface temperature of the CFRP strip uniforms very quickly. This method allows for a rough assessment of the defect size. Distinguishing between voids in the adhesive layer and voids in the concrete cover is possible, but the results are not always clear. In the case of overlapping defects, the differences in measured temperatures were very small, 0.2–0.3 °C, with a maximum measured temperature of 36.5 °C, a difference of approximately 0.68%. Therefore, detecting such a small difference in the real conditions at the site may be difficult or almost impossible. The method allowed the identification of shallow and medium-depth defects ($\Delta T \approx$ 2–3 °C) with high confidence ($\Delta T \gg U_{total}$). However, for deep or small defects, the thermal contrast approached the measurement uncertainty ($\Delta T \approx$ 0.3–0.5 °C), making detection less reliable. These limitations are intrinsic to the thermal diffusion process rather than the camera performance. However, the method offers rapid, contactless inspection and complements ultrasonic and acoustic techniques in comparative diagnostics.
- GPR can be used to detect defects in concrete with externally bonded CFRP strips, but its scope is limited. The 2 GHz antenna allowed for the easy detection of voids in concrete, but imaging delamination at the strip–concrete interface was not possible due to insufficient system resolution and signal overlap with strip reflection. The amplitude decrease observed at the debonding site was very small (approximately 5%). In practical applications, this difference may be insufficient to clearly identify debonding locations. However, the most significant limitation of GPR inspection is the electrical conductivity of CFRP strips, which practically prevents GPR measurements being conducted on their surface. Scanning from the opposite (unreinforced) side, although efficient as indicated in this study, is often impossible in practice because of limited access.

- The results obtained from elastic wave velocity measurements using the Hsu-Nielsen reference source allowed the following conclusions: the method was confirmed to be effective in detecting defects in concrete with the bonded CFRP strip. A decrease in elastic wave velocity was observed in damaged areas compared to areas without defects, averaging 10 to 20%; the magnitude of the difference in elastic wave velocity depends on the size and location of the defect. In the case of local debonding of the CFRP strip (sample D1), a 2% decrease in velocity was observed; a more dense distribution of measurement points is necessary to minimise the risk of underestimation; the velocity values of the Vwof waves passing only through concrete were consistent (3650–3750 m/s), confirming the homogeneity of the material, with the exception of one sample (M2), where the improved concrete compaction may have influenced the observed wave velocity; the type of defect did not significantly affect the recorded wave velocity values. Both adhesive layer defects and concrete cover defects generated comparable velocity drops, suggesting that the method is useful for detecting the presence of damage, not its type; the differences in elastic wave velocity in areas without and with defects are clear and measurable, demonstrating the method's potential in structural diagnostics.

In summary, the choice of the appropriate research method is not unequivocal and depends on the scope and conditions of the survey. The criteria for selecting a diagnostic technique are summarised in Table 8. Given the effectiveness of infrared thermography (IRT) in detecting various defects and even overlapping defects, the authors recommend that it is used first. However, it is important to remember that relying solely on a single NDT method is not entirely objective. To further verify the accuracy of the obtained results, either the GPR method can be used to confirm whether a void exists in the concrete cover or elastic wave velocity measurements can be used to confirm whether a defect is actually present and not an anomaly caused, for example, by contamination of the CFRP strip.

Table 8. Comparison of NDT methods used in the present study.

Criteria	IRT	GPR	WVMM
Damage detectability	Possibility to detect defects both in the adhesive layer (delamination mapping) and voids in the concrete cover, possibility to detect overlapping defects	possible subsurface void detection, not suitable for delamination mapping	Possibility to detect damage in the form of both delamination and concrete loss
Key limitations	Low resolution of thermal imaging cameras, The necessity to heat the sample surface, and susceptibility to disturbances in temperature distribution caused by external factors	inability to detect voids smaller than the antenna resolution (typically about 1 cm), difficulty in visualising the defect, limited usefulness in electrically conductive composites	inability to distinguish the type of damage, measurement possible only in the area where the sensors are applied, it is necessary to place the sensors opposite each other
Equipment	Testo 890 thermal imaging camera (infrared resolution of 640 × 480 pixels)	high frequency antenna (2 GHz or more)	Pocket AE unit and two PK6I resonant sensors with a built-in 26 dB preamplifier with operating frequency range 35–65 kHz
Environmental restrictions	high sensitivity to external conditions (humidity, rain/snowfall, outside temperature, wind), surface contamination, moisture, and the method of sample heating	sensitivity to moisture and chlorides in concrete	maximum operating temperature range of the equipment, shock limit of the sensor, detection very small elastic waves in materials, very noisy environment
Cost	medium	medium	medium

An appropriate database on the presented issue can significantly improve the development of diagnostic methods by using it and computer-based artificial intelligence to create automated solutions and significantly improve the detection, location and recognition of various types of defects, as is currently done in other construction issues [80].

Author Contributions: Conceptualization, P.T.; methodology, P.T., K.B., W.W., M.T., J.T. and S.W.; software, P.T., K.B., W.W., M.T., J.T. and S.W.; validation, P.T., K.B., W.W., M.T., J.T. and S.W.; formal analysis, P.T., K.B., W.W., M.T., J.T. and S.W.; investigation, P.T., K.B., W.W., M.T., J.T. and S.W.; writing—original draft preparation, P.T., K.B., W.W., M.T., J.T. and S.W.; writing—review and editing, P.T., K.B., W.W., M.T., J.T. and S.W.; supervision, P.T. All authors have read and agreed to the published version of the manuscript.

Funding: This research was funded by the Kielce University of Technology within the framework of the statutory work No. 02.0.21.00/1.02.001 SUBB.BKWD.25.001 and No. 02.0.22.00/1.02.001 SUBB.BKKB.25.002.

Institutional Review Board Statement: Not applicable.

Informed Consent Statement: Not applicable.

Data Availability Statement: The original contributions presented in this study are included in the article. Further inquiries can be directed to the corresponding authors.

Conflicts of Interest: The authors declare no conflicts of interest.

References

1. Zhang, J.; Lin, G.; Vaidya, U.; Wang, H. Past, present and future prospective of global carbon fibre composite developments and applications. *Compos. Part B Eng.* **2023**, *250*, 110463. [CrossRef]
2. Klym, A.; Blikharskyy, Y.; Gunka, V.; Poliak, O.; Selejdak, J.; Blikharskyy, Z. An Overview of the Main Types of Damage and the Retrofitting of Reinforced Concrete Bridges. *Sustainability* **2025**, *17*, 2506. [CrossRef]
3. Alsuhaibani, E. Nondestructive Testing of Externally Bonded FRP Concrete Structures: A Comprehensive Review. *Polymers* **2025**, *17*, 1284. [CrossRef]
4. Tworzewski, P.; Alexy, J.K.; Barnes, R.W. Intermediate Crack Debonding of Externally Bonded FRP Reinforcement—Comparison of Methods. *Materials* **2022**, *15*, 7390. [CrossRef]
5. Ke, Y.; Zhang, S.S.; Smith, S.T.; Yu, T. Novel Embedded FRP Anchor for RC Beams Strengthened in Flexure with NSM FRP Bars: Concept and Behavior. *J. Compos. Constr.* **2023**, *27*, 04022093. [CrossRef]
6. Rasheed, H.A.; Zaki, M.A.; Raheem, M.M. Effectiveness of Fiber Anchors in CFRP Flexural Strengthening of RC Girders. *J. Compos. Constr.* **2024**, *28*, e04368. [CrossRef]
7. Michałowska-Maziejuk, D.; Goszczyńska, B. Effectiveness of Strengthening RC Beams Using Composite Materials—An Accelerated Strengthening Method. *Materials* **2023**, *16*, 4847. [CrossRef]
8. Tworzewski, P.; Bacharz, K. Flexural Strengthening of Reinforced Concrete Beams Using Near-Surface Mounted (NSM) Carbon Fiber-Reinforced Polymer (CFRP) Strips with Additional Anchorage. *Materials* **2025**, *18*, 2579. [CrossRef]
9. International Federation for Structural Concrete (FIB). *Externally Bonded FRP Reinforcement for RC Structures*; Bulletin 14; International Federation for Structural Concrete (FIB): Lausanne, Switzerland, 2001.
10. Büyüköztürk, O.; Haupt, R.; Tuakta, C.; Chen, J. Remote Detection of Debonding in FRP-strengthened Concrete Structures Using Acoustic-Laser Technique. In *Nondestructive Testing of Materials and Structures*; Güneş, O., Akkaya, Y., Eds.; RILEM Bookseries: Berlin/Heidelberg, Germany, 2013; Volume 6. [CrossRef]
11. Zatar, W.; Nghiem, H. Detectability of embedded defects in FRP strengthened concrete deck slabs. In *Risk-Based Strategies forBridge Maintenance*; CRC Press: Boca Raton, FL, USA, 2023; pp. 221–234.
12. Kim, H.; Lamichhane, N.; Kim, C.; Shrestha, R. Innovations in Building Diagnostics and Condition Monitoring: A Comprehensive Review of Infrared Thermography Applications. *Buildings* **2023**, *13*, 2829. [CrossRef]
13. Wciślik, S. Thermal infrared mapping of the Leidenfrost drop evaporation. *J. Phys. Conf. Ser.* **2016**, *745*, 032064. [CrossRef]
14. Orzechowski, T.; Wciślik, S. Evaporation of the Leidenfrost Droplet Located on a Thin Metal Foil. *Procedia Eng.* **2016**, *157*, 131–138. [CrossRef]
15. Drobiec, Ł.; Jasinski, R.; Piekarczyk, A. *Diagnostyka Konstrukcji Żelbetowych: Metodologia, Badania Polowe, Badania Laboratoryjne Betonu i Stali*; Publisher PWN: Warsaw, Poland, 2010; pp. 61–115.
16. Sham, F.C.; Chen, N.; Long, L. Surface crack detection by flash thermography on concrete surface. *Insight* **2008**, *50*, 240–243. [CrossRef]

17. Świt, G.; Krampikowska, A.; Tworzewski, P. Non-Destructive Testing Methods for In Situ Crack Measurements and Morphology Analysis with a Focus on a Novel Approach to the Use of the Acoustic Emission Method. *Materials* **2023**, *16*, 7440. [CrossRef]
18. Anwar, M.; Mustapha, F.; Abdullah, M.N.; Mustapha, M.; Sallih, N.; Ahmad, A.; Mat Daud, S.Z. Defect Detection of GFRP Composites through Long Pulse Thermography Using an Uncooled Microbolometer Infrared Camera. *Sensors* **2024**, *24*, 5225. [CrossRef]
19. Jensen, F.; Terlau, M.; Sorg, M.; Fischer, A. Active Thermography for the Detection of Sub-Surface Defects on a Curved and Coated GFRP-Structure. *Appl. Sci.* **2021**, *11*, 9545. [CrossRef]
20. Tang, L. Maintenance and Inspection of Fiber-Reinforced Polymer (FRP) Bridges: A Review of Methods. *Materials* **2021**, *14*, 7826. [CrossRef] [PubMed]
21. Tashan, J.; Al-Mahaidi, R. Detection of cracks in concrete strengthened with CFRP systems using infra-red thermography. *Compos. Part B Eng.* **2014**, *64*, 116–125. [CrossRef]
22. Pan, P.; Zhang, R.; Zhang, Y.; Li, H. Detecting Internal Defects in FRP-Reinforced Concrete Structures through the Integration of Infrared Thermography and Deep Learning. *Materials* **2024**, *17*, 3350. [CrossRef]
23. Yazdani, N.; Beneberu, E.; Riad, M. Nondestructive Evaluation of FRP-Concrete Interface Bond due to Surface Defects. *Adv. Civ. Eng.* **2019**, *2019*, 2563079. [CrossRef]
24. Kuribara, M.; and Nagano, H. Anisotropic Thermal Diffusivity Measurements in High-Thermal-Conductive Carbon-Fiber-Reinforced Plastic Composites. *J. Electron. Cool. Therm. Control.* **2015**, *5*, 15–25. [CrossRef]
25. Terpiłowski, J.; Piotrowska-Woroniak, J.; Romanowska, J. A study of thermal diffusivity of carbon-epoxy and glass-epoxy composites using the modified pulse method. *Arch. Thermodyn.* **2014**, *35*, 117–128. [CrossRef]
26. Talebi, H.R.; Kayan, B.A.; Asadi, I.; Hassan, Z.F.B.A. Investigation of thermal properties of normal weight concrete for different strength classes. *J. Environ. Treat. Tech.* **2020**, *8*, 908–914.
27. Arcone, S.A.; Lawson, D.E.; Delaney, A.J.; Strasser, J.C.; Strasser, J.D. Ground penetrating radar reflection profiling of groundwater and bedrock in an area of discontinuous permafrost. *Geophysics* **1998**, *63*, 1573–1584. [CrossRef]
28. Malla, P.; KhedmatgozarDolati, S.S.; Ortiz, J.D.; Mehrabi, A.B.; Nanni, A.; Dinh, K. Feasibility of Conventional Non-Destructive Testing Methods in Detecting Embedded FRP Reinforcements. *Appl. Sci.* **2023**, *13*, 4399. [CrossRef]
29. Hugenschmidt, J.; Mastrangelo, R. GPR inspection of concrete bridges. *Cem. Concr. Compos.* **2006**, *28*, 384–392. [CrossRef]
30. Wciślik, W. GPR Survey of a Composite Bridge with Prestressed Concrete Beams [in Polish]. In Proceedings of the 56th Scientific Conference of the Committee of Civil and Water Engineering of the Polish Academy of Sciences and the Science Committee of the Polish Association of Engineers and Technicians, Krynica, Poland, 19–24 September 2010; pp. 459–465.
31. Klysz, G.; Balayssac, J.P. Determination of volumetric water content of concrete using ground-penetrating radar. *Cem. Concr. Res.* **2007**, *37*, 1164–1171. [CrossRef]
32. Kaplanvural, I. Volumetric water content estimation of concrete by particle swarm optimization of GPR data. *Constr. Build. Mater.* **2023**, *375*, 130995. [CrossRef]
33. Wciślik, W.; Raczkiewicz, W. Studies on corrosion initiation in reinforced concrete structures using ground-penetrating radar. *Materials* **2025**, *18*, 2308. [CrossRef]
34. Eisenmann, D.; Margetan, F.J.; Ellis, S. On the use of ground penetrating radar to detect rebar corrosion in concrete structures. *AIP Conf. Proc.* **2018**, *1949*, 030009. [CrossRef]
35. Zaki, A.; Johari, M.A.M.; Hussin, W.M.A.W.; Jusman, Y. Experimental assessment of rebar corrosion in concrete slab using ground penetrating radar (GPR). *Int. J. Corros.* **2018**, *2018*, 389829. [CrossRef]
36. Hing, C.L. Nondestructive Evaluation of Fiber Reinforced Polymer Bridge Decks Using Ground Penetrating Radar And Infrared Thermography. Ph.D. Thesis, West Virginia University, Morgantown, WV, USA, 2006.
37. Dolati, S.S.K.; Malla, P.; Ortiz, J.D.; Mehrabi, A.; Nanni, A. Identifying NDT methods for damage detection in concrete elements reinforced or strengthened with FRP. *Eng. Struct.* **2023**, *287*, 116155. [CrossRef]
38. Dutta, S.S. Nondestructive Evaluation of FRP Wrapped Concrete Cylinders Using Infrared Thermography and Groud Penetrating Radar. Master's Thesis, West Virginia University, Morgantown, WV, USA, 2006.
39. Jackson, D.R.; Islam, M.; Alampalli, S. Feasibility Of Evaluating The Performance Of Fiber Reinforced Plastic (FRP) Wrapped Reinforced Concrete Columns Using Ground Penetrating Radar (GPR) And Infrared (IR) Thermography Techniques. In Proceedings of the Structural Materials Technology VI: An NDT Conference, Atlantic City, NJ, USA, 28 February–3 March 2000.
40. Lad, A.A. Qualitative Evaluation Of CFRP-Concrete Bond Using Non-Destructive And Destructive Testing Methods. Master's Thesis, University of Texas at Arlington, Arlington, TX, USA, 2018.
41. Riad, M. Quantitative Non-Destructive Evaluation (NDE) Of FRP Laminate-Concrete Bond Strength. Ph.D. Thesis, University of Texas at Arlington, Arlington, TX, USA, 2018.

42. Riad, M.; Yazdani, N.; Almomani, M.; Gonzales, E. Utilization of ground penetrating radar (GPR) in the non-destructive assessment of FRP laminate-concrete bond strength. In *Advances and Challenges in Structural Engineering*, 1st ed.; Rodrigues, H., Elnashai, A., Eds.; Springer Nature Switzerland AG: Cham, Switzerland, 2019; pp. 244–267. [CrossRef]
43. Ortiz, J.D.; Dolati, S.S.K.; Malla, P.; Mehrabi, A.; Nanni, A. Nondestructive testing (NDT) for damage detection in concrete elements with externally bonded fiber-reinforced polymer. *Buildings* **2024**, *14*, 246. [CrossRef]
44. Li, J.; Liu, C. Noncontact detection of air voids under glass epoxy jackets using a microwave system. *Subsurf. Sens. Technol. Appl.* **2001**, *2*, 411–423. [CrossRef]
45. Yu, T.Y.; Büyüköztürk, O. A far-field airborne radar NDT technique for detecting debonding in GFRP–retrofitted concrete structures. *NDTE Int.* **2008**, *41*, 10–24. [CrossRef]
46. Gholizadeh, S. A review of non-destructive testing methods of composite materials. *Procedia Struct. Integr.* **2016**, *1*, 50–57. [CrossRef]
47. Jodhani, J.; Handa, A.; Gautam, A.; Ashwni; Rana, R. Ultrasonic non-destructive evaluation of composites: A review. *Mater. Today Proc.* **2023**, *78*, 627–632. [CrossRef]
48. Wang, B.; He, P.; Kang, Y.; Jia, J.; Liu, X.; Li, N. Ultrasonic Testing of Carbon Fiber-Reinforced Polymer Composites. *J. Sens.* **2022**, *2022*, 5462237. [CrossRef]
49. Adamczak-Bugno, A.; Lipiec, S.; Adamczak, J.; Vičan, J.; Bahleda, F. Identification of Destruction Processes and Assessment of Deformations in Compressed Concrete Modified with Polypropylene Fibers Exposed to Fire Temperatures Using Acoustic Emission Signal Analysis, Numerical Analysis, and Digital Image Correlation. *Materials* **2023**, *16*, 6786. [CrossRef]
50. Świt, G. *Analiza Procesów Destrukcyjnych w Obiektach Mostowych z Belek Strunobetonowych z Wykorzystaniem Zjawiska Emisji Akustycznej*; Wydawnictwo Politechniki Świętokrzyskiej: Kielce, Poland, 2011.
51. Bacharz, M.; Bacharz, K.; Trąmpczyński, W. The Correlation between Shrinkage and Acoustic Emission Signals in Early Age Concrete. *Materials* **2022**, *15*, 5389. [CrossRef]
52. Bacharz, M.; Bacharz, K.; Trąmpczyński, W. Impact of Early-Age Curing and Environmental Conditions on Shrinkage and Microcracking in Concrete. *Materials* **2025**, *18*, 3185. [CrossRef]
53. Raczkiewicz, W.; Bacharz, M.; Bacharz, K. Experimental verification of the concrete shrinkage strains course according to En 1992-2 standard. *Adv. Mater. Sci.* **2015**, *15*, 22–29. [CrossRef]
54. Goszczyńska, B.; Świt, G.; Trąmpczyński, W.; Krampikowska, A.; Tworzewska, J.; Tworzewski, P. Experimental validation of concrete crack identification and location with acoustic emission method. *Arch. Civ. Mech. Eng.* **2012**, *12*, 23–28. [CrossRef]
55. *PN-EN 1330-9:2017-09*; Badania Nieniszczące—Terminologia—Część 9: Terminy Stosowane w Badaniach Emisją Akustyczną. PKN: Warszawa, Poland, 2017.
56. Angelopoulos, N.; Kappatos, V. An Experimental Assessment Using Acoustic Emission Sensors to Effectively Detect Surface Deterioration on Steel Plates. *Sensors* **2024**, *24*, 6462. [CrossRef]
57. Riccioli, F.; Alkhateeb, S.; Mol, A.; Pahlavan, L. Feasibility assessment of non-contact acoustic emission monitoring of corrosion-fatigue damage in submerged steel structures. *Ocean. Eng.* **2024**, *312*, 119296. [CrossRef]
58. Yu, Y.-H.; Choi, J.-H.; Kweon, J.-H.; Kim, D.-H. A study on the failure detection of composite materials using an acoustic emission. *Compos. Struct.* **2006**, *75*, 163–169. [CrossRef]
59. Sayers, C.M.; van Munster, J.G. Microcrack-induced seismic anisotropy of sedimentary rocks. *J. Geophys. Res. Solid Earth* **1991**, *96*, 16529–16533. [CrossRef]
60. Sayers, C.M.; Kachanov, M. Microcrack-induced elastic wave anisotropy of brittle rocks. *J. Geophys. Res. Solid Earth* **1995**, *100*, 4149–4156. [CrossRef]
61. *ASTM E976-10*; Standard Guide for Determining the Reproducibility of Acoustic Emission Sensor Response. ASTM International: West Conshohocken, PA, USA, 2010.
62. Sause, M.G.R. Investigation of Pencil-Lead Breaks as Acoustic Emission Sources. *J. Acoust. Emiss.* **2011**, *29*, 184–196.
63. Lovejoy, S. Development of Acoustic Emissions Testing Procedures Applicable to Conventionally Reinforced Concrete Deck Girder Bridges Subjected to Diagonal Tension Cracking. Ph.D. Thesis, Oregon State University, Corvallis, OR, USA, 2006.
64. Li, D.; Yang, K.; He, Z.; Zhou, H.; Li, J. Acoustic Emission Wave Velocity Attenuation of Concrete and Its Application in Crack Localization. *Sustainability* **2020**, *12*, 7405. [CrossRef]
65. Hinrichs, R.; Jiang, N.; Beltrán, R.; Krause, T.; Käding, M.; Lange, A.; Schmidt, B.; Ostermann, J.; Marx, S. Analysis of the Repeatability of the Pencil Lead Break in Comparison to the Ball Impact and Electromagnetic Body-Noise Actuator. In Proceedings of the 20th World Conference on Non-Destructive Testing (WCNDT 2024), Incheon, Republic of Korea, 27–31 May 2024. [CrossRef]
66. *Sika® CarboDur® E, Karta InformacyjnProduktu, (Product Data Sheet)*; Sika Poland Sp. z o.o.: Warszawa, Poland, 2025; Version 07.01. Available online: https://pol.sika.com/dam/dms/plcon/s/sika_carbodur_s.pdf (accessed on 1 October 2025).

67. *Sikadur®-330, Karta Informacyjna Produktu, (Product Data Sheet)*; Sika Poland Sp. z o.o.: Warszawa, Poland, 2025; Version 06.01. Available online: https://pol.sika.com/dam/dms/plcon/e/sikadur_-330.pdf (accessed on 1 October 2025).
68. Mostofinejad, D.; Mahmoudabadi, E. Grooving as alternative method of surface preparation to postpone debonding of FRP Laminates in concrete beams. *J. Compos. Constr.* **2010**, *14*, 804–811. [CrossRef]
69. *PN-EN 1542:2000*; Products and Systems for the Protection and Repair of Concrete Structures—Test Methods—Measurement of Bond Strength by Pull-Off. PKN: Warszawa, Polnad, 2000.
70. ACI Committee 440. *Guide Test Methods for FiberReinforced Polymers (FRPs) for Reinforcing or Strengthening Concrete Structures (ACI 440.3R-04)*; American Concrete Institute: Farmington Hills, MI, USA, 2004; p. 40.
71. Annan, A.P. Ground Penetrating Radar Workshop Notes. *Sens. Softw.* **2001**.
72. Patriarca, C.; Lambot, S.; Mahmoudzadeh, M.R.; Minet, J.; Slob, E. Reconstruction of sub-wavelength fractures and physical properties of masonry media using full-waveform inversion of proximal penetrating radar. *J. Appl. Geophys.* **2011**, *74*, 26–37. [CrossRef]
73. *ASTM D 6087-22*; Standard Test Method for Evaluating Asphalt-Covered Concrete Bridge Decks Using Ground Penetrating Radar. ASTM: West Conshohocken, PA, USA, 2022.
74. Elimat, Z.M.; Hamideen, M.S.; Schulte, K.I.; Wittich, H.; de la Vega, A.; Wichmann, M.; Buschhorn, S. Dielectric properties of epoxy/short carbon fiber composites. *J. Mater. Sci.* **2010**, *45*, 5196–5203. [CrossRef]
75. Pocket Ae-2: Protable 2-Channel Acoustic Emission System, Product Bulletin, MistrasGrup. Available online: https://www.physicalacoustics.com/content/literature/multichannel_systems/Pocket_AE-2_Product_Bulletin.pdf (accessed on 1 January 2010).
76. Krautkrämer, J.; Krautkrämer, H. *Ultrasonic Testing of Materials*; Springer: Berlin/Heidelberg, Germany, 2013.
77. PK6I: Medium Frequency Integral Preamplifier Resonant Sensor, Product Bulletin, MIstrasGrup. Available online: https://www.physicalacoustics.com/content/literature/sensors/Model_PK6I.pdf (accessed on 1 January 2010).
78. Teodorczyk, M. Influence of Aggregate Gradation on the Longitudinal Wave Velocity Changes in Unloaded Concrete. *IOP Conf. Ser. Mater. Sci. Eng.* **2017**, *245*, 032084. [CrossRef]
79. Joint Committee for Guides in Metrology (JCGM). *JCGM 100:2008: Evaluation of Measurement Data—Guide to the Expression of Uncertainty in Measurement (GUM 1995 with Minor Corrections)*; BIPM: Sèvres, France, 2008.
80. Rabi, M.; Abarkan, I.; Sarfarazi, S.; Ferreira, F.P.V.; Alkherret, A.J. Automated design and optimization of concrete beams reinforced with stainless steel. *Struct. Concr.* **2025**. [CrossRef]

materials

MDPI

Article

Nonlinear Behaviour of Aluminium and Passive Brackets in Ventilated Facades: Experimental Tests and Numerical Modelling

Łukasz Zawiślak [1,*], Krzysztof Schabowicz [1] and Ołeksij Kopyłow [2]

[1] Faculty of Civil Engineering, Wrocław University of Science and Technology, Wybrzeże Wyspiańskiego 27, 50-370 Wrocław, Poland; krzysztof.schabowicz@pwr.edu.pl

[2] Instytut Techniki Budowlanej, Filtrowa 1, 00-611 Warszawa, Poland; o.kopylow@itb.pl

* Correspondence: lukasz.zawislak@pwr.edu.pl

Abstract: This paper presents the results of experimental tests and numerical analyses of the behaviour of brackets used in substructures of ventilated facades. Two representative solutions were compared: a traditional aluminium bracket and an innovative passive bracket with a composite interlayer. The aim was to assess their load-bearing capacity, deformation and failure mechanisms, and the suitability of the calculation methods used. Laboratory tests were carried out at ITB's accredited Laboratory of Building Elements in accordance with the European Assessment Document (EAD 090034-00-0404). The aluminium bracket was tested under standard environmental conditions. In parallel, finite element (FE) analyses were performed, including elastic–plastic modelling for metallic systems and material and geometric nonlinear analyses for the passive bracket. The results revealed fundamental differences in the behaviour of the two solutions. The aluminium bracket exhibited a predictable plasticisation mechanism, the ability to redistribute stresses, and a gradual loss of capacity. Linear analyses proved sufficient in this case and were consistent with the tests. The passive bracket, by contrast, showed quasi-brittle behaviour, strong temperature sensitivity, and no plastic reserve, resulting in a sudden failure mechanism. For this case, the use of classical linear models leads to unsafe simplifications and underestimated results. The study demonstrates that the development of passive facade bracket technology requires a nonlinear approach and extended long-term testing covering the rheology of composite materials and environmental effects. The findings also reveal a normative gap: current design guidelines and EAD documents focus on metallic solutions while overlooking the specific behaviour of passive brackets. The results constitute an important contribution to knowledge on the safety and durability of ventilated facades and may serve as a basis for developing dedicated design procedures and for updating normative documents.

Keywords: ventilated facade; passive bracket; aluminium bracket; composite insert; FE analysis; material nonlinearity; rheology; structural safety; energy efficiency

1. Introduction and Research Background

Ventilated facade systems are of crucial importance in modern construction, combining energy, functional, and architectural requirements [1,2]. Their popularity in recent decades results from growing expectations placed on building envelopes, regarding not only thermal protection, but also durability, fire safety, and high aesthetic quality. In line with the EPBD [3] directive and the development of nearly zero-energy building (NZEB) standards, (sliding) brackets, which compensate for thermal deformations and eliminate the formation of secondary forces in the system.

In traditional solutions based on steel or aluminium, a bracket can be modelled as a thin-walled cantilever with a known static scheme. Due to the material properties—elasticity in the initial range and the ability to undergo plastic deformation—these elements possess a natural reserve of capacity. In practice, this means that once the elastic limit is exceeded, the structure signals impending failure through large, visible deflections and permanent deformations, which can be observed during service. The failure mechanism is thus continuous and gradual, which constitutes an important safety factor.

Passive brackets with composite inserts behave in a completely different manner. They form a hybrid system (aluminium-composite–aluminium), in which the properties of the polymer insert determine the critical behaviour. Glass-fibre-reinforced polyamide (PA6 GF40) works exclusively within the elastic range, and once the critical strain is reached, it undergoes sudden, brittle failure without a transitional phase and without the possibility of stress redistribution. In practice, this implies the absence of warning signs and the risk of sudden system failure, which fundamentally differentiates this type of bracket from traditional metallic solutions.

These differences are clearly illustrated by stress–strain relationships:

- In the case of aluminium and steel, the curve includes the elastic range, plastic range, and strain hardening, ensuring predictable behaviour under overloads.
- In the case of polymer composites, the curve is limited to the elastic range, and once the critical value is reached, a sudden drop in capacity occurs.

The theoretical and practical consequences are significant:

- For aluminium brackets, the use of linear models is sufficient to predict behaviour within the service range and safe due to the presence of a plastic reserve.
- For passive brackets, linear models lead to errors—underestimation of deflections, failure to capture stiffness degradation mechanisms, and omission of the brittle nature of failure.

An additional factor complicating the modelling of passive brackets is rheological phenomena: creep, stress relaxation, and the influence of humidity and temperature on the properties of the composite. The literature indicates that long-term service may lead to a gradual decrease in stiffness and strength of the inserts, a process not captured by classical calculation schemes [16]. This means that analyses must account for material, geometric, and time-dependent nonlinearities, which significantly increases the computational complexity compared to conventional metallic brackets. A comparison of parameters is presented in Table 1.

Table 1. Comparison of properties of aluminium and passive brackets.

Criterion	Aluminium Bracket	Passive Bracket
Material	Aluminium alloy EN AW 6060 T6/T66	Hybrid system: aluminium + PA6 GF40 insert
Behaviour	Elastic-plastic, capable of plastic deformation	Elastic only, until sudden failure
Failure mechanism	Continuous, plastic, with stress redistribution	Brittle, sudden, without safety reserve
Reserve capacity	Yes—due to plastic deformations	None—capacity loss occurs suddenly
Environmental influence	Relatively minor (aluminium stability)	Significant: creep, stress relaxation, sensitivity to temperature and humidity

Table 1. *Cont.*

Criterion	Aluminium Bracket	Passive Bracket
Computational modelling	Linear models sufficient	Nonlinear models required (material, geometry, rheology)
Design implications	Predictable and highly reliable	Necessity of dedicated calculation and validation procedures

In summary, the differences between aluminium and passive brackets are not limited to thermal aspects. They introduce a new class of engineering challenges, where the decisive factors include the absence of plastic reserve, susceptibility to environmental influences and rheological processes, and the necessity of applying nonlinear calculation models. In the following sections, these issues will be confronted with the results of laboratory tests and FE analyses, enabling a precise evaluation of the actual differences in the behaviour of both solutions.

2. Case Study

For the purposes of comparative analysis, two solutions representative of contemporary ventilated facade substructures were selected: a conventional aluminium bracket and a passive bracket with a composite insert. Both systems were subjected to laboratory testing and subsequently reproduced in numerical FE analyses. The aim was to capture the differences in their mechanical behaviour, load-bearing capacity, and failure mechanisms.

The entire testing procedure was carried out in accordance with the requirements of the European Assessment Document (EAD 090034-00-0404 [8–10]). This gives the obtained results practical relevance, as they may serve as a reference in certification processes and technical assessments of products introduced to the European market.

2.1. Aluminium Bracket—Reference Solution

The first analysed case was a traditional aluminium bracket made of EN AW 6060 T6/T66 alloy with a thickness of 3 mm, working in conjunction with a profile 2 mm thick. The test assembly was mounted to a rigid steel plate 20 mm thick, while the load was applied statically to the extreme point of the profile connected to the bracket. The displacement rate was set at 4 mm/min.

The test setup, along with a view of the mounted specimen, is shown in Figure 1.

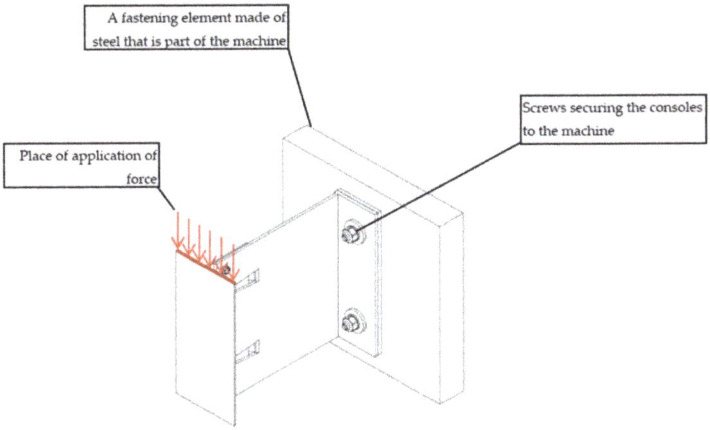

Figure 1. Test of aluminium bracket resistance to vertical load (test setup and view of mounted specimen).

The experiments were carried out under standard laboratory conditions (temperature 20 °C, relative humidity approx. 50%). Forces corresponding to displacements of 0.2% of the cantilever length, 1 mm, 3 mm, and 8 mm were recorded. A total of 10 sets of specimens were prepared. Figures 2 and 3 present dimensioned drawings of the components forming the ventilated aluminium bracket substructure facade.

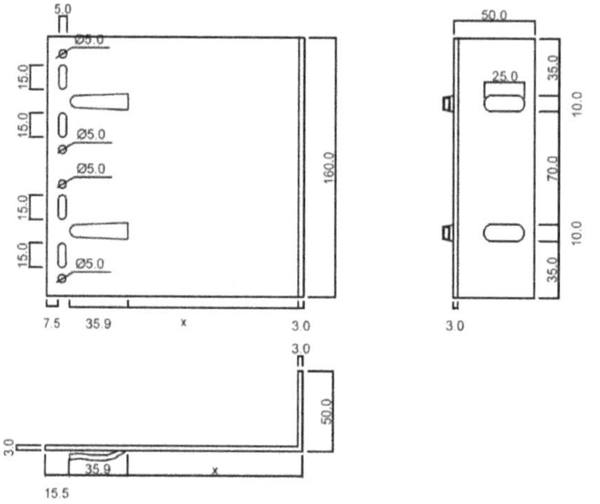

Figure 2. Aluminium bracket—geometry and dimensions [mm].

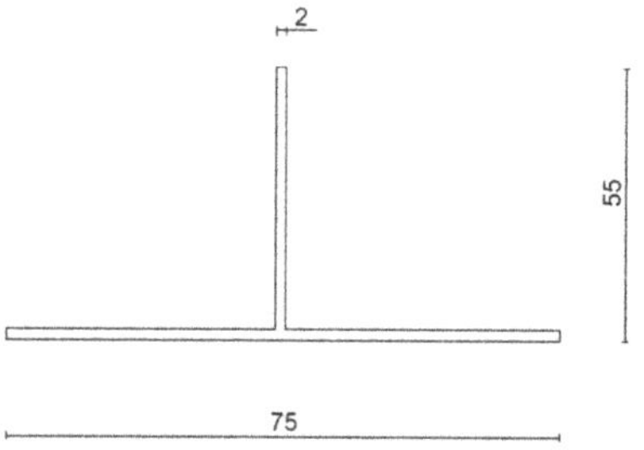

Figure 3. Aluminium profile—geometry and dimensions [mm].

A characteristic feature of the aluminium bracket is its behaviour typical of metallic materials: the ability to undergo plastic deformation, gradual stress redistribution, and a plastic failure mechanism. From the perspective of service safety, this means the presence of reserve capacity and a predictable, well-signalled failure process.

2.2. *Passive Bracket—Innovative Solution*

The second case study concerned a passive bracket incorporating a aluminium insert, an aluminium extension. The element was manufactured from a combination of EN AW 6060 T6 aluminium alloy and a composite interlayer of glass-fibre-reinforced polyamide (PA6 GF40 FR V0).

The test setup, together with a view of the mounted specimen, is shown in Figure 4. The specimens were mounted in accordance with the procedure described in the EAD [8–10], and vertical loads were applied to the end of the cantilever at a rate not exceeding 5 kN/min. Tests were conducted at elevated temperature (+90 °C ± 5 °C) to reproduce environmental effects (exposure to solar radiation and high ambient temperature). Forces corresponding to displacements of 0.2% of the cantilever length, 1 mm, 3 mm, and 5 mm were recorded. A total of nine specimens were prepared and loaded until ultimate values were reached. Figures 5–7 present dimensioned drawings of the components forming the ventilated facade passive bracket substructure.

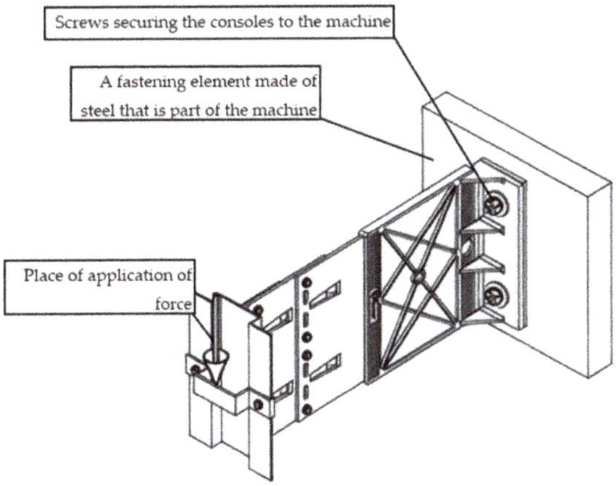

Figure 4. Test of passive bracket resistance to vertical load (test setup and view of mounted specimen) [17] in accordance with the Polish National Technical Assessment ITB-KOT-2022/2265.

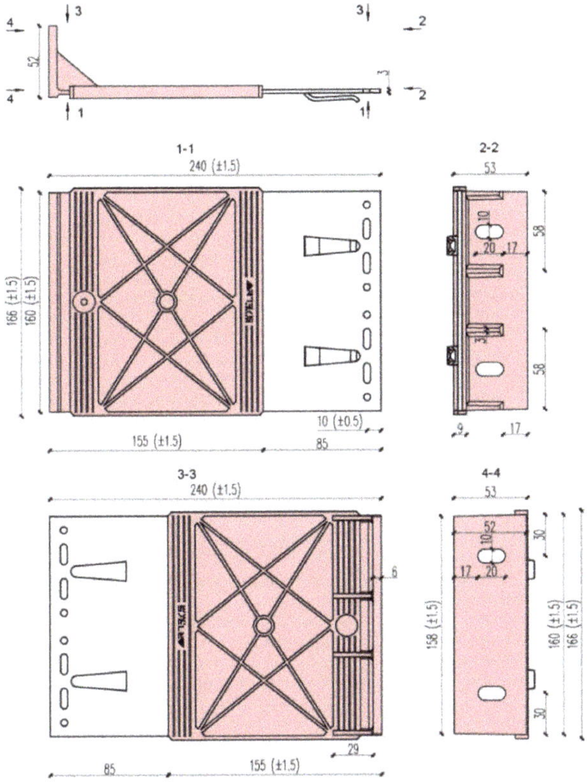

Figure 5. Passive bracket with aluminium insert—geometry and dimensions [mm] [17] in accordance with the Polish National Technical Assessment ITB-KOT-2022/2265.

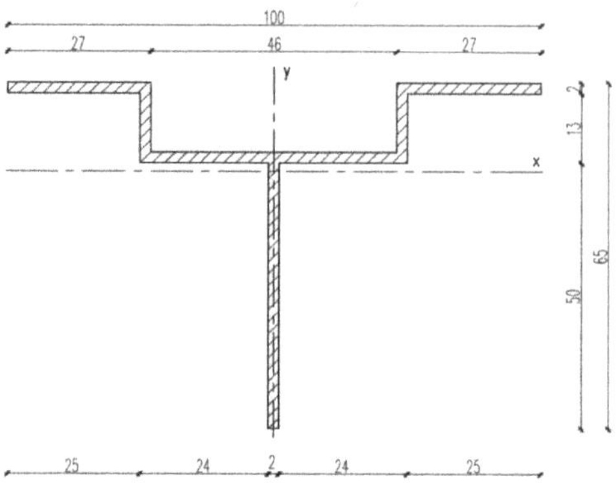

Figure 6. Aluminium profile—geometry and dimensions [mm] [17] in accordance with the Polish National Technical Assessment ITB-KOT-2022/2265.

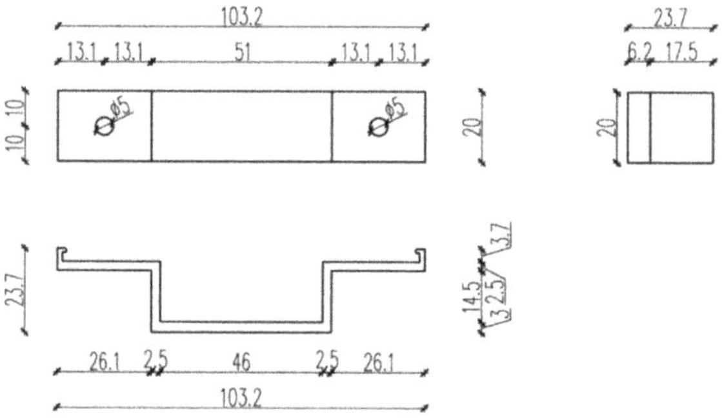

Figure 7. Aluminium holder—geometry and dimensions [mm] [17] in accordance with the Polish National Technical Assessment ITB-KOT-2022/2265.

In contrast to the aluminium bracket, the passive bracket relies on a composite insert integrated with the support, whose behaviour is limited to the elastic range. Once critical values are exceeded, sudden brittle failure occurs without reserve capacity. Temperature sensitivity further exacerbates operational risk, making the assessment of safety more complex and less unambiguous.

2.3. Case Comparison

To facilitate the comparison of both cases, Table 2 summarises the key testing parameters and the characteristic differences in their mechanical behaviour.

Table 2. Comparison of testing and mechanical parameters of aluminium and passive brackets.

Feature	Aluminium Bracket	Passive Bracket
Number of Specimen	10	9
Test temperature	20 °C	+90 °C
Displacement range	0.2%—8 mm	0.2%—5 mm
Failure mechanism	Plasticisation + deformations	Brittle fracture of insert
Failure character	Gradual, signalled	Sudden, without warnings

On this basis, several key differences can be identified:

- Geometry and configuration—The aluminium bracket is a fully metallic system, easily interpreted within a thin-walled model. The passive bracket is a hybrid system requiring nonlinear calculation models.
- Mechanical behaviour—Aluminium allows for plasticisation and signals impending failure, whereas the passive bracket fails suddenly through brittle fracture.
- Testing conditions—The aluminium bracket was tested under standard laboratory conditions, while the passive bracket was tested at elevated temperature.
- Practical significance—The aluminium bracket can be treated as a reference solution, whereas the passive bracket—despite its energy benefits—requires case-by-case validation and distinct design procedures.

2.4. Test Apparatus, Specimen Preparation, and Loading Procedure

The laboratory tests of resistance to vertical load were carried out in accordance with the method specified in Annex H to EAD 090034-00-0404 [8–10]. The specimens, consisting of a bracket connected to a T-profile with two stainless steel self-drilling screws (Ø 4.8 × 19 mm), were mounted to a 20 mm thick steel plate. All specimens (aluminium and passive) were assembled using hex-head bolts M8 × 20 (DIN 933 [18], property class 8.8) together with enlarged M8 washers, 30 mm OD (DIN 9021 [19]). All bolted joints were tightened to a torque of 20 Nm with a calibrated torque wrench.

A static vertical load directed downwards was applied to the extreme point of the T-profile, as schematically shown in Figure 8. The load was increased in steps corresponding to multiples of 100 N at a constant displacement rate of 5 mm/min. At each step, the instantaneous displacement was recorded, after which the specimen was unloaded to 0 N at the same rate and the residual deformation was determined. The cycle was then repeated. To minimise measurement errors due to initial slip in bolted joints, a preliminary preload of 100 N was applied.

Figure 8. Schematic representation of the test setup for vertical load resistance.

The testing machine, of accuracy class 1, was equipped with a force transducer and displacement sensors enabling simultaneous recording of both parameters. The expanded measurement uncertainty at the 95% confidence level (coverage factor k = 2) was ±1% for force, ±0.1 mm for displacement, and ±1 °C for temperature.

Before each series, sensors were zeroed, a "dry run" was performed, and the force transducer characteristics were verified using a reference load (deviation ≤ ±0.5% of the working range). The displacement sensor systematic error was corrected based on calibration curves. During testing, the complete load–displacement or load–time path, as well as the temperature in the node zone (measured with a K-type thermocouple), were continuously recorded. The alignment of the specimens was verified using a feeler gauge and a precision level; permissible misalignment did not exceed 0.2°. In the load application zone, a steel roller was used to distribute the force linearly along the profile edge.

Specimens were cleaned and degreased prior to assembly, and bolted joints were tightened using a calibrated torque wrench. Aluminium specimens were conditioned for 24 h at 20 ± 2 °C and 50 ± 5% RH, while passive specimens were conditioned at 90 ± 5 °C for at least 60 min until stabilisation of thermocouple readings in the node zone.

The sequence of tests was randomised. The minimum interval between successive tests was 5 min for aluminium specimens and 10 min for passive ones (to allow for thermal stabilisation). Sample sizes were $n = 10$ for aluminium brackets and $n = 9$ for passive brackets. For aluminium brackets, displacement control at 4 mm/min was applied, while for passive brackets, force control at 5 kN/min was used, in accordance with EAD 090034-00-0404 [8–10] and the heating chamber manufacturer's recommendations.

For each variable, the mean, standard deviation (SD), and 95% confidence interval (Student's t distribution, $n = 10$ and $n = 9$, respectively) were reported. Differences in mean values at characteristic points (0.2% L; 1 mm; 3 mm; 5/8 mm) were verified using Welch's t-test (two-sided, $\alpha = 0.05$).

3. Summary

The laboratory tests were carried out in an accredited laboratory in accordance with the procedure specified in the European Assessment Document (EAD 090034-00-0404 [8–10]). Each test setup—an aluminium bracket and a passive bracket with a composite insert—was subjected to vertical loading under gradually increasing force. The corresponding displacements were recorded, enabling the development of complete load–deflection curves and the identification of failure mechanisms.

3.1. Aluminium Bracket

Tests of aluminium brackets made of alloy EN AW 6060 T6/T66 were conducted under standard conditions (20 °C, relative humidity approx. 50%). The adopted testing procedure was fully compliant with the requirements of EAD 090034-00-0404 [8–10], giving the results practical value and allowing their direct reference to certification processes.

The load–deflection curves exhibited a typical behaviour for metallic materials: initially linear (elastic range), followed by a clear reduction in stiffness and transition into the plastic range. The bracket demonstrated high initial stiffness—at a permanent deformation of 0.2% of the cantilever length, the force was $F_{r0.2\%}$ = 155 N at a deflection of 0.49 mm.

At subsequent stages, the element showed significant ability to redistribute stresses and carry additional load increments:

- At 1 mm deflection, it carried on average 372 N;
- At 3 mm, 1189 N, and the ultimate load capacity at 8 mm deflection averaged 2606 N.

This represents nearly a sevenfold increase compared to the force corresponding to the first permanent deformations, confirming the presence of a pronounced load-bearing reserve and a plastic failure mechanism.

The full set of results is presented in Table 3, and Figure 9 illustrates a representative example of the load–deflection relationship.

Table 3. Test results of vertical load resistance of substructure elements—aluminium bracket.

Measurement No.	Obtained Force Values (N) Corresponding to Displacements:			
	Lx0.2	1 mm	3 mm	8 mm
1	153	381	1274	2472
2	188	394	1152	2614
3	173	372	1120	2404
4	162	358	1155	2445
5	144	364	1207	2617
6	113	354	1109	2705
7	188	402	1204	2635
8	144	374	1245	2808
9	122	338	1211	2654
10	168	383	1222	2712
Mean value	155.5	372	1189.9	2606.6
Standard deviation	25.4	19.2	53.9	129.0
Design value $F_{u,10}$	106.7	335.1	1086.4	2358.9

$F_{u,10} = F_{mean} - k_n \cdot S$; $k_n = 1.92$.

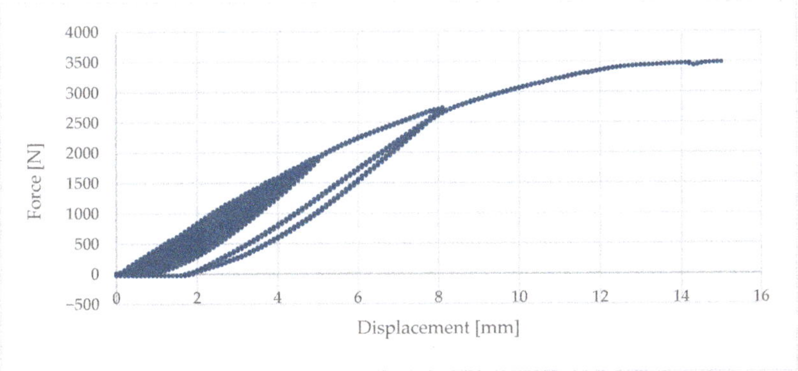

Figure 9. Relationship between displacement and load—aluminium bracket.

The dominant failure mechanism was plasticisation and bending of the bracket arm in the zone of maximum bending moments (at the root). Plastic hinges and permanent deformations were observed, indicating a gradual and well-signalled failure mechanism. In some specimens, additional local displacements were noted in the area of bolted connections (assembly clearances, micro-slips of the profile relative to the bracket); however, these did not determine the ultimate failure.

Compared with the passive bracket, the aluminium solution reached higher force values for the same displacements (e.g., at 3 mm—1189 N vs. 246 N for the passive bracket), due to its superior material properties and shorter reach. This results both from the higher stiffness of aluminium ($E \approx 70$ GPa) and from the material's ability to undergo plastic deformation, which ensures predictability and safety in long-term service.

3.2. *Passive Bracket*

The test of the passive bracket was also carried out in accordance with EAD 090034-00-0404 [8–10] but under more demanding environmental conditions—inside a chamber at +90 °C ± 5 °C. This procedure was intended to simulate the effects of elevated temperatures, solar radiation, and adverse service conditions that may significantly affect the mechanical performance of the composite insert.

Forces corresponding to characteristic displacements were recorded:

- $F_{r0.2\%}$—permanent deformation of 0.2% of the cantilever length (for Lx = 345 mm, deflection 0.69 mm);
- F_{1d}, F_{3d}, F_{5d}—corresponding to displacements of 1 mm, 3 mm, and 5 mm.

The mean values obtained for nine specimens were

- $F_{r0.2\%}$ = 100.53 N;
- F_{1d} = 131.64 N;
- F_{3d} = 246.69 N;
- F_{5d} = 324.54 N.

The standard deviations were relatively high, amounting to 32.05 N, 27.72 N, 38.69 N, and 38.00 N, respectively. The design values Fu9, calculated using Student's factor (k = 1.96, n = 9), are presented in Table 4, while Figure 10 presents a representative example of the load–deflection response.

Table 4. Test results of vertical load resistance of substructure elements—passive bracket.

Measurement No.	Obtained Force Values (N) Corresponding to Displacements:			
	Lx0.2	1 mm	3 mm	5 mm
1	141.16	175.55	293.66	341.54
2	91.82	115.01	256.40	343.48
3	94.10	137.69	258.87	339.09
4	55.34	88.71	181.38	256.62
5	130.50	148.81	267.12	346.31
6	141.02	160.78	269.21	362.71
7	81.74	123.01	280.21	360.19
8	116.42	142.28	234.12	310.66
9	52.64	92.90	179.21	260.30
Mean value	100.53	131.64	246.69	324.54
Standard deviation	32.05	27.72	38.69	38.00
Design value $F_{u,9}$	37.70	77.30	170.85	250.07

$F_{u,9} = F_{mean} - k_n \cdot S$.

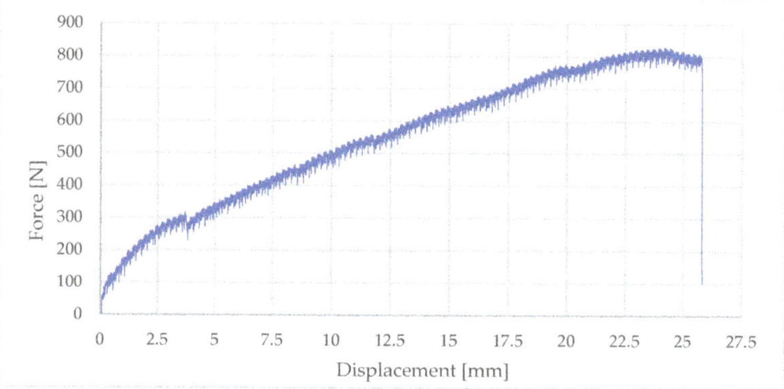

Figure 10. Relationship between displacement and load—passive bracket.

The analysis of results showed that the coefficient of variation decreased with increasing displacement—from 32% at 0.69 mm deflection to 12% at 5 mm. This indicates that the stiffness of the system stabilises only after initial clearances are exceeded and full contact between elements is achieved.

Elevated temperature also had a significant effect, leading to a reduction in effective stiffness in line with the properties of the composite material (PA6 GF40 FR V0). The maximum force value was recorded, followed by a sudden drop in capacity corresponding to quasi-brittle failure mechanisms of the composite insert. The absence of a plasticisation phase and the lack of clear warning signals were characteristic, representing a serious limitation from the perspective of service safety.

4. Finite Element Analysis

The numerical analysis was performed in the ANSYS Mechanical 2024 R2 environment, which enables the modelling of complex structural geometries, meshing with finite elements, and consideration of both material and geometric nonlinearities. Geometric models were prepared in ANSYS Design Modeler based on the actual dimensions of the specimens tested in the laboratory. This allowed accurate representation of the bracket arms, bolted connections, and the composite insert in the passive bracket.

The objectives of the calculations were to

- Verify the consistency of numerical models with experimental results;
- Assess the applicability of a simplified linear approach for both bracket types;
- Identify differences in mechanical behaviour between the aluminium and passive brackets.

4.1. Aluminium Bracket

The aluminium bracket model was prepared analogously, reproducing the actual test setup. It included a load-bearing bracket (EN AW 6060 T6/T66) and profile with a total cantilever length of 245 mm. A finite-element mesh was generated using the hex dominant method, with local adaptivity around fillets and curved features. The target element size was 3 mm (Figure 11).

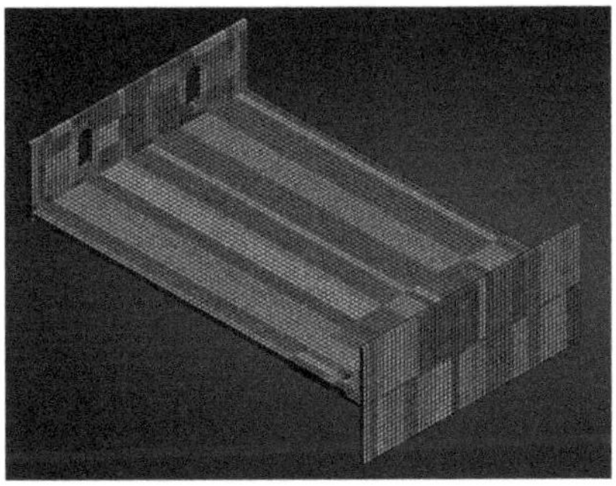

Figure 11. Geometry of the numerical model of the system with the aluminium bracket.

Boundary conditions reflected rigid anchorage in the wall (fixed support at the base plate) and full cooperation of the bracket with the profile (bonded) (Figure 12). The load was applied along the profile edge as a linearly distributed force, consistent with the laboratory testing scheme. The aluminium components (EN AW 6060 T6/T66 bracket and matching profile) were modelled as isotropic, linear-elastic materials. The full set of material parameters used in the model at 20 °C is summarised in Table 5.

The comparison of FEA and laboratory results showed very good agreement:

- For 100 N: model 0.30 mm vs. laboratory 0.49 mm (Figure 13);
- For 350 N: 1.07 mm vs. 1.0 mm (at 335 N);
- For 1100 N: 3.35 mm vs. 3.0 mm (at 1086 N);
- For 2400 N: 7.33 mm vs. 8.0 mm (at 2358 N) (Figure 14).

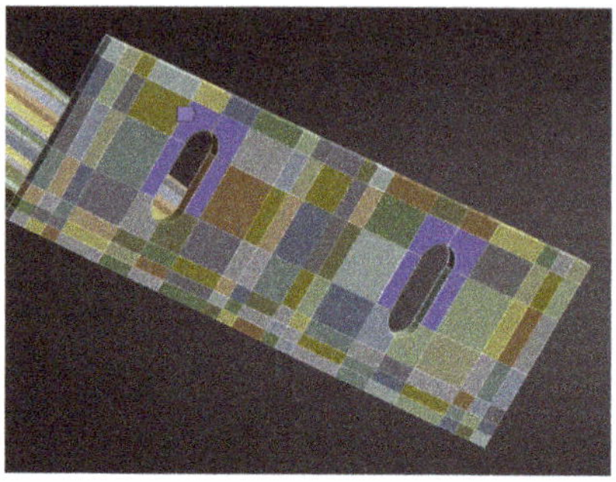

Figure 12. Boundary condition of the aluminium bracket anchorage.

Table 5. Material parameters used in the FE model for the aluminium bracket and profile.

EN AW 6060 T6	
Density	2700 kg/m^3
Young's modulus	69,500 MPa
Poisson's ratio	0.33
Bulk modulus	68,137 MPa
Shear modulus	26,128 MPa
Tensile ultimate strength	190 MPa
Tensile yield strength	150 MPa

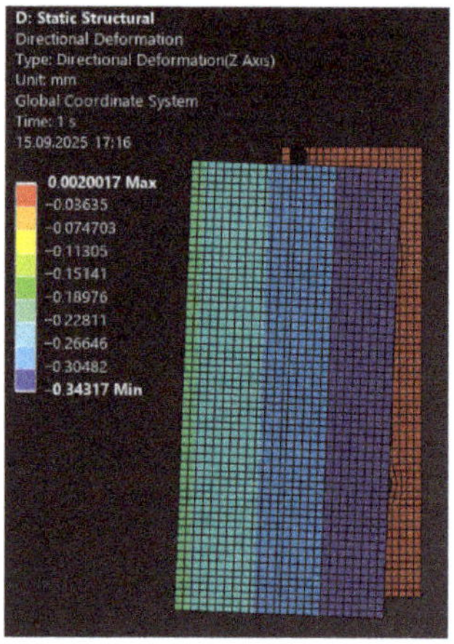

Figure 13. Map of vertical deformations of the aluminium bracket—load 100 N.

Building on the obtained results, Figures 15 and 16 present comparative force–displacement plots for the aluminium bracket and vertical profile. Figure 15 overlays the individual laboratory curves with the FEA prediction, while Figure 16 contrasts the mean laboratory response with the FEA curve, enabling a direct like-for-like comparison.

The high consistency of results indicates that, in the case of the aluminium bracket, linear analysis reliably reproduces its structural behaviour. Minor discrepancies fall within the tolerance limits and do not affect engineering conclusions.

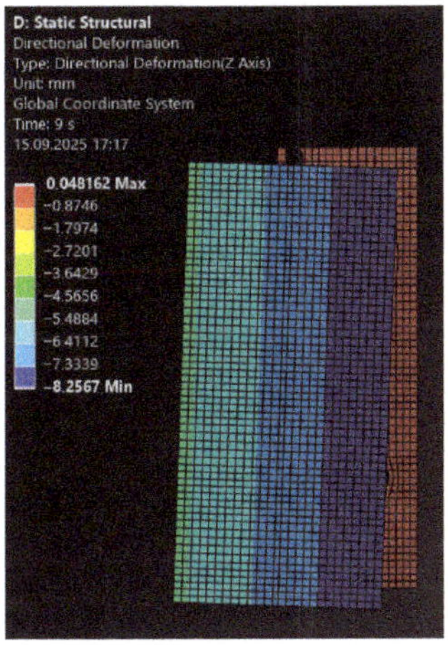

Figure 14. Map of vertical deformations of the aluminium bracket—load 2400 N.

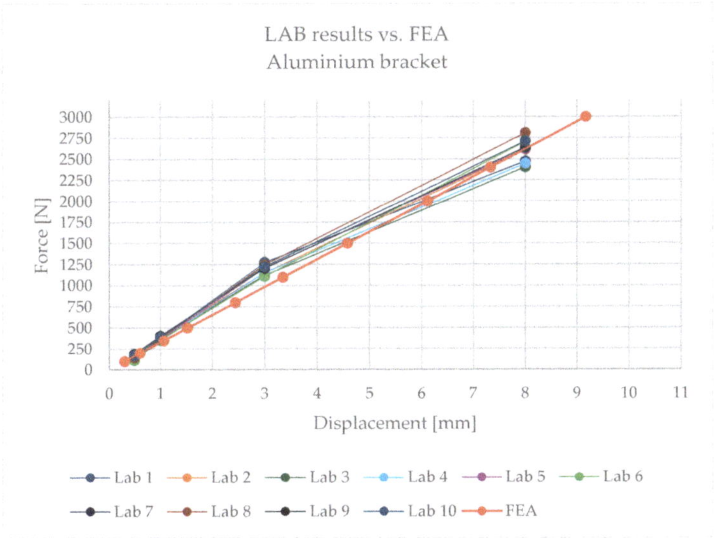

Figure 15. Force–displacement curves for the aluminium bracket and vertical profile): individual laboratory tests overlaid with the FEA prediction.

4.2. Passive Bracket

The numerical model of the passive bracket was reproduced in accordance with the laboratory testing scheme. The system included the aluminium load-bearing arm, extension, "Y"-type profile, hook, and composite insert. The total cantilever length was 345 mm.

A finite-element mesh was generated using the hex dominant method. A dense mesh was particularly important in the region of the polyamide insert and contact zones, where the highest stress gradients were expected. The finite element mesh is shown in Figure 17.

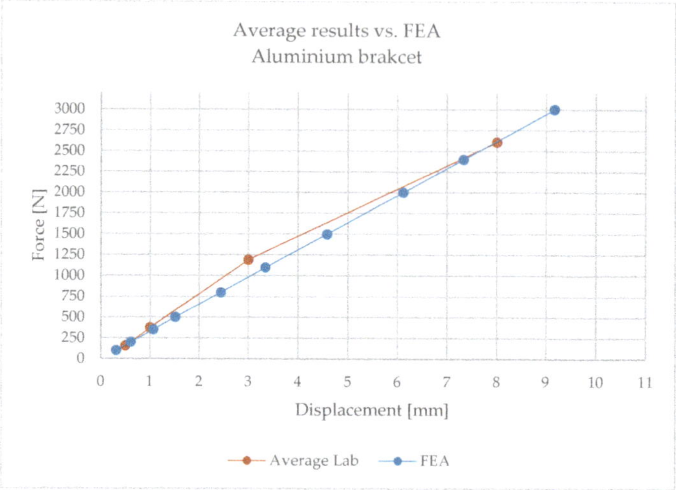

Figure 16. Mean laboratory force–displacement curve for the aluminium bracket and vertical profile compared with the FEA prediction.

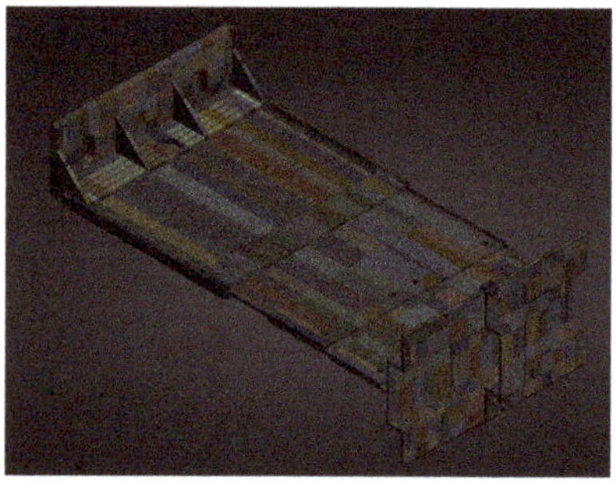

Figure 17. Geometry of the numerical model of the system with the passive bracket.

Boundary conditions reflected the actual mounting: the contact at the anchoring screw was modelled as a fixed support applied to half the circumference of the hole (Figure 18). The remaining contacts between elements were assumed as bonded, corresponding to the rigid cooperation observed in laboratory specimens. The load point was applied to holder, in analogy with the setup of the testing machine.

Figure 18. Boundary condition of the passive bracket anchorage.

All aluminium parts were modelled as isotropic, linear-elastic materials at the relevant test temperature. The composite insert (PA66 GF40) was modelled as an isotropic, linear-elastic material—the elastic constants and density adopted for the insert reflect the passive-bracket test condition (90 °C) and are listed together with the aluminium parameters in Table 6. For clarity of geometry and material layout, Figure 19 includes a photograph of the analysed bracket, showing the polyamide insert region (red part) and the remaining aluminium components (grey part).

Table 6. Material parameters used in the FE model for the passive bracket and profile.

	EN AW 6060 T6	PA6 GF40
Density	2700 kg/m^3	1460 kg/m^3
Young's modulus	69,500 MPa	12,000 MPa
Poisson's ratio	0.33	0.35
Bulk modulus	68,137 MPa	13,333 MPa
Shear modulus	26,128 MPa	4444.4 MPa
Tensile ultimate strength	190 MPa	180 MPa
Tensile yield strength	150 MPa	-

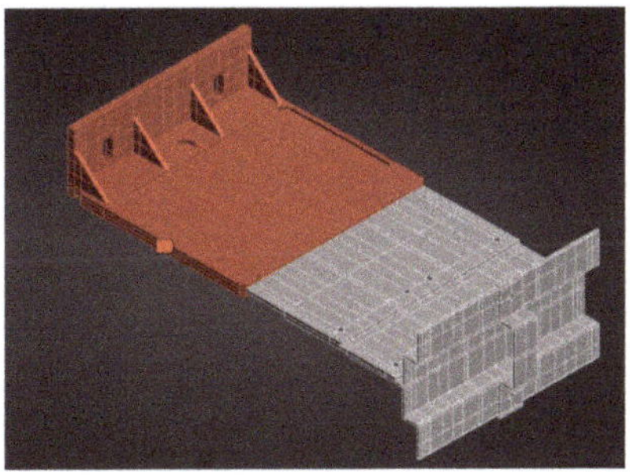

Figure 19. Model view with highlighted composite insert and aluminium parts.

The comparison of FEA and laboratory results revealed significant differences:

- For 100 N, the model predicted a deformation of 0.28 mm, while the laboratory recorded 0.69 mm (Figure 20);
- For 130 N: 0.38 mm vs. 1.0 mm;
- For 250 N: 0.72 mm vs. 3.0 mm;
- For 330 N: 0.95 mm vs. 5.0 mm (Figure 21).

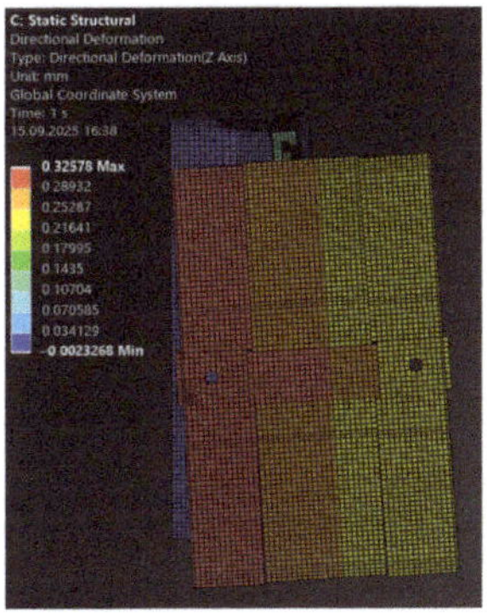

Figure 20. Map of vertical deformations of the passive bracket—load 100 N.

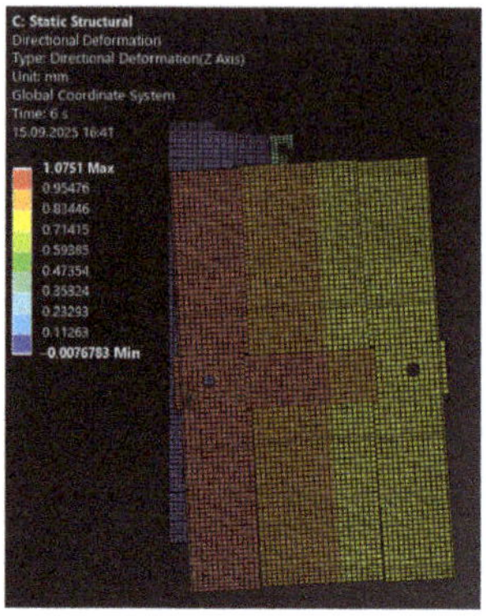

Figure 21. Map of vertical deformations of the passive bracket—load 330 N.

The numerical analysis consistently underestimated displacements, which indicates an overestimation of stiffness in the linear model. This means that the simplified approach does not reflect the actual behaviour of the hybrid element, where the polymer insert with its brittle failure mode plays a critical role.

Following the same procedure, Figures 22 and 23 present comparative force–displacement plots for the passive bracket. Figure 22 overlays the individual laboratory curves with the FEA prediction, while Figure 23 contrasts the mean laboratory response with the FEA curve, enabling a direct like-for-like comparison.

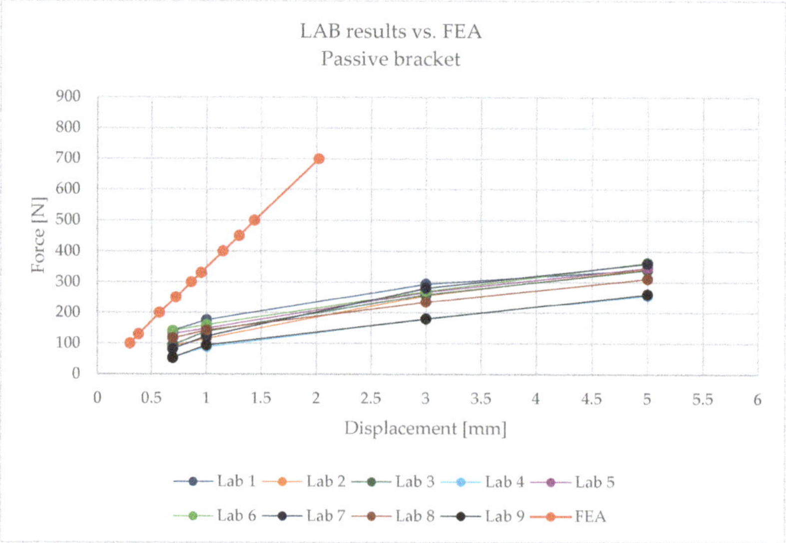

Figure 22. Force–displacement curves for the passive bracket: individual laboratory tests overlaid with the FEA prediction.

The obtained results clearly indicate the necessity of using nonlinear material and geometric models when analysing passive brackets. Only such an approach enables proper representation of stiffness degradation and the sudden nature of failure.

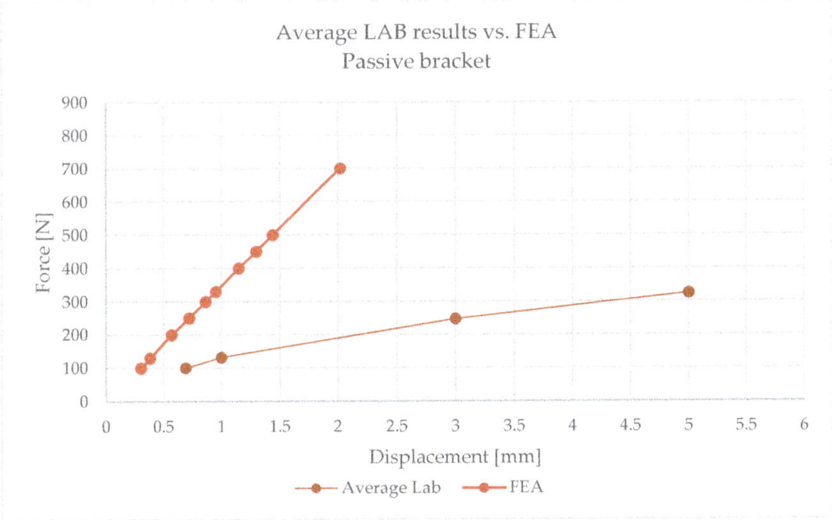

Figure 23. Mean laboratory force–displacement curve for the passive bracket compared with the FEA prediction.

The comparison of laboratory test results and numerical analyses reveals fundamental differences in the mechanical behaviour of aluminium and passive brackets, which have direct implications for engineering practice as well as for the development of normative guidelines. This discussion concerns three key areas: structural behaviour, the reliability of calculation methods, and the implications for the design and operation of ventilated facade systems.

5. Discussion

5.1. Structural Behaviour and Failure Mechanisms

The aluminium bracket behaves in a manner typical of metallic structures: it exhibits a clear elastic range, followed by the ability to undergo plastic deformation and stress redistribution. The failure mechanism is gradual, signalled by increasing deflections and the formation of plastic hinges. Such behaviour provides the user and designer with a "safety buffer"—even in the case of overloading, the element does not fail immediately, and its loss of capacity is predictable and observable.

For the passive bracket, the situation is fundamentally different. The aluminium–composite hybrid system has no plastic reserve. Once critical values are exceeded, the polymer insert undergoes sudden brittle failure without any prior warning signs. From a service perspective, this means a higher risk of sudden failure, difficult to predict and virtually impossible to detect early. Experimental results, where an immediate loss of capacity was observed after reaching maximum force, confirm this hypothesis.

5.2. Reliability of Calculation Methods

Numerical analyses confirmed a high level of agreement for the aluminium bracket: linear FEA models accurately reproduced both the displacement values and the shape of the load–deflection curves. Differences compared with laboratory tests remained within tolerance limits and did not affect engineering interpretation. In practice, this means that classical linear models are sufficient for the safe design of aluminium brackets.

For passive brackets, however, linear analysis consistently underestimated displacements, resulting in overestimated stiffness and incorrect assessment of load-bearing capacity. Such simplifications place the designer on the unsafe side: calculated values may be overly optimistic, while the actual structure is more prone to deformation and failure. This necessitates the use of nonlinear analysis, including

- Material nonlinearity (polymer rheology: creep, stress relaxation, thermal and moisture degradation);
- Geometric nonlinearity (unilateral contact, local displacements, bolt loosening);
- Environmental conditions (temperature, humidity, ageing).

The absence of such models in engineering practice leads to underestimated risks and potentially unsafe design decisions.

5.3. Durability and Environmental Conditions

One of the most important findings is the strong influence of temperature on the behaviour of the passive bracket. Even under laboratory conditions (+90 °C), a significant reduction in effective stiffness and a change in the behaviour of the insert were observed. In real service conditions (exposure to UV radiation, humidity variations, cyclic thermal loads), this effect may intensify, leading to unpredictable degradation over the long service life.

Currently, no data exist on the durability of passive brackets over the 30–50-year service life of a facade. Processes such as creep, thermal ageing, material fatigue, or environmental degradation remain poorly understood and are not covered in normative documents. This means that the operational risk of passive brackets is difficult to assess unambiguously.

5.4. Design and Normative Implications

Current regulations—both Eurocodes and EAD documents—focus primarily on the assessment of entire facade kits, without addressing the specifics of hybrid structural details. For passive brackets, the following are missing:

- Criteria for assessing polymer rheology and long-term degradation;
- Calculation procedures accounting for nonlinear failure mechanisms;
- Guidelines for verifying long-term durability.

In practice, this means that the approval of passive brackets for use relies on individual laboratory tests, which hinders standardisation and comparability. From an engineering perspective, this is unfavourable, as it limits the possibility of routine application of these solutions without additional analyses and testing.

5.5. Practical Significance of the Research

The results of the analyses clearly show that the aluminium bracket can be considered a reference solution, characterised by predictable behaviour, agreement between calculation models and experiments, and a high level of service safety. The passive bracket, in contrast, should be treated as a separate class of structural elements requiring new evaluation and design procedures.

The research also demonstrates that the development of passive bracket technology for facades must be closely linked to the establishment of dedicated calculation and validation methods. Without this, there is a risk that the energy benefits from reducing thermal bridges will be achieved at the expense of mechanical safety and structural durability.

Since the passive bracket was tested at an elevated temperature (+90 °C) and the aluminium bracket under standard conditions (20 °C), the obtained force–deflection (F–δ) values are not directly equivalent. However, this choice of test temperatures was intentional and aimed at highlighting the temperature sensitivity of the hybrid system. Comparative interpretation was therefore based mainly on the relationships between the shapes of the curves and characteristic points (0.2% L, 1 mm, 3 mm), with the explicit reservation that material properties are temperature dependent.

5.6. Limitations

For the aluminium configuration tested at 20 °C, the linear elastic FE model provides qualitatively consistent agreement with the laboratory response in terms of global stiffness trend and overall load–displacement behaviour. Any residual deviations can be attributed to minor fixture compliance, neither of which was modelled explicitly. Within the reported load range, the linear baseline is therefore fit-for-purpose.

For the passive bracket tested at 90 °C, the linear elastic FE model is insufficient to capture the observed behaviour. In particular, for the same applied force, the FE model predicts smaller displacements than those measured experimentally; i.e., the linear baseline is conservative in terms of global deformation. This divergence is consistent with the bracket's temperature-sensitive polymer insert, potential nonlinear contact effects, and brittle failure mechanisms, none of which are represented in a purely linear elastic formulation.

These outcomes delimit the applicability of linear analysis: it can be adequate for the aluminium bracket within the explored range, but a more advanced, temperature-aware, nonlinear model (including contact/friction and, ultimately, time-dependent effects) is required to realistically reproduce the passive bracket's response.

6. Conclusions

The laboratory tests and numerical analyses provided a comprehensive comparison of the mechanical behaviour of aluminium and passive brackets used in ventilated facade substructures. The results clearly indicate significant differences in both structural performance and the requirements for their analysis and design.

6.1. Aluminium Bracket—Reference Solution

1. The aluminium bracket demonstrates a predictable behaviour, with plasticisation and stress redistribution following the elastic phase. As a result, the failure mechanism is gradual, signalled by large deflections, which allows potential service problems to be detected early.
2. The consistency between laboratory results and finite element analyses (FEA) confirms that classical linear models are sufficient for the correct assessment of the capacity and stiffness of aluminium brackets.
    - For example, at a displacement of **1 mm**, the mean laboratory load was **372 N**, while the corresponding FEA result was **327.6 N**.
    - For a displacement of **3 mm**, the mean laboratory load reached **1189.9 N**, compared to **984.6 N** in the FEA model.
    - At **8 mm**, the average laboratory value was **2606.6 N**, and the FEA value was **2618.5 N**.

    This close agreement demonstrates the high reliability of the linear numerical model.
3. Aluminium structural elements can be considered a reference solution in such studies, providing a benchmark for verifying the reliability of both experimental and numerical procedures.

6.2. Passive Bracket—Hybrid Solution

1. The passive bracket, based on a composite insert, is characterised by the absence of a plastic reserve. Once critical values are exceeded, sudden brittle failure occurs without any prior warning signs.
2. Laboratory tests indicate strong sensitivity to elevated temperature, which results in a reduction in effective stiffness and a change in the structural behaviour of the system. Over the long service period, this may lead to material degradation and unpredictable failures.
3. Numerical analyses demonstrated that linear models are insufficient for accurately reproducing the mechanical behaviour of passive brackets. Their application leads to erroneous and overly optimistic estimations of stiffness and load-bearing capacity.
    - For instance, at a displacement of **1 mm**, the mean laboratory value was **131.64 N**, whereas the FEA result was **347.5 N**.
    - For a displacement of **3 mm**, the mean laboratory value was **246.7 N**, while the FEA results for smaller horizontal displacements were significantly higher.

    These discrepancies confirm the need for a more advanced modelling approach.
4. Sound design conclusions require nonlinear analysis, accounting for the rheology of composite materials, creep, stress relaxation, bolt loosening, and environmental effects.

6.3. Design and Normative Implications

1. Aluminium brackets can be safely analysed using linear methods, as the results show a high degree of convergence. For example, for a displacement of 1 mm, the mean laboratory test value was 372.0 N, while the FEM result was 327.6 N. For a displacement of 3 mm, the mean laboratory value was 1189.9 N, compared to 984.6 N from the FEM analysis. Finally, for a displacement of 8 mm, the mean laboratory value was 2606.6 N, while the FEM result was 2618.5 N. This consistency confirms that aluminium brackets exhibit structurally simple and predictable behaviour.
2. Passive brackets should be treated as a separate class of structural elements, requiring the development of new design and normative procedures that account for the specific phenomena associated with hybrid systems.
3. Current normative documents and technical guidelines do not sufficiently address aspects such as polymer rheology, long-term degradation, the effects of temperature and humidity, or contact behaviour in steel-composite systems. This represents a significant gap, the closure of which is essential for the safe implementation of passive brackets in engineering practice.
4. For designers, this means a cautious approach to the use of passive brackets—their implementation should be preceded by additional laboratory tests and numerical validation, and in the case of structures with high safety requirements, by long-term analyses as well.

7. Summary

This paper presented the results of laboratory tests and numerical analyses of brackets used in ventilated facade substructures, covering both traditional aluminium solutions and innovative passive systems based on composite inserts. The comparison of results allowed for the identification of fundamental differences in the structural behaviour of both systems and the formulation of recommendations for their design and evaluation.

Key findings:

1. Aluminium bracket—reference solution
   The aluminium bracket proved to be a reference structural solution, characterised by predictable behaviour and a high degree of agreement between experimental and numerical results.
   Due to the material's ability to plasticise and redistribute stresses, the failure mechanism is gradual and preceded by large deflections, which allows potential service issues to be detected early.
   Quantitative comparison confirmed the reliability of classical linear models:
   - At 1 mm displacement, the mean laboratory load was 372 N, while the FEA result was 327.6 N;
   - At 3 mm: 1189.9 N (laboratory) vs. 984.6 N (FEA);
   - At 8 mm: 2606.6 N (laboratory) vs. 2618.5 N (FEA).

   This close correlation demonstrates that the linear approach is sufficient for accurate assessment of stiffness and capacity, ensuring safe and reliable performance predictions.
2. Passive bracket—hybrid solution
   The passive bracket, incorporating a polymer composite insert, exhibited a fundamentally different behaviour. The lack of plastic reserve led to a sudden and brittle failure once critical stress levels were exceeded, without prior warning signs.
   Significant discrepancies between experimental and numerical results were observed:
   - At 1 mm displacement, the mean laboratory load reached 131.6 N, while the FEA result was 347.5 N;
   - At 3 mm: 246.7 N (laboratory), with considerably higher numerical predictions at small horizontal displacements.

   These results confirm that simplified linear models cannot accurately reproduce the real mechanical response of passive brackets and tend to produce overly optimistic and potentially unsafe design outcomes.
   Furthermore, the tests revealed a strong sensitivity to elevated temperature, leading to stiffness reduction and altered structural response. Over time, this may cause material degradation and unpredictable failures.
3. Design and normative implications
   The current design standards and technical guidelines do not fully reflect the specific behaviour of passive brackets. Aspects such as polymer rheology, long-term degradation, temperature and humidity effects, and the contact behaviour between steel and composite components remain insufficiently addressed.
   Therefore, passive brackets should be treated as a distinct class of structural elements, requiring new design methodologies and verification procedures based on nonlinear analysis, experimental validation, and long-term monitoring.
   For designers, this means adopting a cautious approach: the use of passive systems should be preceded by dedicated testing and numerical verification, and in structures with high safety requirements, by comprehensive long-term analyses.

Practical and scientific significance:

The results indicate that the further development of passive bracket technology is only possible through in-depth experimental research, nonlinear numerical modelling, and long-term studies on the rheological behaviour and degradation of composite materials. Such an approach will provide a robust normative foundation and ensure the safe implementation of these hybrid systems in construction practice.

In a broader context, this study fills an important knowledge gap in the comparison between metallic and hybrid systems used in ventilated facades. The presented findings can serve as a reference framework for future scientific investigations and engineering applications, offering valuable input for upcoming updates to design standards and technical assessment procedures.

Author Contributions: Conceptualization, Ł.Z., K.S. and O.K.; Methodology, Ł.Z., K.S. and O.K.; Software, Ł.Z.; Validation, Ł.Z.; Formal analysis, O.K.; Investigation, Ł.Z. and O.K.; Resources, Ł.Z.; Data curation, Ł.Z.; Writing—original draft, Ł.Z.; Writing—review & editing, O.K.; Visualization, Ł.Z. and O.K.; Supervision, K.S.; Project administration, Ł.Z. All authors have read and agreed to the published version of the manuscript.

Funding: This research was funded by the National Science Centre, Poland, under the MINIATURA programme, grant no. 2024/08/X/ST8/00565, project title "Wpływ podatności węzłów na bezpieczeństwo elewacji wentylowanych stosowanych w budownictwie pasywnym" (The impact of node flexibility on the safety of ventilated facades used in passive buildings) and by grant no. N-001/2025, project title "Ensuring publishing potential at the Institute".

Institutional Review Board Statement: Not applicable.

Informed Consent Statement: Not applicable.

Data Availability Statement: The original contributions presented in this study are included in the article. Further inquiries can be directed to the corresponding author.

Conflicts of Interest: The authors declare no conflicts of interest.

References

1. Bikas, D.; Tsikaloudaki, K.; Kontoleon, K.; Giarma, C.; Tsoka, S.; Tsirigoti, D. Ventilated Facades: Requirements and Specifications Across Europe. *Procedia Environ. Sci.* **2017**, *38*, 148–154. [CrossRef]
2. Schabowicz, K. Ventilated façades. In *Fibre-Cement Board Production Technology and Testing Methods*; Oficyna Wydawnicza Politechniki Wrocławskiej: Wrocław, Poland, 2018. (In Polish)
3. Directive (EU) 2024/1275 of the European Parliament and of the Council of 24 April 2024 on the Energy Performance of Buildings; Recast, Text with EEA Relevance. Available online: https://eur-lex.europa.eu/eli/dir/2024/1275/oj (accessed on 6 November 2025).
4. Petresevics, F.; Nagy, B. FEM-Based Evaluation of the Point Thermal Transmittance of Various Types of Ventilated Façade Cladding Fastening Systems. *Buildings* **2022**, *12*, 1153. [CrossRef]
5. Kopyłow, O.; Chełkowski, F. Problems of using passive brackets in ventilated facades systems. *Mater. Bud.* **2024**, *12*, 34–41. [CrossRef]
6. EOTA. *ETAG 034 Part 1: Ventilated Cladding Kits Comprising Cladding Components and Associated Fixings*; European Organisation for Technical Approvals: Brussels, Belgium, 2012.
7. EOTA. *ETAG 034 Part 2: Cladding Kits Comprising Cladding Components, Associated Fixings, Subframe and Possible Insulation Layer*; European Organisation for Technical Approvals: Brussels, Belgium, 2012.
8. *EAD 090034-00-0404*; Kit Composed by Subframe and Fixing for Fastening Cladding and External Wall Elements. EOTA: Brussels, Belgium, 2016.
9. *EAD 090062-00-0404*; Kits for External Wall Claddings Mechanically Fixed. EOTA: Brussels, Belgium, 2018.
10. *EAD 090097-00-0404*; Kits for External Wall Claddings Glued to the Subframe. EOTA: Brussels, Belgium, 2018.
11. Ujma, A. Ściany zewnętrzne z elewacjami wentylowanymi i ich izolacyjność cieplna. *Izolacje* **2020**, 58–66.
12. *EN 1999-1-1:2007+A1:2009+A2:2013*; Eurocode 9—Design of Aluminium Structures—Part 1-1: General Structural Rules. CEN: Brussels, Belgium, 2013.
13. Cwyl, M.; Prochera, M.; Wojda, J.; Wilga, K.; Radzimirski, P. Badanie nośności wybranych poliamidowo-aluminiowych konsoli nośnych dla okładzin fasad budynków. *Builder* **2023**, *314*, 28–33. [CrossRef]
14. Cwyl, M.; Dmowska-Michalak, I.; Kaczmarczyk, A.; Michalczyk, R. Laboratory tests and numerical analysis of aluminum helping hand brackets with polyamide thermal break. *Arch. Civ. Eng.* **2022**, *68*, 409–426.
15. Kopyłow, O.; Chełkowski, F. Stosowanie narzędzi obliczeniowych do określenia właściwości mechanicznych konsoli do elewacji wentylowanych. *Mater. Bud.* **2023**, *628*, 34–37. [CrossRef]

16. Trejo, D.; Gardoni, P.; Kim, J.J.; Zidek, J. *Long-Term Performance of GFRP Reinforcement: Technical Report*; The Texas A&M University System 2009, Report 0-6069-1; Texas Transportation Institute: College Station, TX, USA, 2009.
17. National Technical Assessment of the Republic of Poland ITB-KOT-2022/2265 wydanie 1. Available online: https://www.google.com.hk/url?sa=t&source=web&rct=j&opi=89978449&url=https://artrys.pl/wp-content/uploads/2025/02/ITB-KOT-2022_2265-wydanie1.pdf&ved=2ahUKEwjR0YO-tICRAxXCFTQIHUJLEwsQFnoECBkQAQ&usg=AOvVaw0yTWDcmjKOA97MR-yIuHeA (accessed on 15 November 2025).
18. *DIN 933*; Hexagon Head Bolts. Deutsches Institut für Normung (DIN): Berlin, Germany, 1987.
19. *DIN 9021*; Plain Washers with Large Outside Diameter. Deutsches Institut für Normung (DIN): Berlin, Germany, 1987.

materials

MDPI

Article

Environmental Life Cycle Assessment of Selected Materials—Building Façades in Poland

Dorota Burchart [1] and Krzysztof Schabowicz [2,*]

1 Faculty of Transport and Aviation Engineering, Silesian University of Technology, Krasińskiego 8, 40-019 Katowice, Poland; dorota.burchart@polsl.pl
2 Faculty of Civil Engineering, Wrocław University of Science and Technology, Wybrzeże Wyspiańskiego 27, 50-370 Wroclaw, Poland
* Correspondence: krzysztof.schabowicz@pwr.edu.pl

Abstract

The use of sustainable building materials is becoming increasingly important in order to reduce their environmental impact. This article draws attention to the lack of life cycle assessment (LCA) of building façades, which would take into account national conditions. The aim of the work is to assess the environmental impact of various building façade solutions. The analysis concerned a ventilated façade on an aluminum substructure with a fiber cement board and external thermal insulation composite system (ETICS) with expanded polystyrene (EPS). The assessed façades differed with regard to the used insulation materials. The study aims to select more ecological façades, while at the same time taking into account national conditions, which is important at the stage of designing a building. The study also aims to fill a gap in the existing literature by providing information concerning the environmental analysis of building façades based on real data. Based on a comparative analysis, it was shown that ETICSs with EPS have higher façade-damage category indicators in all impact categories except for eutrophication, human toxicity (carcinogenic and non-carcinogenic), and resource use related to minerals and metals, for which the ventilated façade shows higher values. Additionally, hot-spots for the analyzed façades were also presented. In the case of a ventilated façade, the determinant is the used insulating material, which is mineral wool. In the case of ETICS, it is the finish coat. For the first time in Poland, the LCA of a ventilated façade and ETICS was presented based on real data. The results of this study can be used as the first step of a full cradle-to-grave LCA for buildings.

Keywords: life cycle assessment; carbon footprint; construction sector; ventilated façade; fiber cement board; external thermal insulation composite system

1. Introduction

The implementation of the European Green Deal, increasing requirements from the sustainable development goals, and the transition into a green economy of the European Union have led to transformations in the construction industry. The construction industry is one of the main sources of greenhouse gas (GHG) emissions. It is responsible for approximately 40% of total energy consumption in the European Union, and for 36% of greenhouse gas emissions related to the energy supply of buildings during their use [1,2]. Paper [3] presented the guidelines and initiatives of the European Commission related to environmental issues for the construction industry, for which managing and reducing the carbon footprint is becoming a key challenge. In order to reduce greenhouse gas emissions in the construction industry, it is necessary to conduct research in order to achieve sustainable buildings. To this end, the first step is to recognize and assess the current condition of buildings in terms of their impact on GHG emissions, while at the same time taking into account not only the building operation stage, but also the construction stage. For this purpose, the life cycle assessment (LCA) technique can be used, which is recommended by the European Commission in the new Directive 2023/1791 of the European Parliament of 13 September 2023 on energy efficiency. To date, many environmental impact assessments based on LCA have been performed for the following building materials: concrete [4,5], constructive solutions [6], cementitious materials [7], prefabricated buildings [8], building materials [9], flat roof systems [10], structural materials [11], innovative building materials [12] and building insulation materials [13,14], and residential building walling [15]. Paper [16] shows the approach needed to evaluate the toxicity of building materials using LCA. It was stated that materials such as insulation or paints have a high human health indicator and ecotoxicity impact. A comparative environmental LCA of road asphalt pavement solutions made of artificial aggregates was shown in paper [17]. There are not a lot of publications concerning the LCA of building façades. The façade, as the first element that protects against undesirable external influences, can contribute to the sustainable development of a building by reducing its impact on the environment. Over the last decade, fiber cement boards have become an interesting architectural material, especially those reinforced with more sustainable fibers. Paper [18] presents an innovative model for assessing the durability index of façade boards, and in particular fiber cement boards. There is currently a wide range of renewable materials that can be used to modernize building façades, including wood, cork, recycled materials, and polymer composites. Life cycle assessment is one of the decision criteria of Multi-Attribute Decision-Making applications in façade design [19,20].

Based on a state-of-the-art review, it was found that several LCA studies have already been carried out regarding the modernization of building façades. Paper [21] presents the potential of abaca fibers to contribute to the circular economy and carbon footprint reducing. Paper [22] shows a LCA for different wall assemblies made of stabilized rammed earth reinforced with natural fibers. The carbon footprint of an insulating material was presented in accordance with the life cycle approach [23]. Other authors focused on the analysis of the environmental impact of the production of lime mortar as a decorative mineral plaster for ETICSs. In paper [24], there are also the results of the analysis of the Multi-Active façade system with ETICS and EPS. There was a lack of research on the analysis of the life cycle of ventilated façades, which would take into account actual data concerning the production of a ventilated façade.

Performing LCA analysis is not yet required in accordance with Polish legal acts; however, taking into account changes in the Directive, as well as the Renovation Wave strategy, the LCA technique becomes an important tool that can help in the analysis of the environmental impact of a building's life cycle. Only a life cycle approach will fully assess and improve the environmental performance of buildings. Until now, there has been no developed LCA methodology for buildings in Poland that covers national conditions. Developing such a methodology and determining the limit indicators for greenhouse gas emissions in a building's life cycle is a complex process. For these reasons, LCA studies covering individual building elements are important. Therefore, this work focused on the analysis of one of the most important elements of a building, which is the façade system.

The aim of this article is to perform an LCA analysis for a building's façade based on actual data, while at the same time taking into account the life cycle of the individual

 https://doi.org/10.3390/ma19040807

materials and elements of the façade. The tests were performed for Polish conditions, which is associated with the fact that insulating materials of greater thickness had to be used in order to meet the thermal insulation requirements and other requirements related to energy saving, A comparative LCA analysis of a ventilated façade with the use of ETICS was performed. In turn, ETICS with EPS was used for comparative analyses, which is due to the fact that it is a system used in buildings in Poland (it has an attractive price). The conducted analysis allowed for the specifying of which system is more ecological, and also showed the hot-spots in each of the façade's systems. Performing an LCA analysis is the first step towards identifying changes and innovations in buildings by relating these changes and innovations to all the factors and materials in the building's life cycle. In the article, the assessment of the life cycle of a ventilated fiber cement board and ETICS was described for the first time in Poland.

2. Materials and Methods

The assessment of the potential impact of a ventilated fiber cement board and ETICS was performed using the life cycle assessment method following the international ISO standards [25,26]. The LCA study includes a definition of its goal and scope, life cycle inventory (LCI), life cycle impact assessment (LCIA), and ends with an interpretation of the results and recommendations [27,28].

2.1. Goal and Scope of the LCA Analysis

The aim of the research was to perform an LCA analysis of building façades that are used in Poland. Both façades were modeled with identical system boundaries, functional unit, harmonized data-collection protocols and the same life cycle impact assessment method. The life cycle assessment was performed using SimaPro Craft 10.3 software with the Ecoinvent database version 3.9 [29]. The "cut-off" system model was adopted, which is consistent with attributional LCA and aligned with the Environmental Footprint (EF 3.1) methodology recommended by the European Commission. The study follows an attributional, process-based approach. Primary data describing façade components and installation processes were directly linked to background processes from Ecoinvent 3.9.

In the construction industry, it is helpful to use EN 15804:2012+A2:2019 [30] and EN 15978:2011 [31] standards for life cycle analysis [32]. For the purpose of the analysis, the rules (PCR—Product Category Rules) for thermal insulation materials were taken as a reference. This is due to the fact that they do not exist specifically for ETICS and ventilated façade systems. For comparative purposes, all analyses were referred to the same functional unit. The used functional unit (FU) was 1 m^2 of a building façade, as in studies [20,33,34]. Based on the analysis of previous LCA studies of various façade systems, the life cycle of a façade system was considered to be 50 years [21,22,35,36].

2.2. The Building Façades' Definition and System Boundary

The aim of this research is to assess the potential environmental impact of two building façade solutions that are used in Polish conditions. The objects of the research included a ventilated façade with a fiber cement board, and ETICS with EPS. The system's boundary was defined, and sets of inputs and outputs were identified as elements of individual system phases. The system boundaries of the study are compliant with ISO 14040:2006 [25] and ISO 14044:2006 [26] standards. The system's boundary covers the scope of processes from cradle to gate: resources used and their respective life cycles, all needed materials and energy production and individual façades' production and installation. The system's

https://doi.org/10.3390/ma19040807

boundaries for individual façades are shown in Figure 1 for the ventilated façade and in Figure 2 for ETICS.

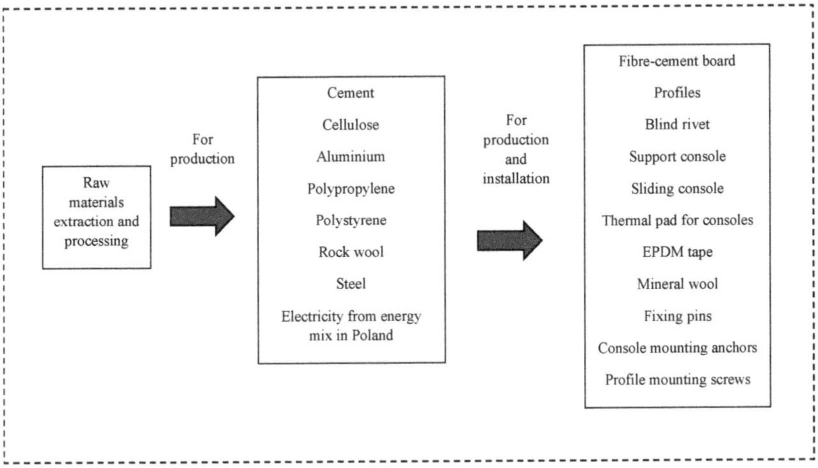

Figure 1. System boundary of the ventilated façade system.

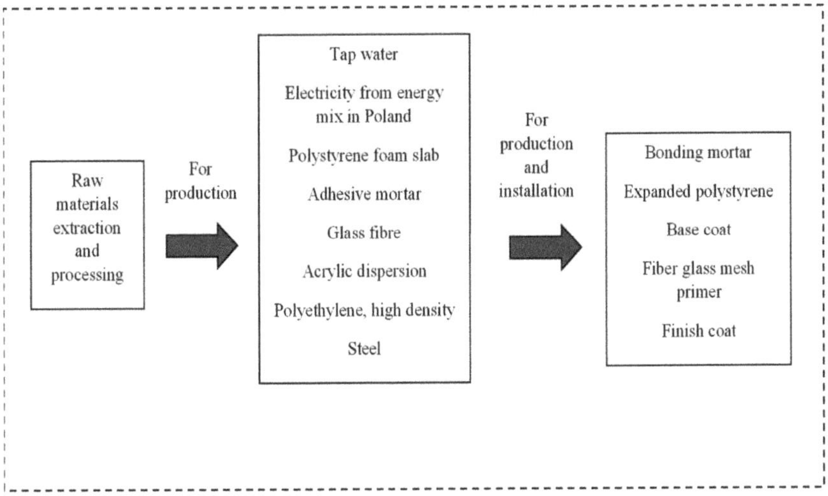

Figure 2. System boundary of the ETICS.

In LCA analyses, in order to demonstrate the reliability of the analysis results, it is important to present the assumptions and limitations of the analysis. Outside of system boundaries are transport and "end of life" stage. Transport of materials is not included due to the difficulty of determining distance, while the "end of life" stage is not included due to the lack of data on the possibility of taking into account the recycling of individual materials after being used for many years. Such cut-off criteria echo an earlier paper regarding LCA analyses for façade systems in other countries. An additional argument for limiting the analyses to the stage of production and installation of the façade system is the purpose and scope of the work, which focuses on comparing the materials used in the production of individual façades. The cradle-to-gate LCA of a building's façade covers both the raw

https://doi.org/10.3390/ma19040807

materials extraction, material and energy production process and the installation stage of a building's façade.

An example of a façade that uses fiber cement boards, a schematic cross-section of the ventilated facade system and a schematic cross-section of the ETICS is available in the Supplementary Materials [Figures S1–S3].

The first type of façade analyzed is a ventilated façade with a fiber cement board.

There is no Polish standard for ventilated façades that considers a whole system. For a given type of ventilated façade that is built on a building, technical approval or technical documentation prepared by the designer is required in order to approve it for individual use [37]. When preparing technical documentation, a designer selects the parameters of individual system components and, on this basis, determines the properties of the entire façade, which must comply with the technical conditions [38,39]. Each component of the façade should be provided with appropriate documents, i.e., a declaration of performance and a CE marking. Ventilated façades are a set of construction products that create an entire system, and should be considered as a whole. Based on the guidelines for European technical approvals, ETAG 034, a technical approval—currently a national technical assessment—was developed for a ventilated façade. These guidelines define the scope of test methods and requirements for a complete façade system. In Poland, technical documents of this type are developed and published by the Building Research Institute [33]. In terms of widespread use, fiber cement boards are the leader among façade claddings and therefore in this study the life cycle assessment of ventilated façade systems with fiber cement boards was performed.

These boards are produced in various varieties. Gray boards are commonly available, but they can be colored according to the color palette offered by the manufacturers. They can also be painted, coated with materials that leave patterns, or can be made of certain materials, e.g., wooden planks. Ventilated façades made of fiber cement boards are very popular in Scandinavia, Russia and North America, although they differ significantly in terms of the used raw materials. In Poland, the use of fiber cement boards is also growing, which is related to their good fire protection and strength parameters. When assessing building façade systems, the environmental aspect is becoming increasingly important. It takes into account the life cycle, starting from the extraction of raw materials, through the production of materials, to the production process of the façade itself.

A ventilated façade is a set of appropriately selected elements that create a complete façade system. This system consists of a substructure called a grid, thermal insulation, a ventilation gap, and the façade's cladding made of fiber cement boards.

An important element of the façade system is the ventilation gap in which air flows between the thermal insulation and the façade's cladding. The flowing air carries away the condensate and moisture that has accumulated in the wall outside the partition. The substructure creates a skeleton, which is used to attach the cladding at a certain distance from the structural wall (taking into account the thickness of the thermal insulation and ventilation gap). The ventilation gap thickness may vary, depending on, among other things, the faults of the wall structure elements, the thickness of the thermal insulation material, and the adopted architectural conditions. The substructure can be made of aluminum, wood, steel or composite material. Thermal insulation is responsible for large part of thermal performance, but only to a minor extent on acoustics. It should also have high vapor permeability and be protected from the outside with a veil or membrane that protects against wind and moisture. The last element of the façade system is the façade's cladding, which primarily gives the facility its appropriate appearance. It must also ensure the drainage of rainwater and protect the thermal insulation against external

https://doi.org/10.3390/ma19040807

environmental and exceptional factors. Façade cladding can be made of various materials, and is available in various colors and formats. One of the more popular solutions used by architects and investors are façade claddings made of fiber cement boards, which were selected for analysis in this study. The growing interest in these types of building façades is also influenced by the perceived shortcomings of the very popular façades that were used in the past and which constitute a complex thermal insulation system with ETICS plasters.

The system boundary of the ventilated façade is depicted in Figure 1.

The system boundary includes all raw materials, materials and energy production and the ventilated façade system production and installation.

The analysis included all the elements of the ventilated façade system, but without the structural wall. The thickness of the fiber cement boards was 8 mm, and the dimensions of the board were 1250 × 3050 mm. The bulk density of the basic board was 15 kg/m^2, and its average bending strength was 24 MPa. Fiber cement boards are non-flammable and have a fire resistance class of at least A2.

The growing popularity of ventilated façades is due to the possibility of using various cladding materials, which are tailored to the requirements for modern façades. This makes it possible to use ventilated façades in various conditions, regardless of the environment and fire requirements, and also with the possibility of obtaining appropriate strength. The structural element of a ventilated façade is the substructure (grid). In the analyzed case, it was made of aluminum.

This solution is currently the most popular due to its large range, simple installation, the possible joining of the façade's cladding (both visible and invisible), and the possibility of using cladding from various materials. Therefore, this solution was used in this research. The grate consists of vertical aluminum profiles and mounting brackets, which are called consoles. The vertical elements of the grid are made of T-sections. The width of the profile provides adequate support for two adjacent façade panels, along with an 8–12 mm wide expansion gap between the panels. Angle bars were used as intermediate support for the panels, and were used in the corners of the façade. The vertical profiles were mounted on the structural wall using consoles made of unequal angle bars. The consoles' reach depends on the distance between the structural wall and the cladding. This distance is necessary to assemble thermal insulation and to provide a ventilation gap of 20 to 50 mm according to ETAG 034 guidelines. Typical console reach is 200–300 mm. The consoles are equipped with special handles (clips that facilitate the installation of aluminum profiles) and a stiffening bracket, which is necessary for greater overhang. The length of the vertical profiles was limited to approximately 3.0 m due to the high thermal expansion of aluminum. Vertical profiles were connected to the consoles using rivets, while taking into account the principle of one fixed fixing point. The remaining joints were slidable, which enables displacements that result from temperature changes to be compensated. It should be remembered that additional stiffening of the grid structure cannot be introduced when designing and assembling the substructure. EPDM spacers were used to enable free movement of the profiles and the façade's panel against each other. The spacer additionally covers the bright aluminum structure that is visible in the joint. Connectors, made of the same material as the joined element of the structure, were used to connect aluminum elements. If the connectors are selected incorrectly, electrochemical corrosion may occur. This phenomenon should be taken into account in aggressive environments or, for example, in seaside conditions. The fixing of the façade cladding should ensure that the cladding does not fall off in the event of a fire, and it should not be below the required fire resistance class of the external wall for the appropriate fire class of the building. The aluminum substructure meets the necessary requirements in terms of fire resistance, which

https://doi.org/10.3390/ma19040807

was confirmed in tests commissioned by the system's manufacturers, including the tests presented in paper [37].

The second type of analyzed façade system was the external thermal insulation system for buildings (ETICS), a complex system of the insulating external walls of a building, which was previously called the seamless thermal insulation system, and even earlier, the light-wet method. The essence of this method comes down to making layers of cooperating and compatible materials on a properly prepared surface (wall), which include thermal insulation and a finished façade layer.

Figure 2 shows the system boundary for the ETICS analysis. As shown in Figure 2, the system boundary includes all raw materials, materials and energy production and ETICS production and installation. This system consists of basic ingredients (adhesive mortar, thermal insulation, mechanical connectors (pins), a reinforcing layer, a finish layer) and additional components (materials for finishing details: skirting boards, protective angle bars, expansion profiles, sealing materials, and other necessary accessories).

Each material used in this building façade system has a different function: thermal insulation ensures adequate thermal insulation and is made of expanded polystyrene board (EPS); adhesive mortar and mechanical connectors (pins) ensure an adequate structural stability of the system; the reinforcing layer (a layer of mortar with an embedded mesh, e.g., made of fiberglass) ensures resistance to damage (e.g., due to impacts) and constitutes a base for the finishing façade layer; and the finishing façade layer (plaster, façade tiles) protects the system layers against weather conditions and aging, and also constitutes a decorative layer.

The selection of a façade system depends on several factors. Generally speaking, it concerns structural safety, fire safety, operational safety, acoustic protection, thermal protection, energy saving, and health issues. Therefore, the approach that states that design is only limited to determining the thickness of the thermal insulation layer to meet the thermal protection requirements and the method of fastening (gluing, doweling, depending on the type of thermal insulation and wear layer) should be considered incorrect. The starting point should be an analysis of the investment's efficiency based on the analysis of operating costs and costs related to the investment. Afterwards, the specificity of the insulated building (the material from which the external walls are made, the shape of the building, the investor's esthetic requirements), its purpose and location, and the financial resources available to the investor need to be analyzed. Fire and acoustic protection requirements must also be taken into account. No less important is the analysis of the correctness of the adopted solution in terms of the requirements of building physics.

The main methodological limitations of this study include the following:

- System boundary is from cradle to gate, excluding transport and end-of-life stages.
- Partial reliance on secondary datasets for certain materials where detailed industrial data were unavailable.
- High uncertainty associated with selected impact categories, particularly water use and human toxicity, due to characterization model sensitivity.

2.3. *Life Cycle Inventory*

Due to the fact that this is the first LCA study for building façades in Poland, an important element was the identification of quantitative data for the needs of analysis, which comes from the authors' own studies of façades and data from the producers of individual materials. It should be emphasized that the performed analyses are innovative. Identified inventory data for the analysis of building façade systems, which take into account national conditions, was presented for the first time. The main sources of the data are manufacturing companies.

https://doi.org/10.3390/ma19040807

The primary data used in this study refer to the year 2024 and are representative of current industrial practice in Poland. Geographical representativeness corresponds to production processes and façade installations carried out within Poland. In the case of the absence of more detailed data on some materials, input and output data were determined based on literature data and the Ecoinvent database.

In the case of the analysis of the building's ventilated façade, all the materials used to develop the ventilated façade with fiber cement board (used as a ventilation cladding) were taken into account. Table 1 presents the inventoried data for the creation of a ventilated façade. All the data were identified and inventoried based on the authors' own research. The data were related to the functional unit.

Table 1. Inventory of materials of ventilated façade system.

Components	Inputs	Amount	Unit
Fiber cement board	Fiber cement	15.273	kg
Profile T T120/70/2	Aluminum	1.776	kg
Profile L L150/70/2	Aluminum	0.778	kg
Blind rivet 4.0 × 20 K14	Aluminum	0.031	kg
EPDM tape 60 mm	Etylen-propylen-Dien-Monomer	0.016	kg
EPDM tape 120 mm	Etylen-propylen-Dien-Monomer	0.281	kg
Support console	Aluminum	0.289	kg
Sliding console	Aluminum	0.397	kg
Thermal pad for consoles	PVC	0.031	kg
Mineral wool	Mineral wool	14.580	kg
Fixing pins	Polipropylen	0.313	kg
Console mounting anchors	Steel	0.413	kg
Profile mounting screws	Steel	0.007	kg
Installation	Electricity from energy mix in Poland	0.08	kWh

The LCA analysis included a ventilated façade with an aluminum substructure. Fiber cement boards with dimensions of 1250 × 3050 × 8 mm were mechanically attached to aluminum profiles with rivets. Thermal insulation was made of mineral wool, which is a thermal insulation material with a low thermal conductivity coefficient. Rock wool was included in this analysis. For the analysis, 14.58 kg/FU of mineral wool with a veil, which was 15 cm thick and had a density of 90 kg/m^3, was considered. Aluminum profiles and rivets were used for the ventilated façade. The supporting console and the sliding console were also made of aluminum. The material used to make the EPDM tape (Firestone Building Products, Zaventem, Belgium) was Ethylene-propylene-Dien-Monomer. PVC was used to produce the thermal pad for the consoles. The console mounting anchors and profile mounting screws were made of steel.

In Poland, when compared to countries such as Portugal, a thicker layer of wool should be used due to climatic conditions and in order to meet the thermal insulation requirements and other requirements related to energy-saving practices specified in the Regulation of the Minister of Infrastructure, regarding the technical conditions to be met by buildings and their locations [40]. The raw materials shown in Table 2 were used to produce fiber cement boards [41].

Detailed requirements for cement, limestone, cellulose fibers and PVA for the production of fiber cement boards were presented in the authors' previous works [33,35].

https://doi.org/10.3390/ma19040807

Regarding the analysis of ETICS with EPS, all the materials used to produce the building façade were taken into account. In the case of the data for the analysis of ETICs, in addition to actual data that takes into account national conditions, secondary data based on literature data were also used. The following components were used to analyze the ETICS: EPS, bonding mortar, base coat, fiberglass mesh, primer, finish coat. The amount of polystyrene foam slab was adjusted to comply with national conditions. Its thickness was assumed to be 15 cm in order to meet the thermal insulation requirements and other requirements related to energy saving specified in the Regulation of the Minister of Infrastructure on the technical conditions to be met by buildings and their locations. The calculations resulted in an amount of 2.7 kg of polystyrene foam slab per 1 m^2 of façade.

Table 2. Comparison of testing [41].

Raw Material	Approximate Contents [a]
Cement	60%
Cellulose (dry)	8%
PVA	2%
Lime	30%
Total	100%
Additives and admixtures	/
Superplasticizer Didecyldimethylammonium chloride (DDAC) or bromide (DDAB)	0.1 dm^3/t
Perlite Mica Microsphere	1 kg/t
Antifoaming agent	0.26 dm^3/t

[a]—Concentrations dm^3/t and kg/t refer to the content in the finished product.

Table 3 presents the inventoried data for the creation of ETICS with EPS, based on the authors' own research. The data analysis included both research-based data and literature-based data. All the data were related to the functional unit.

Table 3. Inventory of materials of ETICS with EPS.

Components	Inputs	Amount	Unit
Bonding mortar	Adhesive mortar	4.500	kg
	Tap water	1.350	kg
	Electricity from energy mix in Poland	0.016	kWh
EPS	Polystyrene foam slab	2.7	kg
Base coat	Adhesive mortar	4.500	kg
	Tap water	1.350	kg
	Electricity from energy mix in Poland	0.016	kWh
Fiberglass mesh	Glass fiber	0.330	kg
Primer	Acrylic dispersion	0.440	kg
	Polyethylene, high density	0.016	kg
	Tap water	1.944	kg
Finish coat	Polyethylene	0.290	kg
	Steel	0.012	kg
	Acrylic dispersion	8.000	kg

The Polish electricity mix was modeled follow "Energy Transition in Poland. 2025 Edition" [42]. The use of the Polish electricity mix ensures geographical consistency of the

https://doi.org/10.3390/ma19040807

modeling framework and reflects national energy conditions relevant for façade production and installation processes.

The exclusion of these elements is due to difficulties in quantifying and inventorying the data. It was also intended to give greater importance to the remaining elements of each system, i.e., the thermal insulation and external cladding.

2.4. *Life Cycle Impact Assessment*

An important stage of the work was the identification and inventory of data regarding the construction of the façade for which the LCA analyzes were performed. The next element of the study involved the assessment of the associated environmental loads, both direct and indirect, and the identification of the main sources of the environmental impact of façades. A quantitative assessment of potential environmental burdens related to the raw materials and the materials used to construct a building's façade was performed. LCA was performed in accordance with ISO 14044:2006 [26].

A cradle-to-gate inventory combined primary data of the chosen building's façades installation with secondary datasets from Ecoinvent 3.9. Sixteen midpoint damage categories were calculated with the European Commission's Environmental Footprint 3.1 (EF 3.1) method [43] and normalized per functional unit. All sixteen impact categories recommended by the European Commission were included: climate change (CC), ozone depletion (OD), human toxicity—cancer (HT-c), human toxicity—non-cancer (HT-nc), particulate matter (PM), ionizing radiation (IR), photochemical ozone formation (POF), acidification (AP), eutrophication terrestrial (ET), eutrophication freshwater (EF), eutrophication marine (EM), ecotoxicity freshwater (EFT), land use (LU), water use (WU), resource use—minerals and metals (RMM), and resource use—fossils (RFs) [28]. The normalization step was applied according to the respective factors provided by the European Commission [43] and no weighting factors have been applied.

3. Results and Discussion

The results from the comparative analysis of the environmental impact categories of the used types of façades in Poland are presented in Table 4 (characterization stage) and damage categories are presented in Table 5 (damage assessment stage).

Based on the LCA analyses performed, it was shown that carbon footprint (which is the most important aspect according to the European Commission guidelines) was higher for the ETICS with EPS than for the ventilated façade with rock wool. Based on the results of analyses, it was also found that it is not possible to clearly determine which façade is more ecological when considering all the impact categories. This is because it depends on the impact category taken into account. The ventilated façade exhibits higher impact indicators in the following damage categories: eutrophication, human toxicity (both carcinogenic and non-carcinogenic), as well as resource use related to minerals and metals. In all remaining impact categories, the environmental burden is higher for the ETICS.

Impact categories for the compared façade systems are shown in Table 4. Based on the performed LCA analyses for the two façade systems, as summarized in Table 5, it was concluded that the ETICS exhibits higher damage category indicators in all impact categories except for eutrophication (freshwater and terrestrial), human toxicity (carcinogenic and non-carcinogenic), and resource use related to minerals and metals, for which the ventilated façade shows higher values. Tables 6 and 7 present the determinants of life cycle assessment for ventilated façades, taking into account damage category. Tables 8 and 9 present the share of individual elements in the life cycle assessment of the evaluated damage categories for the ventilated façade system and ETICS.

https://doi.org/10.3390/ma19040807

Table 4. Comparative LCA for building façades in Poland: Impact category, characterization stage.

Impact Category	Unit	Ventilated Fascade with Rock Wool	ETICS with EPS-PL
Acidification	mol H+ eq	0.2776	0.4375
Climate change	kg CO_2 eq	36.8156	42.5229
Climate change—Biogenic	kg CO_2 eq	0.0617	0.1255
Climate change—Fossil	kg CO_2 eq	36.6420	42.3688
Climate change—Land use and LU change	kg CO_2 eq	0.1118	0.0285
Ecotoxicity, freshwater—part 1	CTUe	85.3324	161.5034
Ecotoxicity, freshwater—part 2	CTUe	13.1038	25.0509
Ecotoxicity, freshwater—inorganics	CTUe	80.8024	160.7112
Ecotoxicity, freshwater—organics—p.1	CTUe	10.9023	3.8792
Ecotoxicity, freshwater—organics—p.2	CTUe	6.7314	21.9640
Particulate matter	disease inc.	2.32×10^{-6}	2.54×10^{-6}
Eutrophication, marine	kg N eq	0.0368	0.0417
Eutrophication, freshwater	kg P eq	0.0176	0.0157
Eutrophication, terrestrial	mol N eq	0.4996	0.4173
Human toxicity, cancer	CTUh	6.47×10^{-8}	5.78×10^{-9}
Human toxicity, cancer—inorganics	CTUh	1.11×10^{-9}	3.50×10^{-15}
Human toxicity, cancer—organics	CTUh	6.36×10^{-8}	5.78×10^{-9}
Human toxicity, non-cancer	CTUh	1.07×10^{-7}	2.93×10^{-8}
Human toxicity, non-cancer—inorganics	CTUh	6.38×10^{-8}	1.61×10^{-9}
Human toxicity, non-cancer—organics	CTUh	4.33×10^{-8}	2.77×10^{-8}
Ionizing radiation	kBq U-235 eq	3.0253	5.0298
Land use	Pt	201.9516	214.6499
Ozone depletion	kg CFC11 eq	1.90×10^{-6}	3.86×10^{-6}
Photochemical ozone formation	kg NMVOC eq	0.1124	0.1552
Resource use, fossils	MJ	403.6371	818.2176
Resource use, minerals and metals	kg Sb eq	0.0055	0.0001
Water use	m^3 depriv.	11.0386	39.1682

Table 5. Comparative LCA for building façades in Poland: Damage category, damage assessment stage.

Damage Category	Unit	Ventilated Fascade with Rock Wool	ETICS with EPS-PL
Acidification	mol H+ eq	0.2776	0.4375
Climate change	kg CO_2 eq	36.8156	42.5229
Ecotoxicity, freshwater	CTUe	98.436303	186.5544
Particulate matter	disease inc.	2.32×10^{-6}	2.54×10^{-6}
Eutrophication, marine	kg N eq	0.0368	0.0417
Eutrophication, freshwater	kg P eq	0.0176	0.0157
Eutrophication, terrestrial	mol N eq	0.4996	0.4173
Human toxicity, cancer	CTUh	6.47×10^{-8}	5.78×10^{-9}
Human toxicity, non-cancer	CTUh	1.07×10^{-7}	2.93×10^{-8}
Ionizing radiation	kBq U-235 eq	3.0253	5.0298
Land use	Pt	201.9516	214.6499
Ozone depletion	kg CFC11 eq	1.90×10^{-6}	3.86×10^{-6}
Photochemical ozone formation	kg NMVOC eq	0.1124	0.1552
Resource use, fossils	MJ	403.6371	818.2176
Resource use, minerals and metals	kg Sb eq	0.0055	0.0001
Water use	m^3 depriv.	11.0386	39.1682

 https://doi.org/10.3390/ma19040807

Table 6. Elements in the life cycle assessment for ventilated façades: damage category.

Damage Category	Unit	Fiber Cement Board with Lime	Mineral Wool with a Thick Veil, 15 cm Rock Wool	Aluminum Profiles	Consoles	The Remainder	Total
Acidification	mol H+ eq	0.0187	0.1619	0.0658	0.0176	0.013	0.2776
Climate change	kg CO_2 eq	8.0991	16.6321	7.1364	1.9168	3.0310	36.8156
Ecotoxicity, freshwater	CTUe	11.1053	39.2803	33.7129	9.0552	5.2824	98.4363
Particulate matter	disease inc.	1.22×10^{-7}	1.25×10^{-6}	5.81×10^{-7}	0	0	2.32×10^{-6}
Eutrophication, marine	kg N eq	0.005	0.0156	0.0105	0.0028	0.0025	0.0368
Eutrophication, freshwater	kg P eq	0.0009	0.0051	0.0075	0.0020	0.0018	0.0176
Eutrophication, terrestrial	mol N eq	0.0594	0.2773	0.1065	0.0286	0.0276	0.4996
Human toxicity, cancer	CTUh	1.55×10^{-9}	5.12×10^{-8}	6.16×10^{-9}	0	0	6.47×10^{-8}
Human toxicity, non-cancer	CTUh	5.89×10^{-8}	7.53×10^{-9}	2.45×10^{-8}	0	0	1.07×10^{-7}
Ionizing radiation	kBq U-235 eq	0.3211	1.9180	0.5158	0.1385	0.1317	3.0253
Land use	Pt	6.9654	139.8213	37.8422	10.1643	7.1583	201.9517
Ozone depletion	kg CFC11 eq	4.15×10^{-8}	1.13×10^{-6}	5.09×10^{-7}	0	0	1.90×10^{-6}
Photochemical ozone formation	kg NMVOC eq	0.0164	0.0464	0.0314	0.0084	0.0096	0.1124
Resource use, fossils	MJ	34.0119	203.9107	82.41632	22.13688	61.16125	403.6371
Resource use, minerals and metals	kg Sb eq	8.14×10^{-5}	4.61×10^{-5}	0.0042	0.0011	0.0001	0.0055
Water use	m^3 depriv.	0.5966	4.0063	4.0387	1.0847	1.3121	11.0386

Table 7. Share of individual elements in the life cycle assessment of the evaluated impact categories for ventilated façade system: damage category.

Damage Category	Fiber Cement Board with Lime	Mineral Wool with a Thick Veil, 15 cm Rock Wool	Aluminum Profiles	Consoles	The Remainder	Total
Acidification	6.74	58.34	23.72	6.37	4.83	100.00
Climate change	22.00	45.18	19.38	5.21	8.23	100.00
Ecotoxicity, freshwater	11.28	39.90	34.25	9.20	5.37	100.00
Particulate matter	5.26	53.88	25.04	0.00	0.00	100.00
Eutrophication, marine	14.37	42.45	28.56	7.67	6.95	100.00
Eutrophication, freshwater	5.34	29.43	42.95	11.54	10.73	100.00
Eutrophication, terrestrial	11.89	55.52	21.33	5.73	5.53	100.00
Human toxicity, cancer	2.40	79.13	9.52	0.00	0.00	100.00
Human toxicity, non-cancer	55.05	7.04	22.90	0.00	0.00	100.00
Ionizing radiation	10.62	63.40	17.05	4.58	4.35	100.00
Land use	3.45	69.24	18.74	5.03	3.54	100.00
Ozone depletion	2.18	59.47	26.79	0.00	0.00	100.00
Photochemical ozone formation	14.62	41.35	27.97	7.51	8.55	100.00
Resource use, fossils	8.43	50.52	20.42	5.48	15.15	100.00
Resource use, minerals and metals	1.46	0.83	75.87	20.39	1.47	100.00
Water use	5.41	36.29	36.59	9.83	11.89	100.00

https://doi.org/10.3390/ma19040807

Table 8. Elements in the life cycle assessment for ETICS: damage category.

Damage Category	Unit	EPS	Bonding Mortar	Base Coat	Fiberglass Mesh	Primer	Finish Coat	Total
Acidification	mol H+ eq	0.0486	0.0512	0.0512	0.0059	0.0146	0.2659	0.4375
Climate change	kg CO_2 eq	11.6153	4.5844	4.5844	0.7836	1.0940	19.8610	42.5229
Ecotoxicity, freshwater	CTUe	6.2530	43.8055	43.8055	0.8448	4.7915	87.0539	186.5544
Particulate matter	disease inc.	5.08×10^{-7}	2.81×10^{-7}	2.81×10^{-7}	4.96×10^{-8}	7.43×10^{-8}	1.35×10^{-6}	2.54×10^{-6}
Eutrophication, marine	kg N eq	0.0076	0.0048	0.0048	0.0011	0.0012	0.0220	0.0417
Eutrophication, freshwater	kg P eq	0.0014	0.0020	0.0020	0.0003	0.0005	0.0093	0.0157
Eutrophication, terrestrial	mol N eq	0.0818	0.0485	0.0485	0.0125	0.0117	0.2140	0.4173
Human toxicity, cancer	CTUh	5.54×10^{-10}	1.30×10^{-9}	1.30×10^{-9}	5.97×10^{-11}	1.37×10^{-10}	2.44×10^{-9}	5.78×10^{-9}
Human toxicity, non-cancer	CTUh	5.21×10^{-9}	3.48×10^{-9}	3.48×10^{-9}	2.33×10^{-10}	8.81×10^{-10}	1.60×10^{-8}	2.93×10^{-8}
Ionizing radiation	kBq U−235 eq	0.5658	0.6599	0.6599	0.1484	0.1564	2.8392	5.0298
Land use	Pt	16.16243	29.53729	29.53729	2.374006	7.150438	129.8885	214.65
Ozone depletion	kg CFC11 eq	2.96×10^{-7}	5.55×10^{-7}	5.55×10^{-7}	7.42×10^{-8}	1.24×10^{-7}	2.26×10^{-6}	3.86×10^{-6}
Photochemical ozone formation	kg NMVOC eq	0.0494	0.0149	0.0149	0.0030	0.0038	0.0690	0.1552
Resource use, fossils	MJ	261.2882	79.2810	79.2810	13.3889	20.0912	364.8871	818.2177
Resource use, minerals and metals	kg Sb eq	2.12×10^{-6}	2.28×10^{-5}	2.28×10^{-5}	4.48×10^{-6}	6.10×10^{-6}	0.000111	0.000169
Water use	m^3 depriv.	8.1827	4.3320	4.3320	0.2472	1.2302	20.8439	39.16824

Elements in the life cycle assessment for ventilated facades and ETICS in the characterization stage according to the EF method and share of individual elements in LCA of the evaluated impact categories for these facades are available in the Supplementary Materials [Tables S1–S4].

The impact category "acidification" shows a higher environmental index for ETICS than for ventilated façades. In the case of ETICS, more than 60% of the acidification impact is associated with the finish coat. For ventilated façades, mineral wool contributes the most to this category, accounting for nearly 60% of the total impact of acidification.

For the impact category of climate change, ETICS represents the less favorable solution, whereas the ventilated façade exhibits lower environmental impacts. In the subcategories "climate change—biogenic" and "climate change—fossil," ETICS also demonstrates higher values. In contrast, for "climate change—land use," the indicator is higher for the ventilated façade. For the climate change impact category—land use and LU change, more than 70% of the environmental burden associated with the ventilated façade is attributable to the production of aluminum profiles, which largely accounts for the elevated contribution observed for this system component.

In the case of ETICS, the highest contribution to the climate change indicator is associated with the finish coat, amounting to 19.86 kg CO_2-eq, followed by the EPS layer, which accounts for 11.61 kg CO_2-eq. In contrast, for the ventilated façade, the largest climate change impact is linked to the production of mineral wool (16.63 kg CO_2-eq), followed by the production of the fiber cement board (8.10 kg CO_2-eq) and aluminum profiles (7.14 kg CO_2-eq).

For all ecotoxicity indicators—except for "ecotoxicity freshwater, organic P1"—the ETICS shows higher ecotoxicity values compared with the ventilated façade. With respect to the contribution of individual material components, in the case of the ventilated façade, the highest share in "ecotoxicity freshwater, organic P1" originates from aluminum profiles, accounting for more than 50% of the total impact. Mineral wool represents nearly 35%

 https://doi.org/10.3390/ma19040807

of the impact in this category. For the indicator "ecotoxicity—organic P2," mineral wool is the dominant contributor, representing 48% of the overall impact, while aluminum profiles account for approximately 30%. For the remaining ecotoxicity categories, the highest impact is likewise associated with the production of mineral wool. In contrast, for the ETICS, the production of the finish coat constitutes the largest contribution to the ecotoxicity indicators.

Table 9. Share of individual elements in the life cycle assessment of the evaluated impact categories for ETICS.

Impact Category	Unit	EPS	Bonding Mortar	Base Coat	Fiberglass Mesh	Primer	Finish Coat	Total
Acidification	mol H+ eq	11.11	11.71	11.71	1.35	3.34	60.78	100.00
Climate change	kg CO_2 eq	27.32	10.78	10.78	1.84	2.57	46.71	100.00
Climate change—Biogenic	kg CO_2 eq	45.05	7.67	7.67	0.71	2.03	36.86	100.00
Climate change—Fossil	kg CO_2 eq	27.28	10.79	10.79	1.85	2.57	46.73	100.00
Climate change—Land use and LU change	kg CO_2 eq	9.03	13.23	13.23	2.57	3.23	58.72	100.00
Ecotoxicity, freshwater—part 1	CTUe	3.22	22.12	22.12	0.42	2.72	49.40	100.00
Ecotoxicity, freshwater—part 2	CTUe	4.19	32.27	32.27	0.63	1.60	29.04	100.00
Ecotoxicity, freshwater—inorganics	CTUe	3.29	21.74	21.74	0.48	2.75	50.00	100.00
Ecotoxicity, freshwater—organics—p.1	CTUe	0.40	31.55	31.55	0.13	1.90	34.47	100.00
Ecotoxicity, freshwater—organics—p.2	CTUe	4.33	34.79	34.79	0.35	1.34	24.39	100.00
Particulate matter	disease inc.	19.97	11.05	11.05	1.95	2.92	53.05	100.00
Eutrophication, marine	kg N eq	18.31	11.60	11.60	2.80	2.91	52.79	100.00
Eutrophication, freshwater	kg P eq	9.45	12.91	12.91	1.99	3.28	59.45	100.00
Eutrophication, terrestrial	mol N eq	19.60	11.64	11.64	3.02	2.82	51.28	100.00
Human toxicity, cancer	CTUh	9.59	22.42	22.42	1.03	2.36	42.18	100.00
Human toxicity, cancer—inorganics	CTUh	8.54	19.58	19.58	1.11	2.67	48.52	100.00
Human toxicity, cancer—organics	CTUh	9.59	22.42	22.42	1.03	2.36	42.18	100.00
Human toxicity, non-cancer	CTUh	17.79	11.88	11.88	0.79	3.01	54.64	100.00
Human toxicity, non-cancer—inorganics	CTUh	8.70	11.74	11.74	3.06	3.38	61.40	100.00
Human toxicity, non-cancer—organics	CTUh	18.32	11.89	11.89	0.66	2.99	54.25	100.00
Ionizing radiation	kBq U-235 eq	11.25	13.12	13.12	2.95	3.11	56.45	100.00
Land use	Pt	7.53	13.76	13.76	1.11	3.33	60.51	100.00
Ozone depletion	kg CFC11 eq	7.66	14.35	14.35	1.92	3.22	58.50	100.00
Photochemical ozone formation	kg NMVOC eq	31.83	9.62	9.62	1.97	2.45	44.51	100.00
Resource use, fossils	MJ	31.93	9.69	9.69	1.64	2.46	44.60	100.00
Resource use, minerals and metals	kg Sb eq	1.26	13.49	13.49	2.65	3.61	65.49	100.00
Water use	m^3 depriv.	20.89	11.06	11.06	0.63	3.14	53.22	100.00

For the impact category "particulate matter," the ETICS demonstrates a higher indicator compared with the ventilated façade. In the ventilated façade, the largest contribution to this category—53.9%—is associated with the production of mineral wool, followed

https://doi.org/10.3390/ma19040807

by aluminum profiles, which account for approximately 25%. In the case of ETICS, the dominant contribution to particulate-matter emissions originates from the finish coat (over 53%), while nearly 20% is attributed to EPS.

Similarly to the particulate-matter impact category, the marine eutrophication indicator is also higher for the ETICS compared with the ventilated façade.

For the impact category of freshwater eutrophication, the ventilated façade shows a higher indicator compared with the ETICS. In the case of the ventilated façade, this impact is primarily associated with the production of aluminum profiles, whereas for ETICS it is mainly linked to the production of the finish coat.

For the terrestrial eutrophication impact category, the indicator is also higher for the ventilated façade compared with the ETICS. In the ventilated façade, more than 55% of the impact is associated with the production of mineral wool, while approximately 21% results from the production of aluminum profiles. In the case of ETICS, terrestrial eutrophication is predominantly driven by the finish coat, which accounts for over 51% of the impact, followed by EPS with a contribution of nearly 20%.

The analysis also demonstrated that, for the impact category "human toxicity," the indicators for all subcategories—both cancer and non-cancer—are higher for the ventilated façade than for the ETICS.

Within the human-toxicity impact category, it was demonstrated that, for the ventilated façade system, the major determinants of the total impact are the fiber cement board, mineral wool, and aluminum profiles. The fiber cement board contributes more than 80% to the subcategory "human toxicity, cancer inorganic" and more than 90% to "human toxicity, non-cancer inorganic." Mineral wool accounts for over 80% of the impact in the subcategory "human toxicity, cancer organic," whereas aluminum profiles contribute more than 55% to "human toxicity, non-cancer organic."

In the case of the ETICS, the finish coat constitutes the dominant contributor to the human-toxicity impact category, accounting for between 40% and 60% of the total impact, depending on the specific subcategory.

For the remaining impact categories—including ionizing radiation, land use, ozone depletion, photochemical ozone formation, resource use (fossils), and water use—higher indicators were observed for the ETICS. In all of these categories, the elevated values for ETICS are primarily driven by the production of the finish coat, which contributes between 45% and 60%, depending on the specific impact category. In the case of the ventilated façade, the impacts in these remaining categories are predominantly associated with the production of mineral wool.

Only in the case of the impact category "resource use—minerals and metals" was a higher indicator observed for the ventilated façade compared with the ETICS. This result is primarily driven by the extensive use of aluminum in the production of aluminum profiles, which accounts for more than 75% of the impact in this category for the ventilated façade. In contrast, for the ETICS, this impact category is mainly influenced by the finish coat, contributing approximately 65%. Uncertainty analysis for the ventilated façade and uncertainty analysis for ETICS are available in the Supplementary Materials [Tables S5 and S6].

One of the limitations of LCA is the uncertainty problem. These uncertainties will directly affect the correctness and reliability of the LCA research conclusions. The Monte Carlo method, as statistical simulation method, is a method of numerical calculation guided by probability statistical theory.

Uncertainty analysis was performed for the ventilated façade system. The Monte Carlo uncertainty analysis provides a quantitative assessment of the robustness and variability of the impact characterization results for the ventilated façade system. Across the evaluated

 https://doi.org/10.3390/ma19040807

impact categories, the magnitude of uncertainty varies substantially, reflecting differences in data quality, model structure, and the inherent variability of underlying life cycle processes. Overall, several impact categories exhibit low relative uncertainty, as indicated by modest standard deviations (SD), narrow 95% confidence intervals, and low coefficients of variation (CV). Categories such as photochemical ozone formation (CV ≈ 6.99%), climate change (CV ≈ 7.10%), and resource use, fossils (CV ≈ 7.09%) demonstrate particularly stable outcomes. Their narrow confidence intervals and low standard errors of the mean (SEM) suggest that the underlying inventory data are consistent and that the model outputs are not highly sensitive to parameter variability. These categories can therefore be interpreted with a high degree of confidence.

In contrast, several categories show substantial uncertainty, with extremely high CV values and wide confidence intervals. The most pronounced example is water use, which displays a CV exceeding 13.500% and an SD two orders of magnitude larger than the mean. Similar issues are observed in Human toxicity, non-cancer and its subcategories, where CV values exceed several thousand percent, reflecting very high sensitivity to parameter uncertainty and potentially low reliability of the available toxicity characterization data.

Uncertainty analyses were also performed for ETICS.

The Monte Carlo uncertainty analysis conducted for the ETICS reveals substantial variation in the robustness of impact characterization results across environmental categories. The magnitude of uncertainty differs markedly between categories, reflecting the heterogeneity of underlying life cycle inventory data, the sensitivity of specific impact pathways, and the methodological limitations inherent to certain characterization models. Several impact categories exhibit relatively low to moderate uncertainty, as indicated by modest standard deviations (SD), narrow 95% confidence intervals, and coefficients of variation (CV) below approximately 15%. This group includes climate change (CV ≈ 9.85%), Climate change—fossil (CV ≈ 9.84%), eutrophication, marine (CV ≈ 10.77%), eutrophication, terrestrial (CV ≈ 10.74%), and photochemical ozone formation (CV ≈ 12.43%). These results suggest that the underlying inventory data for these categories are comparatively consistent and that the model outputs are not highly sensitive to parameter variability. Consequently, these categories can be interpreted with a higher degree of confidence, and the results are considered robust for comparative or decision-support purposes.

The highest uncertainty is observed in categories with very large CV values, wide confidence intervals, and substantial divergence between mean and median values. Notably, eutrophication, freshwater (CV ≈ 47.27%), human toxicity, non-cancer—inorganics (CV ≈ 47.54%), resource use, minerals and metals (CV ≈ 50.66%), and especially water use (CV ≈ 604%) exhibit extreme variability. In the case of water use, the SD (206.17 m^3) is several times larger than the mean (34.12 m^3), and the confidence interval spans from negative to highly positive values, indicating a highly skewed or unstable distribution. Such patterns typically reflect poorly constrained inventory data, the presence of outlier processes with disproportionately large contributions, or methodological limitations in water scarcity characterization models. These categories should therefore be interpreted with significant caution, and their results may not be reliable for comparative decision-making without further refinement of input data.

Based on the performed LCA it was shown that the ETICS exhibits higher damage-category indicators in all impact categories except for eutrophication, human toxicity (carcinogenic and non-carcinogenic), and resource use related to minerals and metals, for which the ventilated façade shows higher values.

The LCA analysis showed that the main determinant of carbon footprint in the case of ventilated façades is the production of mineral wool. Other significant carbon footprint

https://doi.org/10.3390/ma19040807

factors include fiber cement board with lime (where the impact of the board is mainly related to the production of cement) and production of aluminum profiles. It has also been shown that thermal insulation consisting of mineral wool is the main source of influence on acidification, human toxicity-cancer-organics and fossil resource scarcity. The production of fiber cement board was also found to be the factor with the highest impact on human carcinogenic toxicity (inorganics). The impact is primarily related to the production of cement. Clinker production is the main source of negative environmental impact in cement production technology. Human toxicity-cancer-inorganics is caused by the emission of heavy metals from the clinker production process. Human carcinogenic toxicity is mainly connected to the chromium burden generated from transportation, petroleum refinery operation, and coal washing in the clinker production process. Another important factor that causes the highest environmental impacts are aluminum profiles. The profiles are the main determinants of impacts in the impact categories: mineral and metal resource scarcity, ecotoxicity—organics, eutrophication, and climate change, especially a subcategory related to land use. The negative impact of the profiles is primarily related to the aluminum production process, which is very energy intensive.

In the case of the ETICS with EPS assessment, it was shown that the main determinant of the environmental assessment is the finish coat, and especially the use of acrylic dispersion for the finish coat. Other factors that significantly influence the damage categories include the production of EPS and the adhesive mortar that is used in the production of the bonding mortar and base coat. It has been shown that the environmental impact of the ETICS is determined primarily by the acrylic façade paint (acrylic dispersion), and to a lesser extent it depends on the thermal insulation material, i.e., EPS. For this reason, silicone or silicate façade paints are being used more and more often. However, such solutions require further LCAs.

4. Conclusions

In accordance with the guidelines of the European Commission, actions should be taken to reduce the consumption of raw materials and energy, as well as to reduce greenhouse gas emissions during the life cycle of buildings. Due to the fact that the construction industry is responsible for a high consumption of energy and natural resources, a big production of waste, and the fact that it is one of the main sources of greenhouse gas emissions, it is necessary to look for solutions and technologies that will reduce the negative impact of this sector on the environment. This is why research such as this one is important, the aim of which is to demonstrate the environmental impact of the materials used in building façade systems.

The LCA life cycle assessment technique is considered a key method for quantifying and assessing the environmental impact of buildings. The conducted research and the obtained results of LCA analyzes for building façade systems can help decision-makers determine the stages of a façade's life cycle which have the highest impact on the environment. The LCA of a façade that uses insulating materials that are very popular in Poland was conducted during the research.

The article presents the results of the life cycle analysis of individual materials and façade elements based on actual data for Polish conditions. This means that insulating materials of relevant thickness had to be used to meet the requirements of thermal insulation and other requirements related to energy saving, which are specified in the Regulation of the Minister of Infrastructure on the technical conditions to be met by buildings and their locations.

It was shown that the factors that have the greatest impact on the environment, when taking into account the life cycle assessment of the ventilated façade, are mineral wool,

https://doi.org/10.3390/ma19040807

fiber cement board and profiles. However, in the case of the ETICS with EPS assessment, it was shown that the main determinant of the environmental assessment is the finish coat, whereas the impact of the thermal insulation material, i.e., EPS, is lower. In the case of the life cycle assessment of the ventilated façade, it was shown that the most significant impact is the thermal insulation material, which is mineral wool.

The Monte Carlo uncertainty analysis demonstrated that certain impact categories, including climate change, fossil resource use, marine eutrophication, and photochemical ozone formation, show relatively low coefficients of variation and narrow confidence intervals. Differences observed in these categories between façade systems can therefore be considered robust. In contrast, categories such as water use, resource use (minerals and metals), and selected human toxicity indicators exhibit high variability and wide confidence intervals. For these categories, the comparative differences should be interpreted with caution, as the uncertainty ranges partially overlap and reflect limitations of available inventory data and characterization models. Consequently, while the general trend indicates higher environmental burdens for ETICS in most categories under the baseline configuration, not all differences are statistically robust across all impact categories.

The results indicate that, while ETICS shows higher impact values in most categories under the baseline configuration, the statistical significance of these differences varies across impact categories. In some cases, confidence intervals partially overlap, suggesting limited statistical robustness. However, the comparative trend remains consistent across the tested parametric variations.

Performing an LCA analysis is the first step towards identifying changes and innovations in buildings by relating them to all the factors and materials in a building's life cycle.

The article presents, for the first time in Poland, a life cycle assessment of both a ventilated façade made of fiber cement board and the ETICS.

Subsequent research will present new low-energy methods of producing insulating materials, and at the same time will take into account the climatic conditions of individual countries. The results of this study offer a comprehensive environmental analysis of Polish building façade systems, and can be used as the first step in performing a holistic LCA of a building from cradle to grave that includes all the phases of the life cycle.

The authors of the article are currently working on possible alternatives to the determinant contributions and will present the results of their research in subsequent publications.

Supplementary Materials: The following supporting information can be downloaded at https://www.mdpi.com/article/10.3390/ma19040807/s1, Figure S1. Example of a facade that uses fiber-cement boards; Figure S2. Schematic cross-section of the ventilated facade system; Figure S3. Schematic cross-section of the ETICS system; Table S1. Elements in the life cycle assessment for ventilated facades—characterization stage; Table S2. Share of individual elements in the life cycle assessment of the evaluated impact categories for ventilated façade system—characterization stage, Table S3. Elements in the life cycle assessment for ETICS—characterization stage; Table S4. Share of individual elements in the life cycle assessment of the evaluated impact categories for ETICS—characterisation stage; Table S5. Uncertainty analysis for ventilated fascade—characterization; Table S6. Uncertainty analysis for ETICS—characterization.

Author Contributions: Conceptualization, D.B. and K.S.; methodology, D.B. and K.S.; software, D.B.; validation, D.B.; formal analysis, D.B. and K.S.; investigation, D.B. and K.S.; resources, D.B. and K.S.; data curation, K.S.; writing—original draft, D.B. and K.S.; writing—review and editing, D.B. and K.S.; visualization, D.B. and K.S.; supervision, D.B.; project administration, D.B. and K.S. All authors have read and agreed to the published version of the manuscript.

https://doi.org/10.3390/ma19040807

Funding: This research received no external funding.

Institutional Review Board Statement: Not applicable.

Informed Consent Statement: Not applicable.

Data Availability Statement: The original contributions presented in this study are included in the article/Supplementary Material. Further inquiries can be directed to the corresponding author.

Conflicts of Interest: The authors declare no conflicts of interest.

References

1. Hasik, V.; Ororbia, M.; Warn, G.; Bilec, M. Whole building life cycle environmental impacts and costs: A sensitivity study of design and service decisions. *Build. Environ.* **2019**, *163*, 106316. [CrossRef]
2. Nugent, A.; Montano-Owen, C.; Pallares, L.; Richardson, S.; Rowland, M. *EU Policy Whole Life Carbon Roadmap for Buildings*; World Green Building Council: London, UK, 2022. Available online: https://worldgbc.org/article/eu-policy-whole-life-carbon-roadmap-for-buildings (accessed on 19 June 2024).
3. Burchart, D.; Schabowicz, K. Environmental challenges for construction in the context of new requirements of the European Commission. *Przegląd Bud.* **2023**, *11–12*, 88–92. [CrossRef]
4. Colangelo, F.; Petrillo, A.; Cioffi, R.; Borrelli, C.; Forcina, A. Life cycle assessment of recycled concretes: A case study in southern Italy. *Sci. Total Environ.* **2018**, *615*, 1506–1517. [CrossRef]
5. Andersen, J.H.; Rasmussen, N.L.; Ryberg, M.W. Comparative life cycle assessment of cross laminated timber building and concrete building with special focus on biogenic carbon. *Energy Build.* **2022**, *254*, 111604. [CrossRef]
6. Garcia-Ceballos, L.; de Andres-Díaz, J.R.; Contreras-Lopez, M.A. Life cycle study of different constructive solutions for building enclosures. *Sci. Total Environ.* **2018**, *626*, 1167–1174. [CrossRef]
7. Hadj Sadok, R.; Belas Belaribi, N.; Mazouzi, R.; Hadj Sadok, F. Life cycle assessment of cementitious materials based on calcined sediments from Chorfa II dam for low carbon binders as sustainable building materials. *Sci. Total Environ.* **2022**, *826*, 154077. [CrossRef]
8. Gao, H.; Wang, D.; Du, X.; Zhao, Z. An LCA-BIM integrated model for carbon—Emission calculation of prefabricated buildings. *Renew. Sustain. Energy Rev.* **2024**, *203*, 114775. [CrossRef]
9. Chen, W.; Yang, S.; Zhang, X.; Jordan, N.D.; Huang, J. Embodied energy and carbon emissions of building materials in China. *Build. Environ.* **2022**, *207*, 108434. [CrossRef]
10. Botejara-Antúnez, M.; Gonzalez-Domínguez, J.; García-Sanz-Calcedo, J. Comparative analysis of flat roof systems using life cycle assessment methodology: Application to healthcare buildings. *Case Stud. Constr. Mater.* **2022**, *17*, e01212. [CrossRef]
11. Duan, Z.; Huang, Q.; Sun, Q.; Zhang, Q. Comparative life cycle assessment of a reinforced concrete residential building with equivalent cross laminated timber alternatives in China. *J. Build. Eng.* **2022**, *62*, 105357. [CrossRef]
12. Adelfio, L.; La Scalia, G.; La Fata, C.; Giallanza, A. Life cycle analysis of innovative building materials based on circular coffee ground supply chain. *Transp. Res. Procedia* **2022**, *67*, 100–108. [CrossRef]
13. Kumar, D.; Alam, M.; Zou, P.; Sanjayan, J.; Memon, R. Comparative analysis of building insulation material properties and performance. *Renew. Sustain. Energy Rev.* **2020**, *131*, 110038. [CrossRef]
14. Llantoy, N.; Chafer, M.; Cabeza, L. A comparative life cycle assessment (LCA) of different insulation materials for buildings in the continental Mediterranean climate. *Energy Build.* **2020**, *225*, 110323. [CrossRef]
15. Mahlan, S.; Francis, A.; Thumuganti, V.; Thomas, A.; Sadick, A. An integrated life cycle assessment and energy simulation framework for residential building walling systems. *Build. Environ.* **2024**, *257*, 111542. [CrossRef]
16. Rey-Alvarez, B.; Silvestre, J.; García-Martínez, A.; Sanchez-Montanes, B. A comparative approach to evaluate the toxicity of building materials through life cycle assessment. *Sci. Total Environ.* **2024**, *912*, 168897. [CrossRef]
17. Oreto, C.; Russo, F.; Dell'Acqua, G.; Veropalumbo, R. A comparative environmental life cycle assessment of road asphalt pavement solutions made up of artificial aggregates. *Sci. Total Environ.* **2024**, *927*, 171716. [CrossRef]
18. Sadrolodabaee, P.; Hosseini, S.; Ardunay, M.; Claramunt, J.; de la Fuente, A. A New Sustainability Assessment Method for Façade Cladding Panels: A Case Study of Fiber/Textile Reinforced Cement Sheets. In *Fibre Reinforced Concrete: Improvements and Innovations II*; BEFIB 2021, RILEM Bookseries; Serna, P., Llano-Torre, A., Martí-Vargas, J.R., Navarro-Gregori, J., Eds.; Springer: Cham, Switzerland, 2022; Volume 36. [CrossRef]
19. Bianchi, S.; Andriotis, C.; Klein, T.; Overend, M. Multi-criteria design methods in façade engineering: State-of-the-art and future trends. *Build. Environ.* **2024**, *250*, 111184. [CrossRef]
20. Baptista, J.; Kokare, S.; Francisco, A.; Godina, R.; Aelenei, D. A comparative life cycle assessment of ETICS and ventilated façade systems with timber cladding. *Energy Build.* **2024**, *304*, 113842. [CrossRef]

 https://doi.org/10.3390/ma19040807

21. Alcivar-Bastidas, S.; Petroche, D.M.; Ramirez, A.D.; Martinez-Echevarria, M.J. Characterization and life cycle assessment of alkali treated abaca fibers: The effect of reusing sodium hydroxide. *Constr. Build. Mater.* **2024**, *449*, 138522. [CrossRef]
22. Ramezannia, A.; Gocer, O.; Tabrizi, T.B. The life cycle assessment of stabilized rammed earth reinforced with natural fibers in the context of Australia. *Constr. Build. Mater.* **2024**, *416*, 135034. [CrossRef]
23. Tartaro, A.; Mata, T.; Martins, A.; Esteves da Silva, J. Carbon footprint of the insulation cork board. *J. Clean. Prod.* **2017**, *143*, 925–932. [CrossRef]
24. Sattler, S.; Osterreicher, D. Assessment of sustainable construction measures in building refurbishment—Life cycle comparison of conventional and multi-active façade systems in a social housing complex. *Sustainability* **2019**, *11*, 4487. [CrossRef]
25. *ISO 14040:2006+Amd 1:2020*; Environmental Management—Life Cycle Assessment—Principles and Framework. International Organization for Standardization: Geneva, Switzerland, 2020.
26. *ISO 14044:2006+Amd 1:2017+Amd 2:2020*; Environmental Management—Life Cycle Assessment—Requirements and Guidelines. International Organization for Standardization: Geneva, Switzerland, 2020.
27. European Commission; Joint Research Centre; Institute for Environment and Sustainability. *International Reference Life Cycle Data System (ILCD) Handbook: General Guide for Life Cycle Assessment: Detailed Guidance*, 1st ed.; Publications Office of the European Union: Luxembourg, 2010; ISBN 978-92-79-19092-6.
28. Castagnoli, A.; Simi, S.; Pulvirenti, I.; Valese, A. Cradle-to-Grave LCA of In-Person Conferences: Hotspots, Trade-Offs and Mitigation Pathways. *Sustainability* **2025**, *17*, 7604. [CrossRef]
29. Wernet, G.; Bauer, C.; Steubing, B.; Reinhard, J.; Moreno-Ruiz, E.; Weidema, B. The Ecoinvent Database Version 3 (Part I): Overview and Methodology. *Int. J. Life Cycle Assess.* **2016**, *21*, 1218–1230. [CrossRef]
30. *EN 15804:2012+A2:2019*; Sustainability of Construction Works—Environmental Product Declarations—Core Rules for the Product Category of Construction Products. European Committee for Standardization (CEN): Brussels, Belgium, 2019.
31. *EN 15978:2011*; Sustainability of Construction Works. Assessment of Environmental Performance of Buildings. Calculation Method. European Committee for Standardization (CEN): Brussels, Belgium, 2019.
32. Van Gulck, L.; Wastiels, L.; Steeman, M. How to evaluate circularity through an LCA study based on the standards EN 15804 and EN 15978. *Int. J. Life Cycle Assess.* **2022**, *27*, 1249–1266. [CrossRef]
33. Su, X.; Luo, Z.; Li, Y.; Huang, C. Life cycle inventory comparison of different building insulation materials and uncertainty analysis. *J. Clean. Prod.* **2016**, *112*, 275–281. [CrossRef]
34. Hill, C.; Norton, A.; Dibdiakova, J. A comparison of the environmental impacts of different categories of insulation materials. *Energy Build.* **2018**, *162*, 12–20. [CrossRef]
35. Ansah, M.; Chen, X.; Yang, H.; Lu, L.; Lam, P. An integrated life cycle assessment of different façade systems for a typical residential building in Ghana. *Sustain. Cities Soc.* **2022**, *53*, 101974. [CrossRef]
36. Zhang, C.; Hu, M.; Yang, X.; Amati, A.; Tukker, A. Life cycle greenhouse gas emission and cost analysis of prefabricated concrete building façade elements. *J. Ind. Ecol.* **2020**, *24*, 1016–1030. [CrossRef]
37. Schabowicz, K. *Elewacje wentylowane. Technologia produkcji i metody badania płyt włóknisto-cementowych (in Polish) Ventilated Façades. Fibre-Cement Board Production Technology and Testing Methods*; Oficyna Wydawnicza Politechniki Wrocławskiej: Wrocław, Poland, 2018. (In Polish)
38. Gorzelańczyk, T.; Schabowicz, K.; Szymków, M. Application of the Acoustic Emission Method and Artificial Neural Networks to Assess the Damaging Effect of High Temperature on the Structure of Fibre-Cement Boards. *Materials* **2022**, *15*, 6460. [CrossRef]
39. Schabowicz, K.; Gorzelańczyk, T.; Zawiślak, Ł.; Chyliński, F. Influence of Fire Exposition of Fibre-Cement Boards on Their Microstructure. *Materials* **2023**, *16*, 6153. [CrossRef]
40. Regulation of the Minister of Infrastructure of 12 April 2002 on the technical conditions to be met by buildings and their location. *J. Laws* **2002**, *75*, 690. Available online: https://isap.sejm.gov.pl/isap.nsf/download.xsp/WDU20220001225/O/D20221225.pdf (accessed on 5 February 2026).
41. Gorzelańczyk, T.; Schabowicz, K.; Szymków, M. Tests of Fiber Cement Materials Containing Recycled Cellulose Fibers. *Materials* **2020**, *13*, 2758. [CrossRef]
42. Report: Energy Transition in Poland. Edition 2025—Forum Energii. Available online: https://www.forum-energii.eu/en/transformacja-energetyczna-polski-edycja-2025 (accessed on 15 February 2026).
43. European Commission; Joint Research Centre; Institute for Environment and Sustainability. *Normalisation Method and Data for Environmental Footprints*; Publications Office of the European Union: Luxembourg, 2014.

 https://doi.org/10.3390/ma19040807

materials

MDPI

Article

A Model for Comprehensive Material Degradation and Strength Characteristics Assessment of Historic Masonry Structures

Czesław Miedziałowski, Romuald Szeląg and Adam Walendziuk *

Department of Building Structures and Structural Mechanics, Faculty of Civil Engineering and Environmental Sciences, Bialystok University of Technology, Wiejska 45E, 15-351 Białystok, Poland; c.miedzialowski@pb.edu.pl (C.M.); r.szelag@pb.edu.pl (R.S.)
* Correspondence: a.walendziuk@pb.edu.pl

Abstract

Historic buildings are susceptible to damage resulting from material degradation, deterioration of structural properties, changes in loads, overloading, and extraordinary events, which manifest as defects and flaws. These structures are also modernised, rebuilt, and adapted to new functions. In some cases, a rapid assessment of the structural condition is necessary. In practice, it is essential to maintain the safety of structures and their components, which is achieved by conducting an inventory and assessing the technical condition of structural elements and built-in materials. A numerical model was developed using macro-elements, and the practical applications of this model in strength analyses of degraded materials and sections of masonry structures are presented herein. The proposed approach was generalised and extended to include a classification of damaged masonry structures, which can be helpful and applied in indexing and searching databases of building structures, as well as in AI-assisted automatic pattern recognition. This resulted in a coherent and comprehensive approach to assessing the technical and structural condition of historic masonry buildings in use, as well as those undergoing repair, reconstruction, and modernisation.

Keywords: historic masonry buildings; assessment of material strength and structural safety; modelling of damaged heterogeneous materials

1. Introduction

Small-scale elements such as bricks and stones have been used by builders for millennia to erect structures. Today, structures built in past centuries are still in use, retaining their original form. Often, architectural objects are treated in modern times as monuments of great historical value and are subject to special protection. The long-term use of masonry structures contributes to damage as a result of the natural degradation of materials and a reduction in their strength parameters, but also due to changing loads over different periods of use. It is often necessary to carry out ongoing maintenance, repairs and strengthening of such structures.

In this process, in addition to testing the strength of materials incorporated into the structure, it is important to carry out analyses to verify the strength and safety of historic masonry structures and their fragments. The parameters and factors affecting the strength of masonry structures, structural systems and material solutions used in masonry construction are described in studies [1–5], among others.

Historic buildings are also modernised and adapted to new functions, and their great historical value necessitates the preservation of their original features and the requirements defined by a heritage conservator. These processes involve inventorying, monitoring, and non-destructive, semi-destructive, and destructive testing of structural elements [6–10]. In this context, new technologies such as historic building information modelling and structural health monitoring have proven useful for obtaining information about the material and structural condition of buildings and their components [11–14]. These technologies play an important role in assessing the structural condition, identifying existing damage [15], and determining appropriate remedial measures and repair methods [16–19]. In the future, the data collected and analysed will form a resource of technical knowledge regarding the repair methods employed, thereby demonstrating the effectiveness of the solutions adopted and enabling the elimination of methods that have proved ineffective. This is important because most existing historic buildings lack documentation that would allow us to identify past works and structural alterations and determine their impact on the current technical condition and structural integrity of the building. In engineering practice, there are also instances where a rapid and comprehensive assessment of the technical condition of masonry structures and their components is required. This may occur during renovation and modernisation work on structural elements.

This article expands on the content of [20] in the context of the issues discussed above, including the application of the developed numerical model to assess the material properties and strength of historic structures and their components. The aforementioned article addressed issues related to the modelling of masonry structures, as well as details of the strength model developed using macro-elements. Without repeating the information, the authors will merely mention that the strategies used in the modelling of masonry structures aimed at determining the stress–strength state, as well as issues of numerical modelling, can be found, amongst others, in works [21–23]. Further details regarding the proposed method of modelling masonry structures, which will be presented to the extent necessary later in this article, can be found in [20]. The article also includes examples of numerical experiments confirming that the proposed method of modelling masonry structures yields good agreement between calculated and experimental results.

The basis for developing a numerical model is the inventory and spatial and material identification of the historic building, which, in addition to traditional macroscopic measurements, utilises total station surveying, photogrammetry and 3D terrestrial laser scanning [6]. The preparation of documentation using BIM technology creates a database on the historic building, providing consistent documentation that remains accessible throughout its subsequent years of use. The digital model created also enables the development of a structural analysis model of the existing building and the incorporation of any new functional, design and revitalisation assumptions into it. Numerical analysis, combined with stress calculations for the structural elements, aims to avoid the need for demolition and to develop renovation solutions. To illustrate the issue in question, a BIM diagram supplemented with the scope of strength analyses of structural elements in damaged condition was used, as shown in Figure 1.

https://doi.org/10.3390/ma19112250

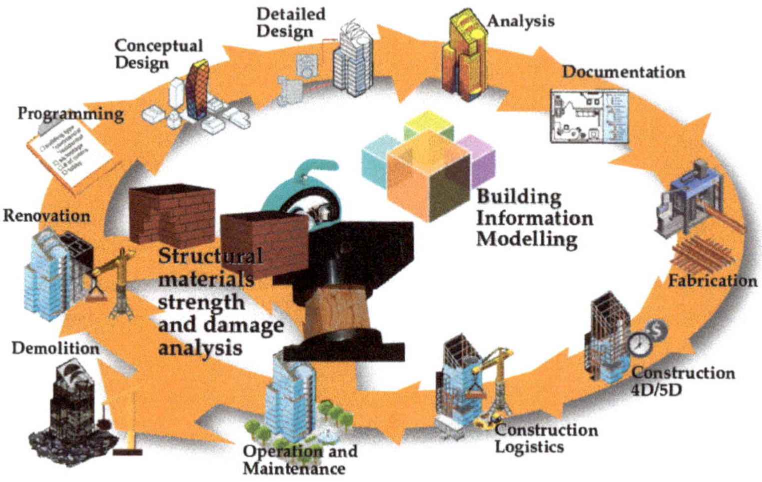

Figure 1. The issues discussed in the context of the general BIM relationship scheme.

2. Assessment of the Technical Condition and Strength of Masonry

2.1. Technical Aspects

The materials used in construction are subject to progressive degradation, resulting in a deterioration of their original properties and parameters over time. The causes of this are physical, chemical, and biological in nature. In addition to mechanical damage to walls, during the operation of a building, the loads acting on it change; some parts of the structure are overloaded, while others are relieved, resulting in a redistribution of stresses in the materials built into the structure. This is important in reconstruction processes when it is necessary to locally replace damaged elements with new ones.

In many regions of the world, an important factor in destruction is the dampness of walls and the processes of freezing and thawing of ceramic materials and mortar. Among the biological causes, factors related to mycology and lichenology, which have a long-term impact, are indicated.

Most often, the degradation of the components of masonry structures is caused by the simultaneous action of many of the above-mentioned factors, resulting in mass losses in the wall structures. The consequence of this process is fragments of the structure with reduced strength and poor technical condition, as shown in Figure 2. The degradation process is inevitable and difficult to stop, occurring in historic buildings over decades. A good illustration of the problem can be found in layered walls (including degraded ones), in the description of which segments in various technical conditions should be distinguished. The segmentation process may result from structural damage to the bricks, which may be accompanied by their disintegration from the structure, or a local cracking process separating continuous structures into local areas due to the occurrence of permanent cracks.

In order to standardise the strength analysis procedure, walls with a regular arrangement of masonry units are considered. The proposed method of description consists in separating segments of a certain size, referred to as macro-elements, in the model of the analysed structure. The macro-elements assume the actual state of degradation, taking into account the variability of physical parameters in the subspaces of the structure and damaged fragments, introducing limits on the variability of material parameters and damage. In this way, it is possible to optimally model the properties of construction materials in a 3D diagram, e.g., using the finite element method.

https://doi.org/10.3390/ma19112250

Figure 2. Various technical conditions of masonry fragments caused by degradation processes.

2.2. Photogrammetric Technique

In the process of locating damage to masonry structures, the representative process of determining areas with different strengths of bricks and mortar, and subsequent classification into the appropriate calculation model, Structure-from-Motion (SfM) photogrammetry has proved to be effective [24–28]. This technology is based on taking a series of photographs at different angles in relation to the structures under investigation in order to obtain a final three-dimensional image that enables the implementation of automation procedures at a later stage. The technology works well for creating images of damaged wall areas, which, due to the degree of degradation, may pose a safety hazard in conditions of direct access [29–32], but it is also recommended for use due to its low cost [33]. The interpretation of the point cloud obtained from the method can be used to divide it into separate regions, which allows for independent analysis of each segment [34]. This approach has been effectively used to prepare data for the calculation of masonry structures in computational areas of walls using the finite element method [35].

In the process of preparing to identify locally different material parameters, it is effective to perform imaging of the analysed structures using SfM technology. This provides a comprehensive image of the entire object and enables material testing in all selected locations without the risk of weak areas being overlooked (Figure 3). Analyses conducted in this manner precede in-situ material tests and are performed after preliminary processing of the results and the creation of a model of structures with similar technical parameters, indicating areas of damage, weld defects, cracks or spalling of the substance, and preliminary classification of wall damage. The correct preparation and selection of research sites is aimed at correctly determining the material parameters of the tested areas, which will later be interpreted as zones with characteristic representative properties. The comprehensive scope of analyses conducted using photogrammetry enables the location of damaged areas for further testing. In this regard, the work carried out aims to determine the type of damage to the walls, the strength parameters of the structures, and, due to the presence of brick

 https://doi.org/10.3390/ma19112250

and mortar losses, individualised calculation models saturated with strength parameters. An example of identifying areas with different characteristics is shown in Figure 4, which was created using photogrammetry.

Figure 3. Masonry structures of the front wall used in material strength analysis obtained using SfM technology. The analysis was performed using 3DF Zephyr Free 8.031 software.

Figure 4. Isolated areas with reduced strength parameters in masonry structures subjected to detailed strength tests.

When SfM procedures are applied, attention is drawn to practical challenges that may arise during the mapping of masonry walls, particularly in homogeneous areas such as window panes. This is due to the difficulty of tracking unique feature points. Furthermore, extended image acquisition under variable lighting conditions, where image sequences may include shadowed wall sections, can cause matching algorithms to incorrectly overlap fragments of the model.

The loss of bricks and mortar, as well as damage to surface layers, indicates the need to develop computational models which, after material testing, will allow an assessment of the impact of masonry discontinuities on local stress increases and damage propagation.

2.3. Material Testing

Brick strength tests were carried out in accordance with the standard [36]. The preparation of elements for testing consists in preparing samples from the tested brick from areas that do not have surface damage and then seasoning them to an air-dry state for a period of 14 days at a temperature of $\geq$15 °C with a relative humidity of $\leq$65%. The initial stage of preparing samples for testing is shown in Figure 5.

https://doi.org/10.3390/ma19112250

Figure 5. Test samples were prepared by cutting cubes from the brick material and then seasoning them in accordance with the provisions of standard [36].

The prepared samples were measured with an accuracy of 0.1 mm and placed in a strength press. The height of the samples was h = 59.0 mm, while the transverse dimension ranged from 50.0 to 50.5 mm. Up to 8 samples for strength testing could be prepared from each brick. The process of testing the strength of one of the samples is shown in Figure 6.

Figure 6. Strength tests were carried out in a strength testing machine, with the load increment set at a speed of 0.3 $(N/mm^2)/s$.

When the strength of a test sample is determined, the shape factor "d" is used to reduce different dimensions to a common comparative value. For the test samples, parameter d equal to 0.90 was used. In masonry structures where there was no surface damage, the normalised strength for samples taken from solid brick was within the range of $f_b = 13.1 \div 15.8\ N/mm^2$, with an average value of $f_b = 14.4\ N/mm^2$; this indicated a reduction in their strength and allowed them to be classified as class 10. Areas where low brick strength was identified were tested using the sclerometric method [37], with a Schmidt hammer of type OS-120PT (Figure 7), in order not to weaken them.

The performance of non-destructive testing requires the derivation of a correlation equation, the coefficients of which are obtained from destructive testing. Alternatively, it is possible to use a standard curve in the form of an equation [38]:

$$f_b(i) = 1.05 \cdot (0.0015R_i^2 + 0.0615R_i - 0.3585) \tag{1}$$

where $f_b(i)$ is the specified strength in $[N/mm^2]$, and R_i is the average measured Schmidt hammer rebound at the measurement point. Measurements carried out on bricks with similar characteristics of damage to the outer layers enabled the test results to be compiled and the strength of the bricks to be determined (Table 1).

https://doi.org/10.3390/ma19112250

Figure 7. Non-destructive strength testing of low-strength bricks was carried out using a Schmidt hammer of type OS-120PT (Proceq, Schwerzenbach, Switzerland). It allows the testing of materials with strength ranging from 1 to 5 MPa.

Table 1. Results of sclerometric tests on wall bricks.

Object	Tenement house	Date of construction								19th century
		Expected grade of bricks								Less than 10 MPa
Element	Front wall. Damaged bricks	Schmidt OS-120PT pendulum								
		Calculation according to standard [39]								
No.	Angle $\alpha = 0°$	Rebound value R 1	2	3	4	5	6	7	8	Rebound value average R_i
1	V—vertical position of the hammer	25	26	25	27	26	25	24	25	25.4
2		29	28	29	30	31	30	31	30	29.8
3		30	31	32	30	29	30	31	30	30.4
4		27	26	30	29	30	29	30	29	28.8
5		29	30	30	28	30	31	30	31	29.9
6		30	30	29	28	30	31	29	30	29.6
7		28	27	26	28	28	29	30	27	27.9
8		27	28	27	28	29	27	29	28	27.9
9		27	28	29	29	27	28	29	28	28.1
10		30	29	31	30	30	30	30	31	30.1
										$\Sigma R_i = 287.8$

The results of parameter testing were prepared in accordance with the ITB instructions [40], specifying:

the average number of rebounds

$$\overline{R_i} = 28.78$$

the standard deviation of the rebound

$$s(R) = 1.523$$

the rebound number variability coefficient

$$v(R) = 5.292\%.$$

the average strength of bricks using the standard function

$$f_b(i) = 2.79\ \mathrm{N/mm^2}.$$

the standard deviation of strength

https://doi.org/10.3390/ma19112250

$$S_R = 0.236\ \text{N/mm}^2,$$

the minimum strength

$$f_{bmin}(i) = 2.40\ \text{N/mm}^2.$$

the coefficient of variation

$$v_R = 8.469\%,$$

and the area homogeneity coefficient

$$k_R = 0.861.$$

The non-destructive testing report leads to the determination of the $f_{bmin}(i)$ strength, which is the basis for the classification of bricks and is a parameter used in the process of creating a calculation model. During the research, it was noticed that in places where the joints did not provide adequate coverage of the bricks, due to their destruction, the bricks were displaced by the impact of the Schmidt hammer. Such places do not ensure proper interaction of materials in wall structures. For bricks with low strength, the sclerometric method using the OS-120PT hammer can be useful for determining their compressive strength. The strength of mortar in masonry can be tested in a similar way to that presented, using a Schmidt hammer similar to the one presented but of type OS-120PM, which has a different tip. The determined strength of the lime mortar was M 0.5 in areas where it retained its technical parameters, and in weakened areas, its strength was close to 0.1 MPa.

2.4. Method of Analysis

In masonry structures, as a result of long-term use accompanied by processes that deteriorate their technical condition, there are fragments that are degraded and damaged to varying degrees. Degradation is understood as a local change in the stiffness of the material, manifested by a loss of mass, cracking, surface flaking and, consequently, changes in the strength parameters of the masonry. Damage causing a break in the continuity of the masonry also excludes the damaged areas or volumes from static operation. It is important to determine the current strength parameters of the tested structures in relation to bricks and mortar. In the process of preparing the calculation model, it is essential to determine the parameters described in Figure 8.

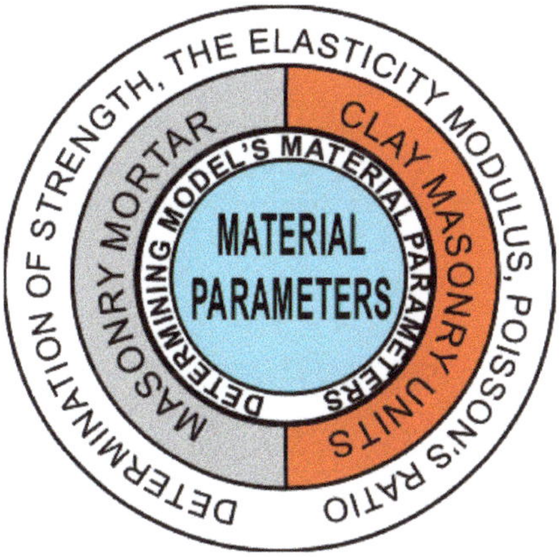

Figure 8. The identification of mortar and brick strength parameters used in the process of creating computational models of masonry structures.

The next stage is the description of degradation and damage analysed on two scales: on the structural scale, an equivalent homogeneous medium is considered, and on the

https://doi.org/10.3390/ma19112250

mesoscale, the complex heterogeneous mesostructure of the wall is taken into account. Reductions in the parameters of the wall mesostructure are systematised, in relation to the degree of damage occurring, with the introduction of an eight-level classification of damage. The D0 ÷ D7 classification proposed in this study indicates the identified level of damage, from D0, corresponding to no damage to masonry structures, up to level D7, which is equated with damage at a level close to 88%. At each level of the scale, the damage increases by 12.5%, which limits the number of macro-elements developed to a closed database of elements. The proposed classification was developed for engineering practice purposes and requires further calibration and validation through research conducted on real-scale structural members. The classification concept prepared in this way is explained in Figure 9, which shows representative macro-elements used in numerical analyses.

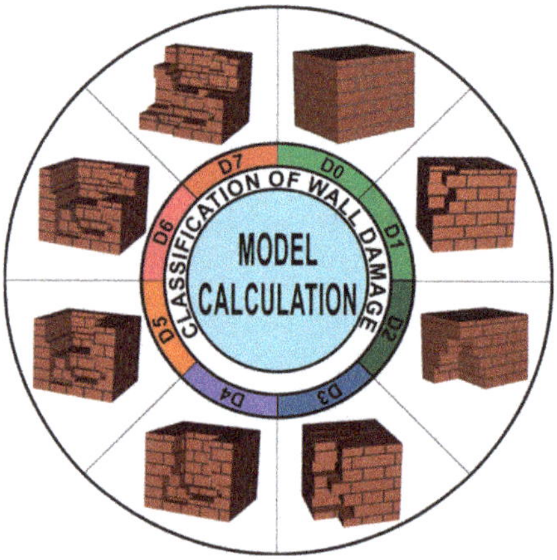

Figure 9. Classification of damaged masonry structures in 20-node macro-elements.

The proposed method of description involves separating segments of a given size, referred to as macro-elements, within the model of the analysed structure. The macro-elements reflect the actual state of degradation, taking into account the variability of physical characteristics in the subspaces of the structure and damaged fragments by introducing limits on the variability of material parameters and damage. Examples of internal structures of macro-elements are shown in Figure 10. This allows for optimal modelling of the properties of structural materials in a 3D diagram, e.g., using the finite element method. The use of macro-elements and stage analyses allows for avoiding computational tasks with a large number of unknowns. In the initial stage, it is important to identify the degree of damage and assign it to the appropriate class, as explained in Figure 11.

(a) (b) (c)

Figure 10. Examples of internal structures of macro-elements: (**a**) undamaged D0; (**b**,**c**) with various forms of degradation or damage, corresponding to class D2.

https://doi.org/10.3390/ma19112250

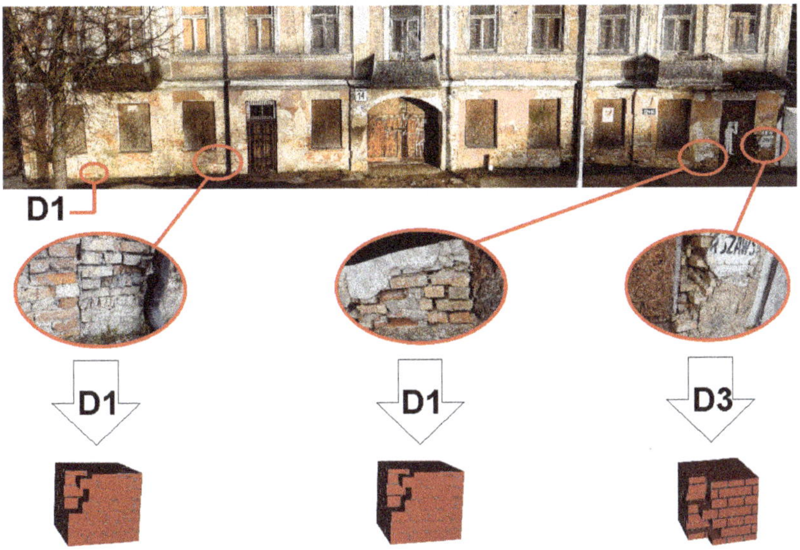

Figure 11. Classification of wall damage to the developed scope; elements of classes D1 and D3 are present.

The location of damage to masonry structures should begin with areas that are particularly exposed to the effects of rainwater and areas of contact with the ground, where the effects of melting snow and water splashes from raindrops can accumulate [41]. If areas of masonry degradation are found, the remaining areas should be examined in detail in order to properly classify the actual masonry structures by assigning them appropriate macro-element models.

The next stage is to apply the developed stiffness matrices of macro-elements $\boldsymbol{K}_e$ obtained by integration in subspaces according to the formula

$$\boldsymbol{K}_e = \int_{V_1} \boldsymbol{B}^T \boldsymbol{E}_1^{(t)} \boldsymbol{B} dV + \ldots + \int_{V_j} \boldsymbol{B}^T \boldsymbol{E}_j^{(t)} \boldsymbol{B} dV + \cdots + \int_{V_n} \boldsymbol{B}^T \boldsymbol{E}_n^{(t)} \boldsymbol{B} dV \quad (2)$$

while the stiffness of the entire structure is described by the set of all macro-elements. The sub-elements of the integration are the material fragments of the actual features used in the masonry structure. In Equation (2), $\boldsymbol{B}$ is the deformation matrix, and the time-varying elasticity matrix of sub-area j (bricks, joints) is written in the form:

$$\boldsymbol{E}_j^{(t)} = \boldsymbol{E}_j = \frac{E_j}{(1+\nu_j)(1-2\nu_j)} \begin{bmatrix} (1+\nu_j) & \nu_j & \nu_j & 0 & 0 & 0 \\ \nu_j & (1+\nu_j) & \nu_j & 0 & 0 & 0 \\ \nu_j & \nu_j & (1+\nu_j) & 0 & 0 & 0 \\ 0 & 0 & 0 & \frac{(1+2\nu_j)}{2} & 0 & 0 \\ 0 & 0 & 0 & 0 & \frac{(1+2\nu_j)}{2} & 0 \\ 0 & 0 & 0 & 0 & 0 & \frac{(1+2\nu_j)}{2} \end{bmatrix} \quad (3)$$

In the designation of the matrix $\boldsymbol{E}_j$ the superscript (t) is used to emphasise the variability of the parameters over time. The elements of the matrix, which depend on the material constants of the Young's modulus E and the Poisson's ratio ν of the brick and mortar, are, in

https://doi.org/10.3390/ma19112250

the general case, variable in the assumed sub-areas j of these materials during the long-term service life of the structure and should be determined in laboratory tests.

The assessment of the safety condition of the structure is proposed to be carried out according to a stepwise algorithm that uses the formulation of the equilibrium equations in the following matrix form:

$$K_i^{(t)} q_i^{(t)} = Q_i^{(t)} \tag{4}$$

where $K_i^{(t)}$ is the stiffness matrix of the modelled medium, $Q_i^{(t)}$ is the load vector, and $q_i^{(t)}$ is the vector of unknown displacements of the model nodes at step i of the iteration. The calculated displacement values are used to determine the deformations and stresses in material structures. The stress values, in turn, are the basis for verifying the strength of bricks and mortar. Failure criteria are used to assess the strength of bricks and mortar in macro-elements. The authors propose the use of the Maxwell–Mohr criterion and, in the case of materials under tension, the Rankine criterion. The formal notation of the failure criteria used is as follows:

$$\frac{1}{2\cos\varphi}(\sigma_i - \sigma_j) + \frac{1}{2}(\sigma_i + \sigma_j)\tan\varphi < f_s, \qquad i,j = 1,2,3 \tag{5}$$

σ_i, σ_j are principal stresses, φ is the angle of internal friction and f_s is the shear strength. At the joints between materials (bricks and mortar), stress and strength are checked using the following criterion:

$$|c - \sigma_n \tan\varphi_m| < f_{sm} \tag{6}$$

where φ_m, f_{sm} are the mortar parameters, σ_n is normal stress at the interface and c is the cohesion. The scheme for conducting the calculations is divided into stages illustrated in Figure 12.

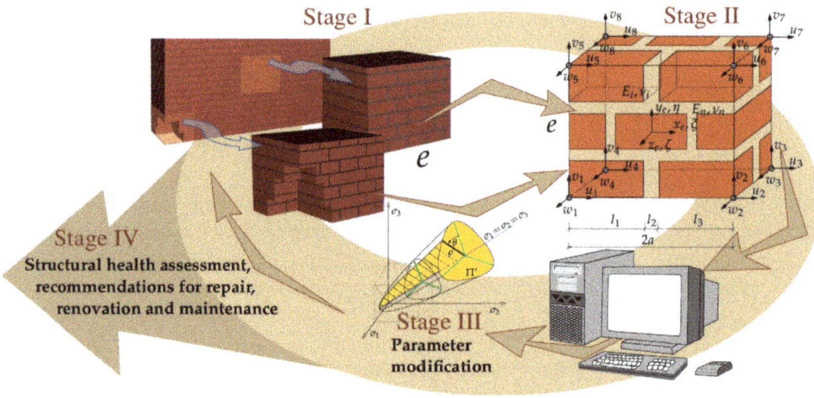

Figure 12. Stages of the calculation process for assessing the current technical condition of a masonry structure.

The procedure used in the calculations comprises four stages of developing a structural model and assessing its technical condition. Stage I involves analysing large surfaces or volumes using photogrammetry and constructing macro-elements. The next stage, Stage II, involves analysing the state of displacement, stress and strain inside the macro-elements. Stage III involves analysing the results obtained and, if necessary, modifying the model parameters and repeating the calculations. The final stage IV involves proposals and recommendations for repairs and maintenance of the structure.

https://doi.org/10.3390/ma19112250

The stiffness matrices used in the numerical analysis of the brick and mortar materials, obtained from the strength parameters tested in the laboratory and tests using the OS-120 Schmidt hammer, are as follows:

$$E_1^{(brick)} = \begin{bmatrix} 4119 & 516 & 516 & 0 & 0 & 0 \\ 516 & 4119 & 516 & 0 & 0 & 0 \\ 516 & 516 & 4119 & 0 & 0 & 0 \\ 0 & 0 & 0 & 2060 & 0 & 0 \\ 0 & 0 & 0 & 0 & 2060 & 0 \\ 0 & 0 & 0 & 0 & 0 & 2060 \end{bmatrix}, \quad (7)$$

$$E_2^{(mortar)} = \begin{bmatrix} 172 & 21 & 21 & 0 & 0 & 0 \\ 21 & 172 & 21 & 0 & 0 & 0 \\ 21 & 21 & 172 & 0 & 0 & 0 \\ 0 & 0 & 0 & 86 & 0 & 0 \\ 0 & 0 & 0 & 0 & 86 & 0 \\ 0 & 0 & 0 & 0 & 0 & 86 \end{bmatrix}, \quad (8)$$

$$E_3^{(brick)} = \begin{bmatrix} 17164 & 2149 & 2149 & 0 & 0 & 0 \\ 2149 & 17164 & 2149 & 0 & 0 & 0 \\ 2149 & 2149 & 17164 & 0 & 0 & 0 \\ 0 & 0 & 0 & 8582 & 0 & 0 \\ 0 & 0 & 0 & 0 & 8582 & 0 \\ 0 & 0 & 0 & 0 & 0 & 8582 \end{bmatrix}, \quad (9)$$

$$E_4^{(mortar)} = \begin{bmatrix} 858 & 107 & 107 & 0 & 0 & 0 \\ 107 & 858 & 107 & 0 & 0 & 0 \\ 107 & 107 & 858 & 0 & 0 & 0 \\ 0 & 0 & 0 & 429 & 0 & 0 \\ 0 & 0 & 0 & 0 & 429 & 0 \\ 0 & 0 & 0 & 0 & 0 & 429 \end{bmatrix}. \quad (10)$$

E_1, E_2—damaged areas of the brickwork; E_3, E_4—brickwork with correct technical parameters. Values are given in units [MPa].

3. Assessment of the Load-Bearing Capacity of Masonry Structures Using Developed Calculation Procedures

Using a database of macro-elements adapted to the damage found in masonry structures, analyses were carried out to verify the correctness of the proposed approach in relation to a section of masonry forming the external façade wall of a historic building. The database should take into account the current condition and information related to the history of the building structure. In addition to the significant advantages of BIM technology in this area, its integration with continuous monitoring provides measurable benefits related to the ongoing updating and expansion of the database. Furthermore, it enables continuous assessment of the building's mechanical performance and safety.

The presented method of assessing the condition of a structure is illustrated by a strength analysis of a 7.42 m long and 51 cm thick section of a solid brick wall, which is part of the 25.36 m long front wall shown in Figure 13.

It is a structural element of a three-storey building without a basement, built of solid brick with lime mortar and cement–lime mortar, with a longitudinal wall structure. In order to assess its current strength, a numerical model of a fragment of the structure was

 https://doi.org/10.3390/ma19112250

developed in the form of a longitudinal wall with geometry corresponding to that of the entire structure. Macro-elements and the proposed method of modelling the structure were used in the construction of the numerical model.

As a result of a phased numerical analysis, results were obtained in the form of stress fields (Figure 14), which allow the current strength condition of the masonry structure to be assessed. It is also important to be able to compare the results of the analyses with the results of previous calculations. Over the years, the presented results will constitute a collection of archival analyses used to identify changes occurring in the wall structures and their impact on the safety of structural systems.

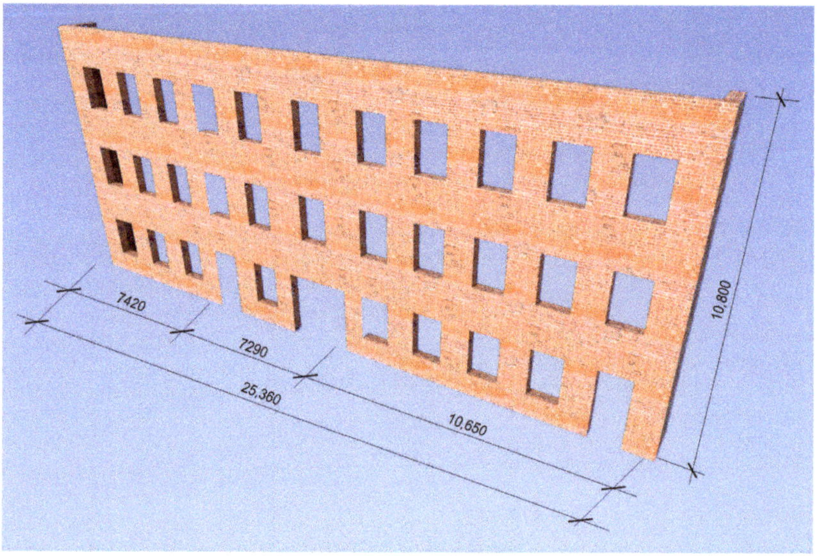

Figure 13. The structure of a brick wall analysed using the developed macro-elements.

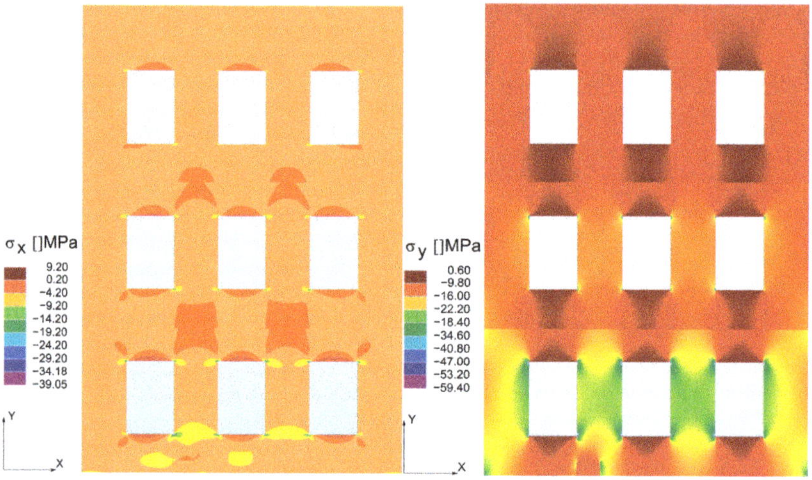

(**a**) (**b**)

Figure 14. Computer model of a fragment of the structure and the results of stress analysis: σ_x (**a**) and σ_y (**b**) in the masonry structure.

https://doi.org/10.3390/ma19112250

4. Conclusions

The continuous development of modern technologies is an undeniable fact. Such technologies are becoming increasingly noticeable and are actively entering many areas of the economy. They include both hardware technologies and software platforms.

In the field of civil engineering, BIM technologies and computational methods such as FEM have enormous potential for the analysis, maintenance and restoration of historic buildings. Thanks to their ability to create detailed digital models that take into account the heterogeneity of materials and their technical condition, and to perform advanced simulations, these techniques can help to better diagnose the static behaviour of historic structures and plan effective measures.

The integration of modern technologies enables the development of a database on the history of structural, repair and conservation changes to buildings, as well as monitoring of their current technical condition.

By combining the collected data on the technical condition [41] with numerical simulations, diagnostics can be expanded to include the technical condition of the structure, and the risk of further damage is minimised. However, the use of other modern technologies, apart from those mentioned in this article, to obtain information useful in assessing the technical condition and safety of structures remains an open question.

The use of photogrammetry significantly speeds up the process of developing macro-elements modelling masonry structures, allowing for the mapping of damage to individual bricks and areas with reduced technical parameters. The implementation of stiffness matrices with individually defined damage parameters enables the prediction of repair effectiveness. This is achieved by replacing damaged areas of the structure with finite macro-elements with actual geometric and strength parameters, thus identifying the local impact of the repairs in relation to the limit state of the entire masonry structure. This article proposes and describes a method for the comprehensive material and strength assessment of masonry structures with varying degrees of degradation. An attempt was made to formalise the technical assessment process for masonry structures, including technologies that are currently becoming standard in diagnostics of existing and newly designed buildings.

The application of a step-by-step diagnostic process based on numerical simulations of stress state changes in macro-elements, taking into account the state of damage, introduces new possibilities for determining the order of repairs in masonry structures characterised by significant damage. The intermediate stage, in which bricks are locally replaced with new ones during the repair process, is simulated using macro-elements with characteristics resulting from the remaining areas in relation to the technical condition of structures that have often been in use for hundreds of years. The development of a model using the proposed solutions speeds up the decision-making process, allows interference with historic structures to be kept to a minimum, and leaves masonry structures in good technical condition for future generations.

Author Contributions: Conceptualisation, A.W. and R.S.; methodology, A.W. and R.S.; software, A.W. and R.S.; validation, R.S. and A.W.; formal analysis, R.S. and A.W.; investigation, R.S. and A.W.; resources, R.S., A.W. and C.M.; data curation, R.S.; writing—original draft preparation, A.W. and R.S.; writing—review and editing, A.W. and R.S.; visualisation, R.S. and A.W.; supervision, A.W., R.S. and C.M.; project administration, C.M.; funding acquisition, C.M. All authors have read and agreed to the published version of the manuscript.

https://doi.org/10.3390/ma19112250

Funding: This research was carried out as a part of the research project No. WZ/WB-IIL/3/2026 at the Bialystok University of Technology and funded by a grant from the Polish Ministry of Science and Higher Education.

Data Availability Statement: The original contributions presented in this study are included in the article. Further inquiries can be directed to the corresponding author.

Conflicts of Interest: The authors declare no conflict of interest.

References

1. Hendry, A.W. *Structural Masonry*, 2nd ed.; Macmillan Press: London, UK, 1998. [CrossRef]
2. Como, M. *Statics of Historic Masonry Constructions*, 3rd ed.; Springer: Berlin/Heidelberg, Germany, 2017. [CrossRef]
3. Shabani, A.; Kioumarsi, M.; Zucconi, M. State of the Art of Simplified Analytical Methods for Seismic Vulnerability Assessment of Unreinforced Masonry Buildings. *Eng. Struct.* **2021**, *239*, 112280. [CrossRef]
4. Kubica, J. Masonry systems: Past, Present and Future. *MATEC Web Conf.* **2024**, *403*, 01003. [CrossRef]
5. Demirel, M.; Topsakal, A.; Altun, M.G. Materials Characterization of Historical Structures: A Review. *Front. Struct. Civ. Eng.* **2025**, *19*, 1461–1477. [CrossRef]
6. Rossi, M.; Bournas, D. Structural Health Monitoring and Management of Cultural Heritage Structures: A State-of-the-Art Review. *Appl. Sci.* **2023**, *13*, 6450. [CrossRef]
7. Jasiński, R. Identification of Stress States in Compressed Masonry Walls Using a Non-Destructive Technique (NDT). *Materials* **2020**, *13*, 2852. [CrossRef]
8. Pallarés, F.J.; Betti, M.; Bartoli, G.; Pallarés, L. Structural Health Monitoring (SHM) and Nondestructive Testing (NDT) of Slender Masonry Structures: A Practical Review. *Constr. Build. Mater.* **2021**, *297*, 123768. [CrossRef]
9. Oikonomopoulou, E.C.; Palieraki, V.A.; Nikolopoulou, V.S.; Vintzileou, E. Application of Non-destructive Techniques for the Investigation of Old Masonry Structures. In *Protection of Historical Constructions. PROHITECH 2021, Lecture Notes in Civil Engineering*; Vayas, I., Mazzolani, F.M., Eds.; Springer: Cham, Switzerland, 2022; Volume 209. [CrossRef]
10. Hoła, J.; Sadowski, Ł. Non-Destructive Testing in Civil Engineering. *Appl. Sci.* **2022**, *12*, 7187. [CrossRef]
11. Knyziak, P.; Spodzieja, S.; Krentowski, J.R.; Pawłowicz, J.A.; Sawczyński, S.; Gil-Mastalerczyk, J. Semi-destructive Testing Methods for Examining the Structural Condition of Historic Buildings. *Eng. Fail. Anal.* **2025**, *171*, 109354. [CrossRef]
12. Ávila, F.; Blanca-Hoyos, Á.; Puertas, E.; Gallego, R. HBIM: Background, Current Trends, and Future Prospects. *Appl. Sci.* **2024**, *14*, 11191. [CrossRef]
13. Pepe, M.; Costantino, D.; Restuccia Garofalo, A. An Efficient Pipeline to Obtain 3D Model for HBIM and Structural Analysis Purposes from 3D Point Clouds. *Appl. Sci.* **2020**, *10*, 1235. [CrossRef]
14. Kavaliauskas, P.; Fernandez, J.B.; McGuinness, K.; Jurelionis, A. Automation of Construction Progress Monitoring by Integrating 3D Point Cloud Data with an IFC-Based BIM Model. *Buildings* **2022**, *12*, 1754. [CrossRef]
15. Soleyman, A.; Jahangir, H.; Nehdi, M.L. Damage Detection and Monitoring in Heritage Masonry Structures: Systematic Review. *Constr. Build. Mater.* **2023**, *397*, 132402. [CrossRef]
16. Valluzzi, M.R.; Modena, C. Masonry. In *Fracture and Failure of Natural Building Stones*; Kourkoulis, S.K., Ed.; Springer: Dordrecht, The Netherlands, 2006. [CrossRef]
17. Wang, C.; Sarhosis, V.; Nikitas, N. Strengthening/Retrofitting Techniques on Unreinforced Masonry Structure/Element Subjected to Seismic Loads: A Literature Review. *Open Constr. Build. Technol. J.* **2018**, *12*, 251–268. [CrossRef]
18. Keshmiry, A.; Hassani, S.; Dackermann, U.; Li, J. Assessment, Repair, and Retrofitting of Masonry Structures: A Comprehensive Review. *Constr. Build. Mater.* **2024**, *442*, 137380. [CrossRef]
19. Angelillo, M.; Lourenço, P.B.; Milani, G. Masonry behaviour and modelling. In *Mechanics of Masonry Structures*; Angelillo, M., Ed.; CISM International Centre for Mechanical Sciences: Udine, Italy; Springer: Vienna, Austria, 2014; Volume 551. [CrossRef]
20. Miedziałowski, C.; Walendziuk, A. Description of Material Properties of Degraded and Damaged Segments of Multi-Leaf Masonry in Analyses of Large Three-Dimensional Structures. *Materials* **2023**, *16*, 4076. [CrossRef]
21. Roca, P.; Cervera, M.; Gariup, G.; Pela, L. Structural Analysis of Masonry Historical Constructions. Classical and Advanced Approaches. *Arch. Comput. Methods Eng.* **2010**, *17*, 299–325. [CrossRef]
22. Asteris, P.G.; Plevris, V.; Sarhosis, V.; Papaloizou, L.; Mohebkhah, A.; Komodromos, P.; Lemos, J.V. Numerical Modeling of Historic Masonry Structures. In *Handbook of Research on Seismic Assessment and Rehabilitation of Historic Structures*; Asteris, P.G., Plevris, V., Eds.; IGI Global: Hershey, PA, USA, 2015; pp. 213–255.

23. D'Altri, A.M.; Sarhosis, V.; Milani, G.; Rots, J.; Cattari, S.; Lagomarsino, S.; Sacco, E.; Tralli, A.; Castellazzi, G.; de Miranda, S. Modeling Strategies for the Computational Analysis of Unreinforced Masonry Structures: Review and Classification. *Arch. Comput. Methods Eng.* **2020**, *27*, 1153–1185. [CrossRef]
24. Verstrynge, E.; Schueremans, L.; Van Gemert, D.; Hendriks, M.A.N. Modelling and Analysis of Time-dependent Behaviour of Historical Masonry under High Stress Levels. *Eng. Struct.* **2011**, *33*, 210–217. [CrossRef]
25. Ferretti, D.; Bažant, Z.P. Stability of Ancient Masonry Towers: Stress Redistribution Due to drying, Carbonation, and Creep. *Cem. Concr. Res.* **2026**, *36*, 1389–1398. [CrossRef]
26. Elkhrachy, I. 3D Structure from 2D Dimensional Images Using Structure from Motion Algorithms. *Sustainability* **2022**, *14*, 5399. [CrossRef]
27. Hoła, J. The Degradation of Historic Buildings Due to Excessive Moisture—Selected Issues. *Bud. Archit.* **2018**, *17*, 133–148. (In Polish) [CrossRef]
28. Gutarowska, B.; Piotrowska, M. Methods of Mycological Analysis in Buildings. *Build. Environ.* **2007**, *42*, 1843–1850. [CrossRef]
29. Seaward, M.R.D. Lichens as Agents of Biodeterioration. In *Recent Advances in Lichenology*; Upreti, D., Divakar, P., Shukla, V., Bajpai, R., Eds.; Springer: New Delhi, India, 2015. [CrossRef]
30. Armesto, J.; Arias, P.; Roca, J.; Lorenzo, H. Monitoring and Assessing Structural Damage in Historic Buildings. *Photogramm. Rec.* **2008**, *23*, 36–50. [CrossRef]
31. Biscarini, C.; Catapano, I.; Cavalagli, N.; Ludeno, G.; Pepe, F.A.; Ubertini, F. UAV Photogrammetry, Infrared Thermography and GPR for Enhancing Structural and Material Degradation Evaluation of the Roman Masonry Bridge of Ponte Lucano in Italy. *NDT E Int.* **2020**, *115*, 102287. [CrossRef]
32. Lluisi i Ginovart, J.; Toldrà, J.M.; Costa, A.; Coll, S. Close Range Photogrammetry and Constructive Characterization of Masonry gothic Vaults. *Rev. Constr.* **2014**, *13*, 47–55. [CrossRef]
33. Loverdos, D.; Sarhosis, V.; Adamopoulos, E.; Drougkas, A. An Innovative Image Processing-based Framework for the Numerical Modelling of Cracked Masonry Structures. *Autom. Constr.* **2021**, *125*, 103633. [CrossRef]
34. Hain, A.; Zaghi, A.E. Applicability of Photogrammetry for Inspection and Monitoring of Dry-stone Masonry Retaining Walls. *Transp. Res. Rec.* **2020**, *2674*, 287–297. [CrossRef]
35. Riveiro, B.; Caamaño, J.C.; Arias, P.; Sanz, E. Photogrammetric 3D modelling and mechanical analysis of masonry arches: An Approach Based on a Discontinuous Model of Voussoirs. *Autom. Constr.* **2011**, *20*, 380–388. [CrossRef]
36. *EN 772-1:2011+A1:2015*; Methods of Test for Masonry Units—Part 1: Determination of Compressive Strength. European Standard: Brussels, Belgium, 2015. Available online: https://www.en-standard.eu (accessed on 18 May 2026).
37. Bajno, D.; Schabowicz, K.; Grzybowska, A. Material Testing of Historic Bricks and Mortars in Degraded Masonry Structures. *Materials* **2024**, *17*, 3192. [CrossRef] [PubMed]
38. *Schmidt OS-120 Operating Instruction*; Proceq SA: Schwerzenbach, Switzerland, 2017. Available online: https://media.screeningeagle.com/asset/Downloads/Schmidt%20OS-120_Operating%20Instructions_English_high.pdf (accessed on 18 May 2026).
39. *EN 12504-2:2021-12*; Testing Concrete in Structures—Part 2: Non-Destructive Testing—Determination of Rebound Number. European Standard: Brussels, Belgium, 2021. Available online: https://www.en-standard.eu (accessed on 18 May 2026).
40. *Instructions for the Use of Schmidt Hammers for Non-Destructive Quality Control of Concrete in Structures Manual 210*; Building Research Institute: Warsaw, Poland, 1977. (In Polish)
41. Stryszewska, T.; Kańka, S. Forms of Damage of Bricks Subjected to Cyclic Freezing and Thawing in Actual Conditions. *Materials* **2019**, *12*, 1165. [CrossRef]

 https://doi.org/10.3390/ma19112250

MDPI AG
Grosspeteranlage 5
4052 Basel
Switzerland
Tel.: +41 61 683 77 34

Materials Editorial Office
E-mail: materials@mdpi.com
www.mdpi.com/journal/materials